P. W. West

Growing Plantation Forests

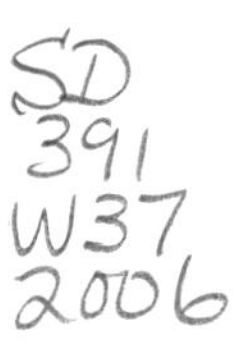

P. W. West

Growing Plantation Forests

With 36 Figures and 10 Tables

Professor Philip W. West
School of Environmental Science and Management
Southern Cross University
Lismore, NSW 2480
Australia

and

SciWest Consulting
67 Gahans Road
Meerschaum Vale, NSW 2477
Australia

Cover photo: A plantation of messmate stringybark *(Eucalyptus obliqua)*. This species is native to southeastern Australia, but in this photo it is being grown as an exotic species in an experimental plantation in southwest Western Australia. It has been found subsequently that Tasmanian blue gum *(E. globulus)*, another species native to southeastern Australia, grows rather better than messmate stringybark in Western Australia and is now planted widely there.

Library of Congress Control Number: 2006923374

ISBN-10 3-540-32478-X Springer Berlin Heidelberg New York
ISBN-13 978-3-540-32478-2 Springer Berlin Heidelberg New York

Springer is a part of Springer Science + Business Media
springer.com

Editor: Dr. Dieter Czeschlik, Heidelberg, Germany
Desk editor: Anette Lindqvist, Heidelberg, Germany
Cover design: *design & production*, Heidelberg, Germany
Production: LE-TEX Jelonek, Schmidt & Vöckler GbR, Leipzig, Germany
Typesetting: Camera ready by author
31/3100/YL – 5 4 3 2 1 0 – Printed on acid-free paper

To My wife and our companions,
who make life so pleasant

Preface

So varied are plantation forests around the world and so varied are the practices necessary to grow them successfully, it is impossible to consider them completely in a single volume. However, there is a set of scientific principles which underly what is done, wherever successful plantation forests are grown. It is these principles this book describes, so readers may glean some idea of how their plantations will behave and the problems they will face wherever they are in the world.

I have attempted to reach a wide range of readers, from landholders and farmers, who have no formal forestry education, through undergraduate students of forestry, field foresters and, finally, forest scientists involved in research. This has meant dealing with topics from basic plant biology, through to advanced concepts of forestry science. To maintain the interest of all readers, I have tried wherever possible to illustrate the concepts with practical examples drawn from plantation forests throughout the world.

The decisions as to which topics to include or exclude were not easy. In general, I have tried to cover as much as I feel can be taught reasonably in a one semester undergraduate university course on plantation forestry. No individual has the breadth of knowledge and expertise to ever hope to cover all the topics discussed here with complete authority. I am certainly no exception to that rule; inevitably, experts in any particular topic will see where my strengths and limitations lie.

Several colleagues, Humphrey Elliott, Mickie Fogarty, Graeme Palmer, Libby Pinkard, John Turner and Peter Volker gave generously of their time in reviewing all or part of the manuscript and I thank them deeply for their efforts and suggestions.

P.W. West
March 2006

Contents

1 Plantation Forests

As the population of the world increases and its economies grow, so does the amount of wood people use, for firewood, for building, for paper and for many other purposes. Large areas of native forests (that is, naturally occurring forests) are cleared every year and converted to other uses. Conservationists are becoming increasingly concerned about this. They are insisting that the native forests that remain be preserved and used much less for wood harvest.

Plantations are forests created by man, by planting seeds or seedlings, usually at a regular spacing. Most contain a single tree species, although plantations of a mixture of species are important in some parts of the world. Plantation forestry is often much like any other agricultural enterprise, aiming to grow highly productive forests on relatively small areas of land, so the land is being used most efficiently. To do so requires that much attention be paid to the silviculture of plantations, that is, to the tending of the trees to achieve some desired objectives. Silvicultural techniques are the principal topic of this book.

There is now a global trend to encourage the establishment of plantations; eventually, it is likely that they will produce a large proportion of the wood used around the world. As well, plantations are being grown to provide various environmental benefits and may or may not produce wood for consumption at the same time.

1.1 Plantation Forests Around the World

The most recent figures available on the area of plantation forests are given in Table 1.1. They were collated by the Food and Agricultural Organisation of the United Nations in 2000 (FAO 2001). They show that about 30% of the land surface of the world (excluding frozen parts such as Antarctica) is covered by forests and about 5% of those are plantations. The area of native forests has been declining; over the decade 1990–2000, an average of 16 million ha was cleared annually. By contrast, plantation areas have been increasing, by an average of about 3 million ha/year over the same period.

Table 1.1. The total area (millions of hectares) of land in various parts of the world, together with the areas covered by forests and forest plantations in 2000 (Source—FAO 2001)

Location	Land	Forest	Plantations
Africa	2,980	650	8
Asia	3,080	550	116
Europe	2,260	1,040	32
North and Central America	2,140	550	18
South America	1,750	890	10
Oceania	850	200	3
Total	13,060	3,870	187

About half the plantations of the world are being grown to produce wood for industrial use, that is, to be processed principally into building products or paper (Sect. 3.3). About one quarter of the plantations are for non-industrial uses, that is, for firewood, environmental protection and other products (including animal fodder, fruits, leaf oils and tannins). The purpose of the other quarter of the plantations around the world could not be determined when this information was collated. The largest areas of industrial plantations are in China (37 million ha), the USA (16 million ha) and India (12 million ha). Of non-industrial plantations, the largest areas are in India (21 million ha), China (8 million ha) and Indonesia and Thailand (4 million ha each).

Plantations already make a substantial contribution to the supply of wood throughout the world. It has been estimated (ABARE–Jaakka Pöyry Consulting 1999) that nearly 4 billion m^3 of wood would have been used throughout the world in 2000, just under half of which would have been for industrial use. The rest would have been firewood, largely for domestic use, principally in Asia and Africa. Plantations supply about one third of this industrial wood; the rest comes from native forests. This relatively large production from plantations, which make up a relatively small proportion of all forests (Table 1.1), emphasises that the productivity of plantations is much greater generally than that of native forests. This is a result of the attention paid to plantation silviculture.

Two groups of tree species dominate plantations around the world, the pines (scientifically speaking, members of the plant genus *Pinus*) and the eucalypts (the genus *Eucalyptus*). Pines make up about 20% of the plantations of the world and eucalypts about 10% (FAO 2001). The pines are softwoods and are native to a number of countries of the northern hemisphere. Botanically speaking, a softwood is any tree species (not just pines) which produces its ovules (eggs in common parlance) and its pollen in cones; after the ovules have been fertilised by pollen, the scales of the cone protect the seeds as they develop. The eucalypts are hardwoods and are native to Australia, almost exclusively. A hardwood is any tree species which produces flowers; their seeds grow within, and are protected by, the ovary of the flower.

Not all softwoods have particularly soft wood, nor hardwoods particularly hard wood (the balsa tree, the species *Ochroma lagopus*, is a hardwood renowned for its soft, light wood). However, the properties of the wood of the two groups differ in various respects, which are important for their usefulness; this is discussed in Sect. 3.3.

Of course many tree species other than pines and eucalypts are used in plantations around the world. In one way or another, I make reference in this book to 81 different species used in plantations, 17 each of pines and eucalypts, 14 other softwoods and 33 other hardwoods. This is only a small proportion of all species used in plantations.

1.2 Purposes of Plantation Forests

On all the continents, there are large plantation areas grown to produce wood for industrial use. They often cover tens of thousands of hectares and are major commercial enterprises. Some are publicly owned and some are privately owned. They are important economically to the countries concerned, often earning export income and providing employment, both in their growing and in the processing of the wood they produce. Many such plantations are grown for only 10–15 years before harvesting (in forestry, the period from planting to final harvest of a plantation is known as a rotation). Usually, these produce wood for papermaking. Others are grown on 20–30 year rotations, by which time the trees will be large enough for their stems to be sawn to produce timber.

An emerging type of industrial plantation is being grown to produce bioenergy (Evans 1997; Fuwape and Akindele 1997; Kuiper et al. 1998; Toivonen and Tahvanainen 1998; van den Broek et al. 2000; Updegraff et al. 2004; Weih 2004; Andersen et al. 2005; Hoffman and Weih 2005). The wood from such plantations is used to make energy, either through fermentation to produce ethanol (the chemical name for alcohol) for use as motor fuel or through burning the wood to generate electricity.

The primary stimulus for the establishment of bioenergy plantations comes from concerns about the global warming of the atmosphere, which is believed to be caused by the release of greenhouse gases. One of the most important of these is carbon dioxide, which is released when fossil fuels (coal, oil and natural gas) are burnt. Greenhouse gases trap energy from sunlight, energy which would otherwise be reflected back into space from the surface of the earth; this leads to warming of the atmosphere.

It is argued that bioenergy plantations are carbon dioxide neutral. As will be discussed in Sect. 2.1, plants take in carbon dioxide from the air (about 50% of the structural matter of plants consists of the element carbon, derived from carbon dioxide from the air). If the wood grown in bioenergy plantations is burnt ultimately as fuel, it will release back to the atmosphere exactly as much carbon dioxide as the plantations took up. This contrasts with fossil fuels, where the carbon they contain was stored millions of years ago in the plant material from which they derived; burning them today adds additional carbon dioxide to that which is present normally in the atmosphere (carbon dioxide makes up about 0.035% of the atmospheric gases). Bioenergy plantations are usually grown on very short rotations (perhaps of 3–5 years); the small size of their trees at harvest is of no importance, because they are simply going to be burnt or used for fermentation.

Although they can be considered industrial plantations, bioenergy plantations also provide an environmental benefit by reducing the overall emissions of carbon dioxide to the atmosphere. Plantations are being used around the world to provide various other environmental benefits as follow:

- Clearing land of forests, for agriculture, for mining or for other reasons, can lead to soil erosion by wind or water. Tree roots hold the soil and plantation establishment on cleared land can prevent soil loss, siltation of waterways or can otherwise rehabilitate cleared sites (Costantini et al. 1997a; Mazanec et al. 2003; Coates 2005; Udawatta et al. 2005).
- Clearing forests for agriculture can allow the soil water table to rise, bringing salt close to the soil surface. High salinity can prevent growth of agricultural crops; this is becoming a serious problem in agricultural areas in various parts of the world. Re-establishment of trees on salt-affected sites can draw down the water levels in the soil, taking the salt with them and rehabilitating the soil (Lambert and Turner 2000; Theiveyanathan et al. 2004; Marcar and Morris 2005).
- Plantations can be used to dispose safely of sewage waste and some other forms of industrial waste (Myers et al. 1996; Labrecque et al. 1997; Rosenqvist et al. 1997; Hasselgren 1998; Weih 2004; Mercuri et al. 2005; Morris and Benyon 2005; Rosenqvist and Dawson 2005a). These are often rich in various nutrient elements (Sect. 2.1.4), which the trees take up; their disposal otherwise, often in streams or the sea, can lead to serious environmental damage.

- Scattering plantation areas throughout a region, which has been cleared largely for agriculture, can enhance biodiversity (the variety of plants and animals that live within a region) (Keenan et al. 1997, 1999; Bonham et al. 2002; Hobbs et al. 2003; Lindenmayer et al. 2003; Weih 2004; Kanowski et al. 2005). Plantations increase the variety of habitats available for animals and other plants, although generally not to the same extent as native forests.

Of course, when plantations are established principally to gain environmental benefits, it is often hoped they will provide wood for sale also. In this case, it can be just as important as with industrial plantations to use appropriate silvicultural practices to ensure wood is produced efficiently.

Finally, plantations have an important role as part of the mix of activities undertaken on farms; the combination of agriculture with forestry is known as agroforestry. In some cases, plantations are simply established on parts of a farm not being used for agricultural purposes. In others, the trees are grown in combination with other activities; other plant crops might be afforded protection by being grown between the trees, livestock might be permitted to graze between widely spaced trees or the leaves of the trees might be fodder for livestock. So diverse are agroforestry systems around the world that it is impossible to summarise them here. A number of excellent texts and papers describe them in detail (e.g. MacDicken and Vergara 1990; Jarvis 1991; Knowles 1991; Nair 1991, 1993; Huxley 1999).

1.3 About This Book

In writing this book, my principal problem was to deal with the enormous diversity of plantation forests. Different species are grown in different parts of the world, the products they yield differ, the climates in which they grow differ and the soil types on which they grow vary enormously. Unless there has been previous experience with plantations in a particular part of the world, it is impossible to say just how they should be managed there.

However, over many years forest scientists have gained considerable understanding of how plantations grow and develop and what problems are encountered with them. It is the scientific principles they have developed which this book attempts to describe. It concentrates on plantations where the principal objective is the commercial production of wood. Through an understanding of the science of plantation forestry, I hope plantation growers anywhere in the world will gain some appreciation of why their plantations behave as they do and what types of remedy might be necessary if things go wrong. I have tried to choose examples from plantation forests all over the world to illustrate the issues that are discussed.

It should be appreciated that when problems are encountered with a plantation forest, the remedy may be neither quick nor easy. For example, suppose a new disease (Chap. 11) appears which threatens the health of a plantation. It may require a very large research effort, over many years and involving many scientists in various specialised disciplines, to identify the disease, assess how damaging it is and find possible remedies, which are neither prohibitively expensive nor damaging to other parts of the environment. Such an effort is far beyond the resources of a small plantation grower; only governments, universities, forest research organisations and large forestry corporations have the staff and funds available to deal with such issues.

The smaller plantation grower faces special problems in establishing a successful plantation. It is not particularly difficult to plant tree seedlings on what appears to be a suitable piece of land and hope they will eventually produce an attractive forest. Often they will. However, if the objective is to grow them rapidly, to yield a valuable wood product in a sufficiently short time to gain a worthwhile financial return, the advanced silvicultural practices discussed in this book will have to be used. To be commercially successful, the small grower will often need to obtain outside, expert advice.

In writing this book, I have tried to make it accessible to a wide audience, from the farm forester who has no specialised training in forestry or the biological sciences, through undergraduate students of forestry, the professional forester and, finally, the forest scientist. For scientists in particular, I refer to recent scientific literature, so they will be able to access more advanced aspects of a particular topic. Less advanced readers will often use grey literature (less formally published material) to obtain specific information about plantation forestry in their region; this type of material is often produced by research organisations and through government agencies. I hope this book will make it easier to read and understand that more local material.

To meet the needs of this wide audience has required covering topics from quite basic through to advanced levels. For example, Sect. 2.1 introduces some basic principles of tree and forest biology, principles well known to readers with a biological education. Then, Sects. 2.2–2.4 cover more advanced material, which describes our present knowledge of how plantations grow and develop during their lifetime. I have tried to minimise the use of biological and forestry jargon. Nevertheless, many terms will be unfamiliar and a glossary has been included as Appendix 1. Words in the glossary are shown in **bold** type when they are first encountered in the text.

The book is structured so that Chaps. 2 and 3 give information about the biology and growth of **plantation** forests and the properties and uses of the **wood** that is harvested from them. That information is referred to constantly in later parts of the book. From Chap. 4 on, the specifics of plantation **silviculture** are discussed. It follows plantations from the planning stages, through their initial establishment and right through to final harvest.

The metric system of weights and measures has been used throughout. A table of metric–imperial conversion factors has been included as Appendix 2, together with some of the relationships between units in the imperial system.

2 Biology of Plantation Growth

Growers want their plantations to be healthy and to grow as rapidly as possible. To appreciate how plantation silviculture can achieve this, it is necessary first to understand how plantations grow and what resources they need to do so.

This chapter describes the growth of normal, healthy plantations. It concentrates on monocultures, that is, plantations of a single **tree** species, since these are the most common types of plantations grown commercially. **Mixed-species** plantations, that is, those which contain two or more tree species, behave differently in some respects from monocultures and are discussed in Chap. 13. Much of the information in this chapter will be referred to in later parts of the book, as the various aspects of plantation silviculture are discussed in detail.

2.1 Basic Plant Biology

For readers without any detailed biological knowledge, this section describes briefly the biology of plants in general and trees in particular. Standard texts on plant biology can be consulted for more details. More advanced texts, such as the excellent volume by Atwell et al. (1999), provide much more scientific detail of plant biology.

2.1.1 Tree Requirements and Characteristics

In common with all plants which grow on land, trees have certain fundamental needs which they must get from the environment around them. In particular, they need:

- Sunlight
- Carbon dioxide from the air
- An appropriate air and soil temperature
- Water and **nutrients** from the soil.

Each of these will be discussed in more detail later. As long as they are all available, trees have anatomical, physiological and metabolic characteristics which allow them to live and grow. For the present discussion, the most important of these characteristics are:

- Leaves, which take in carbon dioxide from the air and convert it chemically to food for the tree, using energy from sunlight in the process
- Branches, which support the leaves high in the air to intercept sunlight
- Roots, which anchor the tree in the ground and take up water and nutrients from the soil
- A stem, which supports the tree upright, and through which water and nutrients are transported up from the roots to the leaves and down which food is transported from the leaves to the roots.

Other land plants have similar characteristics, but what distinguishes trees is that they are tall. The tallest in the world, the redwoods (*Sequoia sempervirens*) of California and the eucalypts, mountain ash (*Eucalyptus regnans*) and Tasmanian blue gum (*E. globulus*), of southern Australia (Potts and Reid 2003), may grow to over 100 m. Most tree species grow to much lesser **heights** and the definition of what is then a tree and what is a shrub becomes rather arbitrary.

The key to the great height of trees is that they have massive stems made of wood. Wood is a strong material, its strength coming from the particularly thick walls of the microscopic plant **cells** of which it is made. Wood consists mainly of dead tissue. That is, the cells have been emptied of their living contents, so the tree needs no longer to supply them with food. Not only do tree stems contain wood, but branches and large roots do also.

Wood serves two purposes. It provides strength and is also the pathway through which water is transported from the roots to the leaves. The dead, empty wood cells can be thought of as a system of interconnecting pipes, through which water passes up the whole length of the tree, a process known as **transpiration**. The tissue which transports water in plants in general, not just trees, is known as **xylem** and wood is one such tissue.

A thin, outer layer of living tissue (known as **phloem**) surrounds the wood of roots, stems and branches. Food is transported down through the phloem from the leaves to the tiny, living fine roots, at the extremities of the root system. Between the phloem and the wood is a very thin layer of tissue called the **cambium**. When cells in the cambium divide, they form new wood cells towards the inside of the stem or new phloem cells towards the outside. Outside the phloem is the **bark**, which consists also largely of dead tissue and serves to protect the thin layer of living tissue beneath it; bark is often 2–3-cm thick.

Speaking ecologically, the great height of trees allows them to carry their leaves high in the air. This gives them an advantage in that they can receive the sunlight they need and deny light to the smaller, shaded plants below. A tree stem can be considered as an engineering structure, in fact simply as a tall, tapered pole. That pole must be strong enough to support both its own weight and the weight of the branches and leaves it carries. It must also be strong enough to resist the **stresses** to which it is subjected as the wind blows on the tree **crown** (a term used for the foliage and branches of an individual tree). Engineering theory shows that the taller a pole, the larger must be its diameter at its base for it to remain upright. So it is for tree stems (King and Loucks 1978; King 1981, 1986; Osler et al. 1996b) and very tall trees must have very large stems. The tallest trees have girths at the base of their stems of 20–30 m.

2.1.2 Photosynthesis and Water Use

The surfaces of the leaves of plants which grow on land are covered by microscopic holes called **stomata**. These have a special structure, which allows them to open and close. When they are open, carbon dioxide may enter the leaves from the air. Then, within specialised cells in the leaf, a complex sequence of chemical reactions occurs. Using energy from sunlight, these reactions chemically convert carbon dioxide and water to sugars, which are energy-containing food for the plant. Sucrose (the chemical name for table sugar), glucose and starch are all sugars produced by plants; both sucrose and starch can be converted to glucose, which is the form of sugar which plants use ultimately as food.

This whole process of food production by plants is known as **photosynthesis**. A by-product of photosynthesis is oxygen, which is released from the leaves into the air, through their stomata. Animals then breathe this oxygen; animal life as we know it on earth is possible only because oxygen is released by plants through photosynthesis.

The ultimate result of photosynthesis, and use by the plant of the food it produces, is that plants grow and increase their **biomass**. The word biomass means the weight of a living organism. It may include the water in the organism, when it is referred to as fresh biomass. Usually about half the weight of plants is made up of water. However, since plants take up water from the soil, they do not have to produce it chemically; all the other tissue of which plants consist has to derive ultimately from food produced through photosynthesis. Because of this, the **oven-dry biomass** of plants is referred to commonly in biological science. This is the tissue weight after the water has been removed by drying; it is a measure of what the plant has actually produced through its **metabolism**.

Unfortunately for plants, not only does oxygen escape from leaves through their stomata, but water does also, as water vapour. The living tissue of the leaves needs water to stay alive, but the presence of the stomatal holes in the leaves means that a lot of water is evaporated from them. On hot, dry days or during droughts, plants close their stomata to prevent excessive water loss. However, as long as their stomata are closed, they cannot take in carbon dioxide from the air and are unable to produce food through photosynthesis. As will be discussed in more detail later, the availability of water from the soil is often the most crucial environmental factor which determines how well plants grow on any particular **site**.

2.1.3 Temperature

As temperature varies from season to season and from time to time during any day, it affects the rate of metabolism of plants, that is, the rate of the chemical reactions within cells which provide the energy and the materials for their growth, maintenance and reproduction. If the temperature is too low or too high, the chemical reactions cannot proceed at all and the plant stops metabolising.

Within this range from too low to too high temperatures, a plant has some temperature at which its metabolism and, hence, its growth is maximised. Both the temperature at which this maximum occurs and the temperature range within which any growth occurs vary from plant species to plant species. Some species are adapted to grow better in cooler climates, whilst others grow better in warmer climates. Plants do not tend to occur naturally on earth in places where the annual average air temperature is outside the range of about 5–45°C.

2.1.4 Nutrients

Nutrients are chemical elements which play a wide variety of roles in the metabolism of plants; without them, plants cannot survive. There are 15 nutrient elements believed to be essential for plants. Table 2.1 lists them and the minimum **concentrations** (averaged over plants generally) at which they need to be present in actively metabolising leaf tissue for plants to function normally (the concentration of something is the proportion it makes up of the whole of which it is part). For nutrient elements in plants,

Table 2.1. The chemical elements considered essential nutrients for plants. The chemical symbol by which each is known is shown, together with the minimum concentration (averaged over plants generally) at which each needs to be present in actively metabolising leaf tissue for plants to function normally. The concentrations are shown as milligrams of the chemical element per kilogram of oven-dry weight of plant tissue. The elements required in higher concentrations are known as macronutrients and those required in lower concentrations as micronutrients (Source—Atwell et al. 1999)

Element	Chemical symbol	Concentration (mg/kg)
Macronutrients		
Nitrogen	N	25,000
Potassium	K	10,000
Magnesium	Mg	2,000
Phosphorus	P	2,000
Calcium	Ca	2,000
Sulphur	S	1,000
Micronutrients		
Sodium	Na	500
Chlorine	Cl	100
Iron	Fe	100
Manganese	Mn	20
Zinc	Zn	20
Boron	B	12
Copper	Cu	3
Nickel	Ni	0.1
Molybdenum	Mo	0.1

their concentration is usually expressed as a weight of the element per unit weight of the oven-dry biomass of the plant.

It is obvious from Table 2.1 that the amount of each nutrient required for plant metabolism varies enormously, from 25,000 mg/kg of nitrogen to 0.1 mg/kg of nickel and molybdenum. Because the first six nutrients are required in much larger amounts than the others, they are referred to commonly as macronutrients. The last nine are called micronutrients.

Plants take up virtually all their nutrient requirements from the soil. They do so through fine roots, which are found at the extremities of the root system. As their name implies, fine roots are small and thin. They consist of living tissue and have not developed wood. For nutrient elements to be taken up by fine roots, they must be dissolved in the water which fills the spaces between the particles which make up the soil. Both nutrients and water are then taken up together by fine roots.

The uptake of water and nutrients is aided by a symbiotic relationship between fine roots and certain types of **fungi** (Sect. 11.1), known as **mycorrhizas**. In a symbiotic relationship, both organisms involved derive benefit from their association. In the case of a plant and a mycorrhiza, the plant benefits from improved water and nutrient uptake, whilst the mycorrhiza is provided with food by the plant.

There are many types of mycorrhizal fungi and they are associated with a large proportion of tree (and other plant) species throughout the world. There are two principal groups of mycorrhizas, ectomycorrhizas which form a sheath of fungal tissue around fine roots and vesicular arbuscular mycorrhizas which grow within the root. Both extend an extensive fine web of filaments (known as hyphae) beyond the roots into the soil; effectively, they increase greatly the surface area of the plant root system for uptake of water and nutrients. Many plant species are unable to grow and develop adequately unless they have a mycorrhizal association with their roots. Various texts give more information about mycorrhizas and their importance in forests (e.g. Read et al. 1992; Vogt et al. 1997; Smith and Read 1997).

Once water and nutrients have been taken up by the fine roots, both are transported into the woody roots, then into and up through the wood of the stem and branches, where, finally, they reach the leaves. This transport of water and nutrients from the roots to the top of even the tallest trees is powered directly by energy from sunlight. The sunlight evaporates water from the leaves, through their stomata, and a continuous stream of water is literally pulled right up the length of the tree from its fine roots to its leaves. Thus, trees do not have to use any energy from the food they have produced through photosynthesis to raise water and nutrients to their tops.

2.2 Principles of Plantation Growth

Consider a newly established forest plantation. Tree seedlings have been raised in a nursery (Sect. 5.2) and planted out (Sect. 5.3). Usually, the trees will have been planted in rows, typically with 2–3 m between each row and 2–3 m between each tree in a row (Sect. 7.2). The soil has been prepared for planting by some form of cultivation (Sect. 5.1). **Weeds**, which might compete with the seedlings, have been controlled (Sect. 5.4). Various other treatments may have been applied also to give the seedlings their best chance to survive and grow rapidly.

Because of their small size at the time of planting, the seedlings are very vulnerable. They face hazards such as hot, dry weather, insect infestation (Sects. 10.2, 10.3), browsing by larger animals (Sect. 10.4), frost or competition from vigorous weed growth (Sect. 5.4). Often, it is accepted as normal in plantation **forestry** that 5–10% of the seedlings will die from one or other of these causes over the first year or so after planting out.

Assuming it survives these early hazards, each seedling will then start to grow. During the first year or so, seedlings are so small that their requirements for nutrients and water from the soil are correspondingly small. As the trees continue to grow, the biomasses of their leaves, branches, roots and stems all increase. As well, the trees increase in height and their crowns increase both in length and width. Their root systems grow deeper into the soil and spread in width. As the biomass of the principal living tissues, the leaves and fine roots, of each seedling increases, the amount of water and nutrients the seedling requires to keep it alive increases also; the expansion of its root system will allow it to take up the extra water and nutrients it needs.

Eventually, the root systems and crowns of the individual trees expand to such an extent that they meet those of neighbouring trees. The individual tree crowns then form a closed **canopy** over the whole plantation, that is, a more or less continuous layer of leaves and branches covering the whole area. Below ground, the root systems will be spread also, more or less continuously over the entire area.

On some sites, the canopy may become sufficiently dense, that it intercepts nearly all of the sunlight falling on it, leaving the ground heavily shaded. When this happens, there would obviously be no advantage to the trees to increase their leaf biomass any further, because there would be no additional sunlight for those leaves to intercept. This will happen only on sites which are so plentifully supplied with water and nutrients that there is more of both available than the trees can use.

On drier or less fertile sites, the root systems of the trees will continue to expand until they can gather all of the water and nutrients available to them from the soil. Once this happens there will be no opportunity for the trees to increase their leaf and fine-root biomasses any further. Under these conditions, the canopy will be less dense than on sites where there is more water and nutrients available than the trees can use.

It follows from this discussion that there will be a limit to the biomasses of the principal living tissues which can be supported on any site. This limit will be determined by the availability of sunlight, water or nutrients. In fact, whichever of those three is in least supply will determine the limit for a particular site. From a forestry point of view, this limit is extremely important. It will determine the amount of photosynthesis that can occur on a site and so, ultimately, the amount of wood available for harvest.

One of the vital resources from the environment required for plant growth, carbon dioxide (Sect. 2.1.1), has not been mentioned as one of the factors which could limit plantation growth. The amount of carbon dioxide available does not depend on the properties of a site, because carbon dioxide is obtained from the air. Fresh supplies are brought continuously to any site with the wind. The last environmental factor which affects plant growth, temperature (Sect. 2.1.3), is most important in determining the *rate* at which plants grow, rather than the total amount of growth which can occur on a site.

2.3 An Example of Plantation Growth

The principles of plantation growth established in Sect. 2.2 can be illustrated with an example. This is taken from an experimental plantation of flooded gum (*E. grandis*) in southeastern Queensland, Australia. Flooded gum is an important plantation species in subtropical, eastern Australia and elsewhere in the world. The experimental details and some results from the first few years of growth were given by Cromer et al. (1993a, b). To provide the information for this example, I used those results with a forest growth modelling system (adapted from the system of Running and Coughlan 1988 and Running and Gower 1991) to predict how the experimental plantation would have grown over 20 years.

The results are shown in Fig. 2.1. Note that they are shown as **stand** results, that is, as the weight of oven-dry biomass of all the trees per unit ground area they occupy (the units used are tonnes of oven-dry biomass per hectare of ground area). Stand is a peculiarly forestry term, which refers to a more or less homogeneous group of trees in a forest in which an observer might stand and look about him or her.

2.3.1 Leaf Development

Figure 2.1 shows that, over the first 3 years, the stand leaf biomass increased steadily until it reached a maximum approaching 8 t/ha. This was the period during which the small planted seedlings were growing and expanding their crowns, until they formed a continuous canopy over the entire plantation area. Thereafter, the leaf biomass stayed more or less constant from year to year, with an average biomass of about 7.6 t/ha. That is, after 3 years of age, the canopy of the plantation had reached its limit of leaf biomass. That limit will have been determined by the availability at the site of water, nutrients or sunlight (Sect. 2.2). In this case, it was probably sunlight; the plantation was growing in a region with a high annual rainfall (1,440 mm/year) and had been heavily fertilised to ensure there was an adequate supply of mineral nutrients.

This development of leaf biomass was matched by a corresponding development of the surface area of the leaves. This is illustrated in the bottom section of Fig. 2.1, which shows the change with age in **leaf area in-**

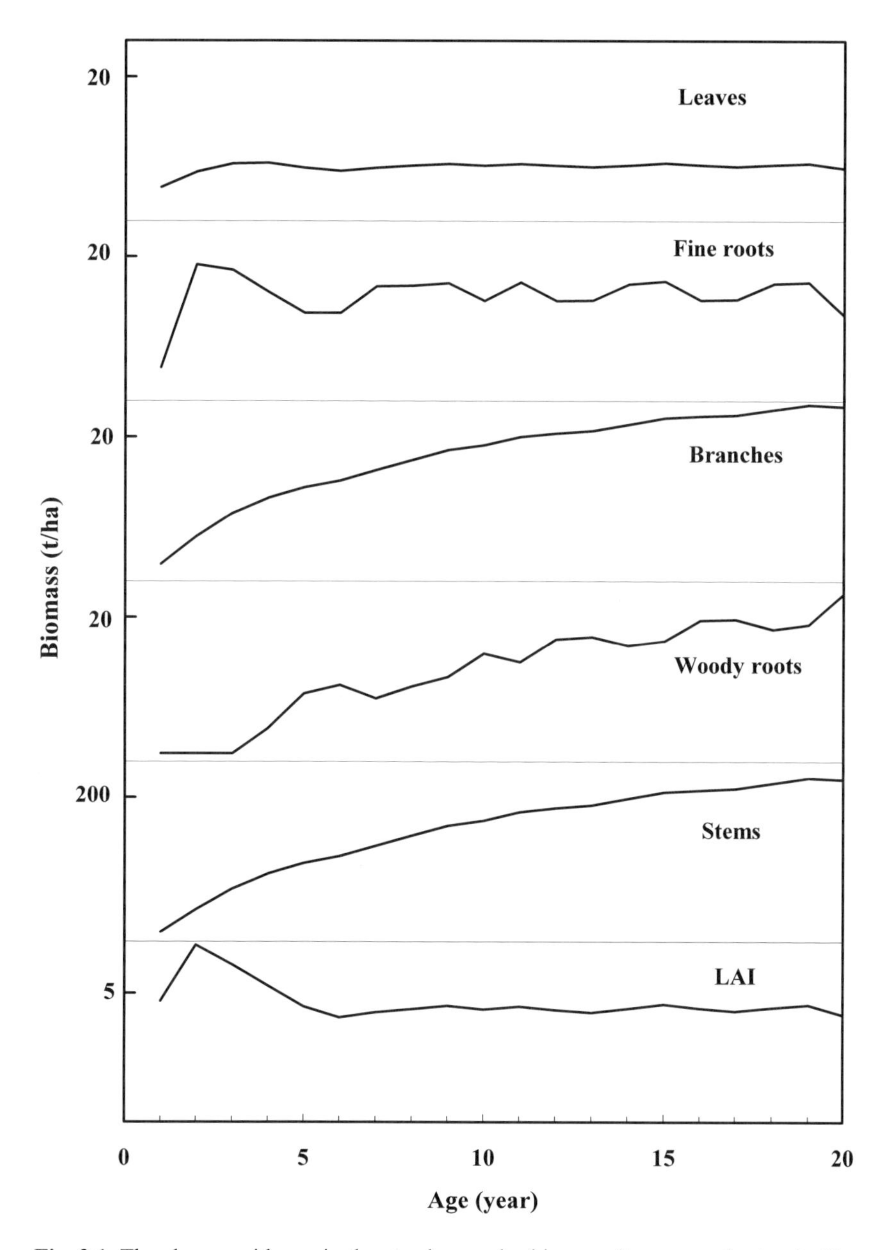

Fig. 2.1. The change with age in the stand oven-dry biomass (tonnes per hectare) of leaves, fine roots, branches, woody roots and stems and of the stand leaf area index (*LAI*, square metres per square metre) of an experimental plantation of flooded gum (*Eucalyptus grandis*) established in southeast Queensland, Australia (derived by the present author, based on results of Cromer et al. 1993b for plantations which had been fertilised heavily)

dex (often abbreviated to LAI) of the stand. Leaf area index is the area of the leaves of the canopy expressed per unit area of the ground they cover (the units used in Fig. 2.1 are square metres of leaf area per square metre of ground area covered by the canopy). Leaf area is defined as the area of the shadow which the leaves would cast if they were laid flat and lit vertically from above; by using a shadow area, scientists can define readily the areas of leaves with either flat surfaces or needle shapes. There are various instruments and techniques available to measure the leaf area index of plantations at any time during their life. Texts on forest measurement should be consulted to learn more about these methods (e.g. West 2004).

As can be seen in Fig. 2.1, the leaf area index of the plantation rose to a peak of 6.9 m^2/m^2 at 2 years of age, then declined to a more or less constant value, which averaged 4.4 m^2/m^2 over 5–20 years of age. A long-term, constant leaf area index of 4–6 m^2/m^2 is a figure typical for highly productive eucalypt plantations (Beadle 1997; Whitehead and Beadle 2004). Leaf biomass did not show a similar peak. The peak in leaf area index reflects the fact that leaves of very young eucalypt seedlings are often much thinner than leaves of more mature trees (Doley 1978; Linder 1985; Beadle and Turnbull 1986; Cromer and Jarvis 1990; West and Osler 1995; Whitehead and Beadle 2004); the balance between increasing leaf biomass with age and increasing thickness of the leaves leads to the peak in leaf area at 2 years of age in Fig. 2.1. Such a peak often seems to occur in plantations at about the age the individual tree canopies meet to form a closed canopy over the whole plantation (Pinkard and Beadle 2000).

Results have been shown here for leaf area index and leaf biomass because both are important in understanding how the canopy of a plantation develops. Leaf biomass represents the amount of living leaf tissue in which plant metabolism can occur. Leaf area index is important because it is the surface area of the leaves which intercepts the sunlight falling on the canopy. The larger the leaf area index of a stand, the more sunlight will the leaves be able to intercept and, hence, the more photosynthesis will they be able to undertake. In fact, considerable research has shown that the rate of growth of forests is directly proportional to the amount of sunlight their leaves intercept (Atwell et al. 1999, p. 405 et seq.; Landsberg and Gower 1997, p. 136). This fact has been used as the basis of many mathematical modelling systems which have been, and continue to be, developed to predict how forests grow (Bartelink et al. 1997; Le Roux et al. 2001); it was an important part of the model system which I used to obtain the results in Fig. 2.1.

The leaf area indices reached by plantations of different species vary quite widely. Beadle (1997) compiled a list, from research studies throughout the world, where leaf area indices of different **softwood** and **hardwood** plantations varied over the range 2–11 m^2/m^2. The variation between different species occurs because it is not just the area of the leaves which determines the amount of sunlight they will intercept, but it is also the way they are distributed in the canopy. In some species, the leaves are held rather more in clumps, whilst in others they are spread evenly throughout the canopy. In some species the leaves hang more vertically than in others. These leaf arrangements will affect the amount of sunlight they can intercept, which, in turn, will determine the leaf area index required to intercept the sunlight available (Fleck et al. 2003; Niinemets et al. 2004). Thus, it is the arrangement of leaves in the canopy as well as the availability of water, nutrients and sunlight which will determine the eventual leaf area index achieved by any particular plantation. In later parts of this book, there will be more discussion of leaf area index and its importance for plantation silviculture.

2.3.2 Root, Branch and Stem Growth

It can be seen from Fig. 2.1 that the biomass of the fine roots followed a pattern similar to that of leaf biomass. Fine-root biomass eventually reached a more or less steady amount, of nearly 15 t/ha, after about 5 years of age. This is consistent with the earlier discussion (Sect. 2.2) that fine roots will reach a steady, long-term stand biomass so they can continue to supply water and nutrients to the steady, long-term stand leaf biomass.

By comparison with the leaves and fine roots, the stand biomasses of woody roots and branches continued to increase continuously with age, well after the stand biomasses of leaves and fine roots had become more or less steady. Their need to do this follows from the fact that some trees die as the plantation grows (Sect. 2.4). Despite the loss of those trees, the biomasses of both the leaves and the fine roots in the plantation will remain the same; their amounts will continue to be determined by the availability of sunlight, water and nutrients from the site.

To make up for the deaths, each surviving tree will need to increase its leaf and fine-root biomass. As well, it will need to spread further both its crown and its woody root system, to support the additional leaves and fine roots. The spread will occur into gaps left by the dead trees. Because of the spread, both branches and woody roots will have to become longer. Engineering theory shows that, as this happens, they will need to become disproportionately larger in diameter, hence biomass, to maintain the strength they need to support the weight of the leaves or to ensure the tree remains anchored securely in the ground. The increase in branch and woody root biomass across the whole plantation will more than offset the corresponding biomass lost through tree deaths. Thus, both branch and woody root stand biomasses will continue to increase with time.

The stand biomass of stems also continued to increase continuously with time. This occurred because trees grow continuously in height. The diameter of their stems, and hence their biomass, will have to increase also, or the stem will have insufficient strength to support the tree upright (Sect. 2.1).

2.3.3 Growth Variations and Leaf and Root Turnover

There are several other features of the results of Fig. 2.1 which are worth noting. Firstly, it is obvious that the long-term leaf and fine-root biomasses were not exactly constant from year to year. Nor were the increases in branch, woody root and stem biomasses consistent from year to year.

These variations reflect the fact that weather conditions vary from year to year. If one year is slightly warmer than another, it might be expected that growth rates might be a little different in that year than in the other. If rainfall was particularly low in one year, there might be a shortage of water available from the soil, at least for some part of the year. The plantation would respond to the lack of water by losing some of its leaves and reducing its leaf area index to a value consistent with the reduced water availability; fine-root biomass would be expected to change accordingly in that year, to correspond with the change in leaf biomass. Almeida et al. (2004) have given an interesting example for flooded gum plantations in Brazil, which illustrates how variable plantation growth can be from year to year as weather conditions vary.

Secondly, it should be realised that the leaves and fine roots which make up the leaf and fine-root biomass are not the same leaves and fine roots all the time. After the canopy has reached its maximum size, leaves are shed from its more shaded base and progressively replaced by new leaves at its better lit top. There are several reasons for this continual turnover of leaves as follow:

- On highly productive sites, such as the one considered in this example, the leaves absorb most of the sunlight falling on the canopy. Thus, shaded leaves near the base of the canopy no longer receive sufficient sunlight for them to carry out photosynthesis. As they then no longer have any use, they are shed by the tree and replaced by new leaves near the well-lit top of the canopy.
- On less productive sites, where the availability of nutrients from the soil is relatively low, trees may recycle nutrients from the more shaded leaves near the base of the canopy to better-lit leaves near the top of the canopy. Those leaves near the top are then positioned better

to carry out photosynthesis than shaded leaves. The now nutrient deficient leaves near the base of the canopy would then be shed.

- Living tissue, such as leaves, has a limited lifespan. For leaves this is often around 2–3 years, but may be as long as 10–12 years (Muukkonen and Lehtonen 2004; Harlow et al. 2005; Muukkonen 2005), and they are shed and replaced by new leaves after that time. The turnover of leaves may vary seasonally also, so that the total biomass of leaves on a tree will differ from season to season during a year (Sampson et al. 2003; Roig et al. 2005). Of course, in deciduous forests leaves have a lifespan of only 1 year and are all shed and replaced annually.

Fine roots too have a limited lifespan, which can be as short as a few months (Santantonio and Santantonio 1987; Fahey and Hughes 1994; Rytter and Rytter 1998; Mäkelä and Vanninen 2000; Baddeley and Watson 2004; Kern et al. 2004; Tingey et al. 2005). Thus, they are continually turned over, as some die and are replaced by new ones. Because the lifespan of fine roots is generally much shorter than that of leaves, a much higher proportion of the food produced by trees from photosynthesis is used in continually turning over their fine roots than is used in continually turning over their leaves.

After they have been turned over and replaced, the dead leaves and fine roots rot away in the soil. As they do so, the nutrients they contain are returned to the soil. The living fine roots then take up those turned-over nutrients from the soil and make them available to new leaves and new fine roots as they develop. The importance of this process of nutrient cycling will be discussed in Sect. 6.3.

2.4 Growth of Individual Trees

The discussion in Sect. 2.3 concentrated on the growth of plantations as a whole, that is, when the total plantation growth is considered over all the individual trees which make up the plantation. From a forestry point of view, this is obviously important: on sites where more of the resources necessary for growth are available to the trees, plantations will produce more wood than on sites where lesser amounts of resources are available.

However, it is not only the total amount of wood that is produced by plantations which is important to forestry. Wood is sold from plantations as logs, which have been cut from the stems of the individual trees. There are certain minimum sizes logs must have before they are large enough to be sawn to produce any of the various types of sawn **timber** (or lumber as it is termed in the USA) used in building and for many other purposes

(Sect. 3.3). Larger logs are able to produce larger timber sizes, which generally attract higher prices at sale. Hence, it is not only the total amount of wood produced by a plantation that is important in determining its value, but also the sizes of the stems of the individual trees in the plantation. This section describes how individual trees grow in plantations and how their sizes are determined.

After planting, each tree seedling in a plantation starts to grow using the sunlight, water and nutrient resources available in its immediate vicinity (Sect. 2.2). However, different seedlings will grow at somewhat different rates for two reasons. Firstly, each seedling has its own **genetic** characteristics which will determine its inherent growth capability (von Wuehlisch et al. 1990). Secondly, the availability of water and nutrients from the soil may vary quite appreciably from metre to metre across the site (Thomson 1986; Phillips 2001; Guo et al. 2004; Hutchings and John 2004; Phillips and Marion 2004; Roy et al. 2004). These micro-site variations occur for several reasons:

- Small-scale variation in the topography of a site influences how water moves through the soil and just how much is available to a tree at any particular point.
- Variation in the soil parent material (the underlying rock from which soil is formed) may influence nutrient availability at any spot.
- Vegetation which grew on the site before the plantation was established may have affected the site, the effects of which will vary from spot to spot.

By the time the seedlings have grown sufficiently large that their canopies and root systems have spread to make contact with neighbouring trees, these differences in individual growth rates will ensure that some are already taller and larger than others. At that stage, the trees start to interact with each other and to compete for the sunlight and soil resources available for growth (Sect. 2.2).

Considerable research has been undertaken to understand how trees compete with each other in monoculture plantations. Above ground, the principal competitive process is for taller trees to shade smaller trees and deny them sunlight for growth (Weiner and Thomas 1986; Hara 1986a, b; West et al. 1989; Schwinning and Weiner 1998). This type of competition is known as asymmetric competition; since taller trees can shade smaller ones, but not vice versa, the taller plants obtain a disproportionately large share of the available sunlight (Schwinning and Weiner 1998; Park et al. 2003). Below ground, there is symmetric competition between the trees for water and nutrients; since the roots of each tree occupy a volume of the soil which is proportional to the size of its root system, each tree can take up amounts of water and nutrients which are proportional to its size and, hence, its metabolic needs.

The result of the asymmetric competition for sunlight is that taller trees are able to grow disproportionately faster than smaller trees. This will lead to an ever-increasing range of tree sizes within the plantation. Eventually, some of the smaller trees will be shaded so heavily that they will be unable to continue to grow and will die.

The effects of this asymmetric competition are illustrated in Fig. 2.2. It shows results, for a plantation of flooded gum in subtropical eastern Australia, of the frequency distribution of individual tree stem diameters at **breast height** over bark (forest scientists conventionally measure tree stem diameter at breast height, which is 1.3 m above ground, or 1.4 m in some countries). Results are shown at three different ages of the plantation (it is not the same plantation as that in Fig. 2.1, but the results were determined using other information available to me for such plantations).

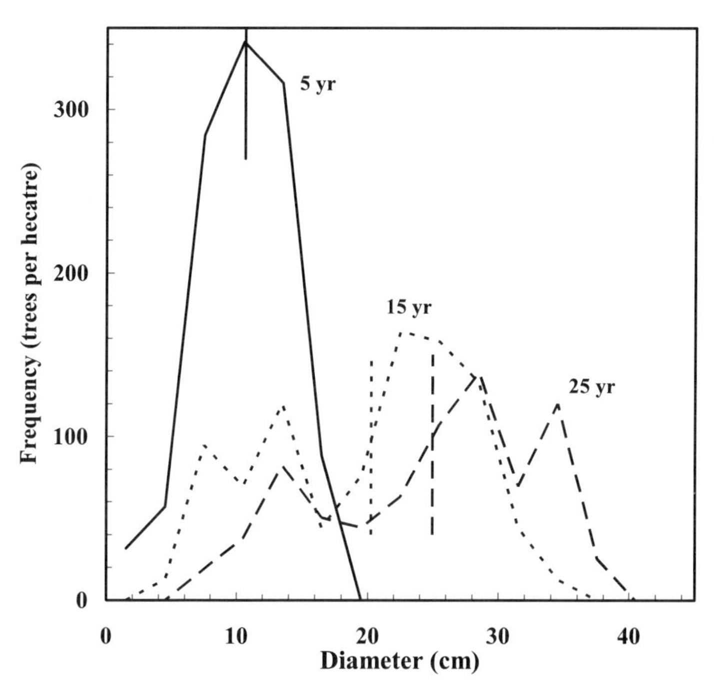

Fig. 2.2. The frequency distribution, at 5 (——), 15 (- - -) and 25 (— —) years of age, of stem diameter at breast height over bark in a plantation of flooded gum (*E. grandis*) growing in subtropical, eastern Australia. The *short, vertical lines* show the average diameter of the trees in the plantation at each age (derived using an unpublished plantation growth and yield model developed by the present author)

Frequency distributions are used generally in science to show the distribution of sizes amongst a group of things (stem diameters amongst a group of trees in this case). Tree stem diameter is shown here because it is used commonly in forest science to represent the overall size of any tree; research by forest scientists over many years has shown that there is a close correlation between the overall biomass of trees and their stem diameters (West 2004).

The results of Fig. 2.2 illustrate how the competitive processes in a monoculture plantation affect the development of individual trees. On average, the trees increased in size as the plantation grew; their average diameter increased from 11 cm at 5 years of age, to 20 cm at 15 years of age and then to 25 cm at 25 years of age. At 5 years of age, the plantation contained 1,119 trees per hectare. As larger trees suppressed smaller ones through asymmetric competition, and some of the smaller trees died, this **stocking density** was reduced to 929 trees per hectare by 15 years of age and then to 759 trees per hectare by 25 years of age.

What is most striking in the figure is how the spread of diameters was affected as the plantation grew. At 5 years of age, the range of diameters was 1–17 cm. This had spread to 6–34 cm by 15 years of age and, by 25 years of age, even further to 7–39 cm. This reflects the disproportionately larger growth rates of the taller trees as they shaded the smaller ones.

3 Growth Rates and Wood Quality

Substantial effort and expense are involved in the establishment and **management** of forest plantations. There would be no point in growing them unless they provided something of value to the plantation grower, either as a financial return or otherwise. As discussed in Sect. 1.2, the value of plantations may arise from the wood products they produce or the environmental benefits they confer. However, in keeping with the general approach of this book, this chapter will concentrate on wood production.

From this perspective, several things determine the financial return which an investor can expect from a forest plantation:

- The rate at which it grows—the faster it grows, the earlier can the plantation be harvested and the larger will be the quantities of the products which can be harvested from it.
- The quality of the wood it produces—that is, the better suited is the wood for the products to be produced ultimately from it (Sect. 3.3), the more readily will the market be willing to purchase it and the higher the price will it pay.
- The costs involved in establishing and managing it—these are determined by the labour market in the region where the plantation is being grown, the prices of the things needed for establishment and management (seedlings, fencing, fertiliser and so on) and the costs of the research and development the grower has undertaken.
- The prices received for the products which are sold from it—these are determined both by their quality and by the general economic circumstances of the market for timber within which the plantation operates.

The economic issues which surround the costs and returns involved in plantation enterprises, as alluded to in the last two of these dot points, are more properly the realm of text books on forest **economics** or management (e.g. Davis et al. 2001). This chapter will not consider these economic issues further, but will concentrate on the biological issues of the growth rates of plantations and on the quality of the wood which is produced from them.

To assess growth rates, a plantation owner will need some way of measuring and expressing them. Section 3.1 discusses the ways in which this is done in forestry. An owner will usually wish to assess how well a plantation is growing in relation to other plantations in the region, or indeed to other plantations elsewhere in the world; this will tell him or her how well the plantation is performing in comparison with the best practices used elsewhere. Plantation growth rates around the world are discussed in Sect. 3.2. Many different characteristics of wood determine its suitability for different uses. Section 3.3 describes those characteristics which are most important in determining its quality. These issues of growth rate and wood quality, and how they are influenced by plantation silviculture, will be referred to continuously in later parts of this book.

3.1 Expressing Growth Rates of Plantations

The rate of growth of trees in a plantation depends on the species planted, the environmental circumstances of the site on which it is planted and the silvicultural practices which are employed. Much of the substance of this book is concerned with these three issues.

Certain conventions are used in forestry science to describe forest growth rates and these will be described here using an example, again for flooded gum (*Eucalyptus grandis*) plantations in subtropical eastern Australia (Sects. 2.3, 2.4). Figure 3.1 is reproduced from my textbook on forest measurement (West 2004). Figure 3.1a shows the change with age in the wood volume contained in the stems of the trees of a typical flooded gum plantation stand. The shape of this growth curve is typical for forests and, indeed, for many other biological organisms; the growth rate increases with time initially (up to about 15 years of age in the figure) and declines steadily thereafter.

Given a growth curve like that of Fig. 3.1a, stand growth rate is usually expressed in one of two ways. The first is called **current annual increment** (often abbreviated as CAI and also termed periodic annual increment, or PAI). It is the immediate growth rate of the stand at any age. It changes with age and is illustrated by the solid line in Fig. 3.1b. It shows the increase in growth rate, to a maximum at about 15 years of age as mentioned in the preceding paragraph, with a progressive decline thereafter.

The pattern of change with age in current annual increment evident in Fig. 3.1b is typical for plantation forests anywhere in the world (and indeed for many **native forests**). The period whilst the growth rate is increasing, up to 15 years of age in the example, is the period when the trees are growing to the stage that the leaf and fine-root stand biomasses become more or less constant (Sects. 2.2, 2.3). The steady decline in growth rate after the maximum is reached is rather difficult to explain. It is obviously

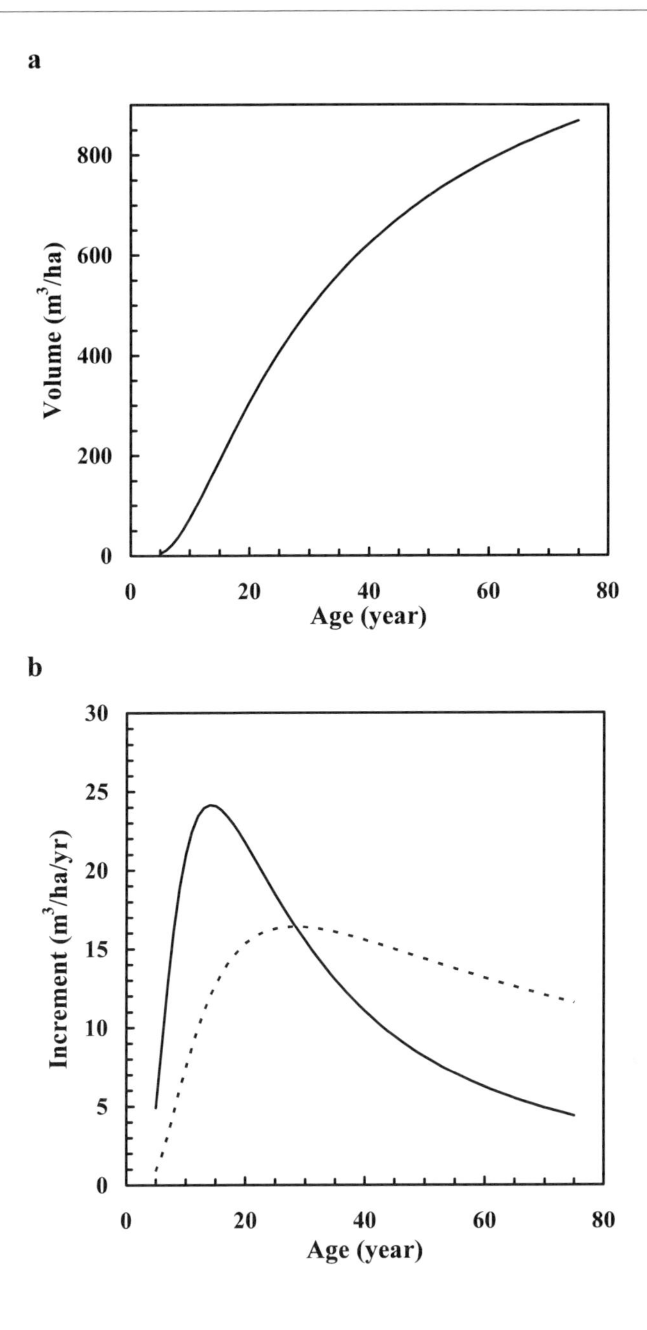

Fig. 3.1. a Change with age in the wood volume contained in the stems of the trees of a typical flooded gum (*Eucalyptus grandis*) plantation stand in subtropical eastern Australia. **b** Change with age in current annual increment (——) and mean annual increment (- - - -) in stand stem wood volume for the stand shown in **a** (reproduced from Fig. 8.3 of West 2004)

of great interest to forestry. If it did not happen, if growth continued at a constant rate instead of declining, forests would yield much more timber at much earlier ages.

One theory to explain the decline in growth rate is that, as trees become larger with age, the path for water transport from the roots to the leaves becomes longer and more tortuous. This is due both to the trees growing taller and to some of the pipes (Sect. 2.1.1), which transport water up the roots, stem and branches, becoming blocked, so water must detour through other pipes. Both these things lead to a higher level of water stress in the leaves, which then reduce their degree of stomatal opening. In turn, this reduces the amount of photosynthesis a tree can carry out and, hence, its growth rate.

This theory remains controversial (Gower et al. 1996; Ryan et al. 1997; Murty and McMurtrie 2000; Smith and Long 2001; Zaehle 2005). An alternative has been proposed by Binkley (2004). He suggested that competitively less successful trees continue to use resources from the site (light, water and nutrients) for their growth and maintenance. However, because they are shaded by more successful competitors (Sect. 2.4), they are unable to use those resources to produce new biomass as efficiently as the more successful competitors (but see Reid et al. 2004). Over a stand as a whole, this means that less biomass will be produced per unit of resource available to produce it; in consequence, stand growth rate will decline. Alternatively again, Niinemets et al. (2005) have suggested that leaves in the canopies of older trees of some species are arranged to absorb sunlight less efficiently than in younger trees, hence reducing their photosynthetic capability and growth rate. Much further research is necessary to investigate further the causes of the decline in growth rate of trees with advancing age.

A second method used to describe stand growth rate in forestry is called **mean annual increment** (often abbreviated as MAI). This is the average rate of production to any particular age of the stand. It is determined simply as the growth (stand stem wood volume in the example) divided by the age at which the growth is measured. It is probably the most popular measure used by foresters to indicate how fast a forest grows. Mean annual increment changes with age during the life of the forest; it is illustrated by the dashed line in Fig. 3.1b.

Because both current and mean annual increments change with age, it is important to mention what age is being referred to when using them to describe the growth rate of any plantation. It is common also when referring to growth rates, that foresters will refer to the amounts of wood which can be harvested and sold from a forest (often called the merchantable volume), rather than to the total stem wood volume. These amounts may even be subdivided further into particular log size classes (Sect. 2.4). If this is being done, it is obviously important that the log class sizes be defined carefully, so it is quite clear what wood volumes are being referred to when expressing a plantation growth rate. Of course growth rates of things other than stem wood volume, things like stand biomass, are often considered by forest scientists; the same conventions are used to describe their growth rates.

3.2 How Fast Do Plantations Grow?

Perhaps surprisingly, there is a rather limited amount of published information available to give a very reliable answer as to just how rapidly plantations around the world can grow. Pandey (1983) collated a large amount of growth data from the tropics. No other major data collation seems to have been attempted more recently, although individual reports on particular species in particular parts of the world have been published since then.

Figure 3.2 shows results from both Pandey's work and some other reports. Only measurement data from rapidly growing plantations of hardwood species have been included there. The figure shows how mean annual increment in stand stem wood volume of live trees (inclusive of any volume removed during the life of a stand by **thinning** the stand, that is by removing some of the trees from the stand—Chap. 8) varies with age in those data. In all cases, the data were obtained from plantations which had been planted with a stocking density more or less normal for commercial plantations grown for wood production, say in the range 800–2,500 trees per hectare. Some of the plantations had been measured on several occasions and the trajectories of growth of those are shown as solid lines. Some had been measured once only and their results are shown as single points. The results include data from plantations in Africa, India, Asia (Indonesia, Philippines), Oceania (Australia, New Zealand), Hawaii and South America. Most of the data are for eucalypt species, with some from other hardwood species (*Acacia dealbata*, *Falcataria moluccana*, *Gmelina arborea* and *Tectona grandis*). Figure 3.3 illustrates one of the plantations.

The results suggest that the mean annual increment of the fastest growing hardwood plantations increases rapidly to a maximum of about 60 m^3/ha/year at about 6–7 years of age, then declines steadily to just over 30 m^3/ha/year at 30 years of age. I positioned the dashed line in Fig. 3.2 to represent an upper limit to hardwood plantation growth rates around the world. It might be considered as an approximate world maximum, by which the growth rate of a hardwood plantation can be judged. Results for some particularly fast-growing species are identified specifically in the figure; those shown for flooded gum (*E. grandis*) in Australia (Cromer et al. 1993a) are from the example plantation used in Sect. 2.3.

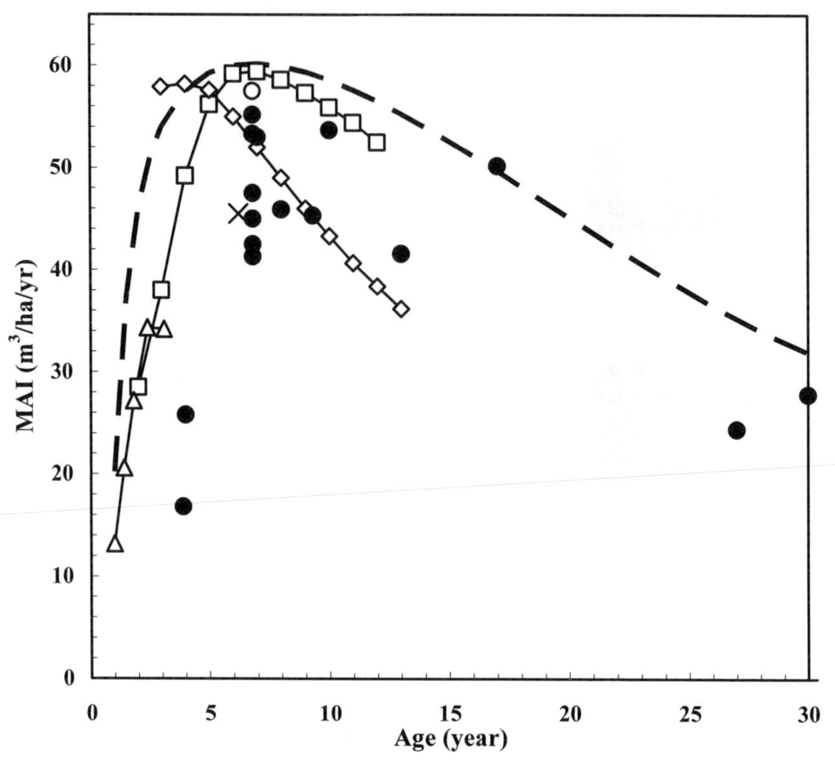

Fig. 3.2. For fast-growing hardwood plantation forests around the world, a scatter plot of published values of mean annual increment (*MAI*) in stand stem wood volume of live trees, inclusive of any volume removed at thinnings, against age. Plantations for which information was reported at several ages are shown as *continuous lines* and those with information for a single age as *filled circles*. Specifically identified plantations are (×) *E. grandis* in Hawaii, USA (see Fig. 3.3), (O) *E. dunnii*, *E. saligna* and *E. smithii* in Africa (each with the same result) (Schönau and Gardener 1991), (△) *E. grandis* in Australia (Cromer et al. 1993a), (□) *Falcataria moluccana* (formerly known as *Albizia falcataria*) in Indonesia (Pandey 1983) and (◇) *Gmelina arborea* in Africa (Pandey 1983). The *dashed line* was positioned by eye by the present author as an approximate world maximum for hardwood plantations (Sources—Bradstock 1981; Pandey 1983; Frederick et al. 1985a, b; Beadle et al. 1989, 1995; Schönau and Gardener 1991; Birk and Turner 1992; Cromer et al. 1993a; J.B. Friday 2005, personal communication)

Figure 3.4 shows similar data, but for softwood plantation species. The results include data from plantations in Africa, India, Asia (Indonesia), Central and South America and Australia. Most of the data are for species of pine (*Pinus*) with some from two other softwood species (*Cryptomeria japonica* and *Cupressus lusitanica*).

Fig. 3.3. An example of a hardwood plantation forest which is growing relatively rapidly by world standards. It is a 6-year-old plantation of flooded gum (*E. grandis*) growing in Hawaii, USA; it is shown as the symbol (×) in Fig. 3.2. The stand contained 1,210 trees per hectare. Their average diameter at breast height over bark was 18 cm and their average height was 20 m (measured by J.B. Friday, University of Hawaii) (Photo—P.W. West)

The world maximum for softwood plantations, shown by the dashed line in Fig. 3.4, suggests that the growth pattern for fast-growing softwood plantations is somewhat different from that for hardwoods. Early growth of softwoods rises to a maximum mean annual increment of nearly 50 m^3/ha/year at about 15–17 years of age, 9–10 years later than the age at which the maximum for hardwood plantations is reached. It then declines steadily to just over 35 m^3/ha/year at 30 years of age.

These world maxima should be useful guides for plantation growers to indicate how well their plantations are performing in relation to the fastest growing plantations in the world. However, a number of things must be borne in mind when using them:

- They refer to plantations being grown for timber production and established at stocking densities of about 800–2,500 trees per hectare. Plantations grown under other circumstances may have higher or lower growth rates. For example, one of the highest values of mean annual increment reported for plantations in the scientific literature is a value of about 100 m^3/ha/year, to 3 years of age, for the stem wood volume of a small experimental plantation of swamp gum (*E. ovata*) growing in New Zealand (Sims et al. 1999). This is a growth rate far

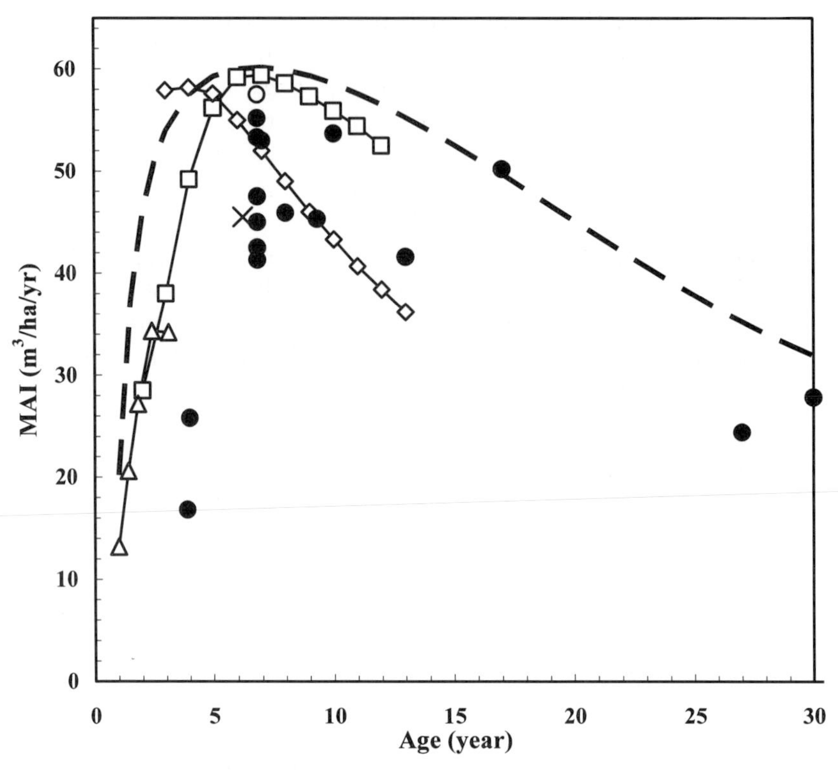

Fig. 3.4. Results, similar to those of Fig. 3.2, for fast-growing softwood plantation forests around the world. Specifically identified plantations are (O) *Cupressus lusitanica* in Africa (Pandey 1983), (◇) *Pinus elliottii* in South America (Pandey 1983), (□) *P. patula* in Africa (Pandey 1983) and (△) *P. radiata* in Australia (Myers et al. 1996, 1998) (Sources—Lewis et al. 1976; Pandey 1983; Myers et al. 1996, 1998; Toro and Gessel 1999)

in excess of the value of about 54 m^3/ha/year, to 3 years of age, for the hardwood world maximum shown in Fig. 3.2. This was a plantation being grown for **bioenergy** production and there are several reasons why its growth rate was so much higher than those shown in Fig. 3.2; these are discussed in more detail in Sect. 5.5.1. No doubt world maxima for growth rates of bioenergy plantations will be developed as more and more growth data are published from them; they are likely to be well above the maxima shown in Figs. 3.2 and 3.4.

- Different species will have different patterns of growth rate over their lifetimes. Some will grow rapidly early in their lifetime, perhaps approaching the world maximum for some years, then may slow their growth rate and fall below it. This growth pattern is illustrated by the results in Fig. 3.2 for *Cupressus lusitanica* in Africa. Other species may grow more slowly early in their lifetime, but then maintain a relatively high growth rate later in life and reach the world maximum at later ages. The results in Fig. 3.2 for *Pinus patula* in Africa illustrate that trend. The growth rate pattern that different species adopt can be very important when choosing which species is to be planted at a particular site. If the plantation is to be grown to produce only small trees, say for paper-making or for bioenergy, a species which grows rapidly early in its lifetime would be preferred, so that as much as possible of the desired product would be available as early as possible. However, if the plantation is being grown to produce much larger logs for sawing into timber, a species which maintains a higher growth rate later in its lifetime might be preferred.

- Plantations would be expected to grow at rates as fast as the world maxima only under very special circumstances. The temperature regime at the site would have to be close to an optimum which maximises the metabolic activity of the species concerned (Sect. 2.1.3). The site would need to have a plentiful rainfall (or perhaps even be irrigated), together with a soil capable of storing adequate moisture to ensure the trees always have a more than adequate water supply (Sect. 2.1.2). The soil would have to be very fertile (and perhaps supplemented by fertiliser), to ensure the trees have a more than adequate supply of nutrient elements (Sect. 2.1.4). Lastly, appropriate silvicultural practices must have been applied at the site to ensure the trees can avail themselves of all the resources available at the site. Few sites anywhere in the world have all these desirable characteristics and so most plantations can be expected to have growth rates somewhat below the world maxima.

3.3 Wood Quality

World markets for industrial wood (Sect. 1.1) are becoming increasingly discriminating of the quality of the wood that is sold to them. The principal industrial products produced from wood are (Bowyer et al. 2003):

- Timber—this consists of boards sawn from logs which have been cut from tree stems. Timber has a multitude of uses, for furniture, flooring, panelling, framing, boxes, pallets, railway sleepers and many other things. When cut fresh from the log, wood contains a lot of water. If it is put into service without drying, it will shrink and warp. To avoid this, usually it is dried before it is sold, either by being left out in the air for some months or by being heated for some hours in a kiln.

Wood is a rather variable material, which requires a high level of skill and advanced technology to ensure it is both sawn and dried successfully.

Larger pieces of timber can be manufactured from smaller pieces cut from smaller logs. These can be finger-jointed and glued together, to make longer boards, or many can be glued together to make very large laminated beams.

- Plywood—this is a panel product, manufactured by gluing together several layers of thin veneer. Plywood has many and varied uses, especially in building. Veneers of wood with attractive grain are often glued to panels to make feature wall panelling. Veneer is produced either by peeling, with a large blade, a thin layer from a log mounted in a lathe or by slicing strips from the log.
- Particle-based panels and timber—in manufacturing these, wood waste or small logs are broken down into small pieces which are then glued and pressed into large sheets or boards. The small pieces can be chips, flakes, shavings, sawdust or slivers, all of which can be used to make boards with different properties. Often these products can be made from the waste left after producing timber.
- Other fibre products—these include hardboards, insulation boards and others, where, in effect, pulp (see paper, below) is pressed and dried to form large panels.
- Round wood products—these include poles, posts, piles and others, where the tree stem is simply cut into lengths. Often these products are impregnated with preservative chemicals; these prevent rotting and decay when the products are in use, especially where they are sunk partly into the ground or are used in water. Other forms of timber which are to be exposed in use are often treated also with preservatives.
- Paper—where wood is first macerated mechanically or chemically to separate its individual fibres (Sect. 3.3.1) to make pulp. The pulp is then ground further and mixed with water to form a thin mat. The water is drained from the mat, which is pressed, rolled and dried to form paper. There is an enormous variety of papers, from strong cardboards for packaging to very high-quality papers for fine printing.

The properties which make wood most suitable for paper production, say, are rather different from those required of wood to be sawn to produce timber. The study of wood, its properties and its conversion to wood products is a specialised field of study, far too large to consider in detail in this book. Texts such as Desch and Dinwoodie (1996) and Bowyer et al. (2003) will provide introductions to the topic. However, the various silvicultural practices used in plantation forestry, which *are* the subject of this book, can affect appreciably the quality of the wood produced. Some of these effects are discussed in later chapters.

The remainder of this section will describe briefly some of the properties of wood which are important in determining its quality for industrial end uses. Only some of the more important properties are described; fuller discussion can be found in the texts just mentioned. To ensure that wood of appropriate quality is being produced in a plantation, its assessment in trees before they are felled is becoming increasingly important. Downes et al. (1997) have discussed in detail how this is done; although their book concentrates on eucalypts, its concepts will have application to many other plantation species.

3.3.1 Tracheid and Fibre Length

About 90–95% of the wood (xylem) of the stems of softwood trees is made up of cells (Sect. 2.1.1) known as tracheids. These are long (averaging about 3–4 mm, which is long by wood standards), thin (diameter about 0.025 mm) cells arranged with their long axes vertically in the stem. Like most cells of wood (Sect. 2.1.1), they are dead cells which have lost their cell contents and retain only their cell walls. They are joined one to the other by holes through their walls, known as pits. They provide the structural strength to softwood tree stems and allow water to be transported up the stem, passing from tracheid to tracheid through their pits (making the pipes referred to in Sect. 2.1.2). Softwood xylem does contain some living cells, known as parenchyma, which act as food-storage tissue for the tree and may produce resin. Some softwoods also contain relatively large, hollow resin ducts (Desch and Dinwoodie 1996).

The principal structural support of hardwood tree stems is provided by cells known commonly as fibres. These are anatomically similar to, and have much the same diameter as, tracheids in softwoods; however, they are much shorter (usually a little under 1-mm long). Depending on the species, fibres make up 15–60% of the wood of hardwoods. However, water transport through hardwood tree stems is principally through hollow cells known as vessels. Vessels are shorter than fibres (around 0.2–0.5-mm long), but have much larger diameters (varying over the range 0.02–0.4 mm). They are often large enough in diameter to be seen with the naked eye in a cross-section cut from a hardwood tree stem.

Vessels are commonly arranged one above another in the stem to make long pipes up which water can pass. Depending on the species, vessels make up 20–60% of the wood of hardwoods. The presence of excessive numbers of vessels in wood can affect the quality of paper (Higgins 1984;

Downes et al. 1997) and particle-based boards (Macmillan 1984), because they sometimes come loose from otherwise smooth surfaces. Hardwood stems also contain living parenchyma and may also have resin or gum ducts.

Figure 3.5 illustrates the relative sizes of vessels and fibres in hardwoods and tracheids in conifers. Fibre or tracheid length varies substantially between different tree species and also varies within the stem of individual trees, both vertically up the stem and horizontally across its cross-section (Higgins 1984; Hillis 1984; Downes et al. 1997; Raymond et al. 1998; Mäkinen et al. 2002c). The length of the tracheids or fibres is particularly important in paper-making. It affects the extent to which they weave together to form the final paper and, hence, its strength. Because softwood tracheids are longer than hardwood fibres, papers which require strength (such as packaging papers, cardboards or newsprint) tend to be made with high proportions of softwood pulp. Very high-quality printing papers often contain more hardwood pulp. Various other wood characteristics, not just fibre length, are also important in paper manufacture and some of these will be alluded to in the following subsections. However, paper-making is a complex process and many factors, not just wood quality, determine the final paper quality. Because other fibre products (hardboards, insulation boards and others) are produced from pulped wood, fibre or tracheid length can affect their properties also (Macmillan 1984).

3.3.2 Microfibril Angle

Microscopic examination of the walls of wood cells shows that they are made up of long strands called microfibrils, which are so small they can

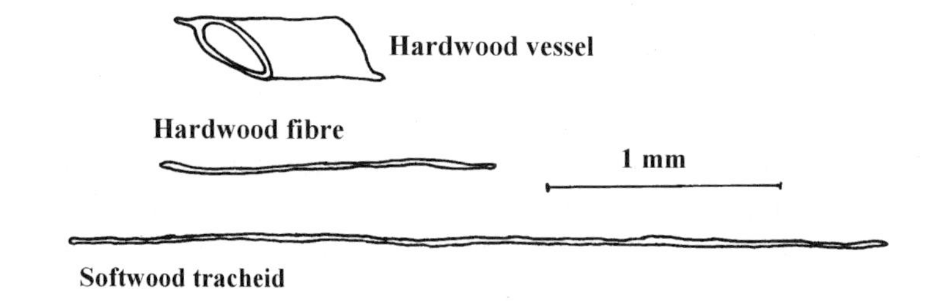

Fig. 3.5. Relative sizes of principle types of cells which make up the xylem of hardwoods and softwoods. Note that in the living tree, these cells are arranged parallel to the axis of the stem, branch or woody root (after Bowyer et al. 2003)

be seen only with the very high magnification of an electron microscope. In turn, microfibrils consist of many strands of a substance called **cellulose**. Chemically speaking, cellulose is a complex carbohydrate, consisting of many sugar molecules strung together in a long sequence. Microfibrils are bound together in the cell wall by a substance called lignin. Lignin has a complex chemical structure, which has not yet been fully determined. In effect, lignin can be thought of as being like cement and microfibrils as steel reinforcing rods embedded in it (Downes et al. 1997). Cellulose and lignin are the two most abundant organic chemicals (that is, chemicals relating to or derived from living organisms) on earth, because about 50% of the dry biomass of plants is composed of them.

The microfibrils are wound spirally around the cell wall. The angle, from the long axis of the cell, at which they are wound determines the strength of fibres and tracheids, particularly their resistance to bending; the smaller the angle, the stronger the cell. In tree stems, the angle is often greater nearer the centre of the stem and declines towards the outer parts of the stem cross-section (Hillis 1984; Evans et al. 2000; Jordan et al. 2005). Wood with a high microfibril angle tends to have appreciably less strength when sawn and suffers more distortion when it is dried (Desch and Dinwoodie 1996; MacDonald and Hubert 2002).

3.3.3 Wood Density

The density of wood (its weight per unit volume) is the characteristic used perhaps most frequently to assist in the assessment of wood quality. It is usually expressed as **basic density**, which is the oven-dry weight of wood per unit volume of the undried (fresh) wood.

The dimensions of the cells which make up wood, that is, the thickness of their cell walls and the size of the empty space the walls surround, determine the **wood density** at any point in the stem. In hardwoods, the relative frequency of larger-diameter vessels when compared with smaller-diameter fibres can affect wood density. The average basic density of the wood in stems varies greatly from tree species to tree species around the world, over a range of about 200–1,200 kg/m^3, the range being appreciably wider in hardwoods than in softwoods (Desch and Dinwoodie 1996).

Within the stems of individual trees, basic density can vary considerably, both with height in the stem and across the stem cross-section. In some species, density increases from the inner to the outer part of the stem section, perhaps by as much as 100–150 kg/m^3. In other species, the reverse occurs, and in yet others, there is little change across the section.

Woodcock and Shier (2002) have advanced the interesting hypothesis that increases in wood density across stem sections tend to occur in species which grow rapidly in full sunlight. The reverse occurs in species which are adapted to grow more slowly, in the shade below the canopy of other species. They suggested that the rapidly growing species produce low density wood initially, until they become tall enough to be at risk of blowing over or snapping off in the wind. They then produce denser wood on the outer parts of the stem to increase its strength, hence its ability to resist wind damage. The trees which grow more slowly in shade will already have much larger stems before they become tall enough to be at risk of wind damage and it is unnecessary for them to produce denser wood in the outer parts of the stem.

The species used for major plantation forestry developments around the world are commonly those which grow rapidly in full sunlight and, hence, are mainly species in which basic wood density tends to increase towards the outer parts of the stem. Good examples of this have been given for Sitka spruce (*Picea sitchensis*) in Great Britain (MacDonald and Hubert 2002), Scots pine (*Pinus sylvestris*) in Finland (Kellomäki et al. 1999), radiata pine (*Pinus radiata*) in Chile (Zamudio et al. 2002) and shining gum (*E. nitens*) in New Zealand (Evans et al. 2000).

These works show also how density varies with height up the tree stem in these species. The stem cambium (Sect. 2.1.1) is progressively younger further up the stem (simply because trees grow progressively in height) and younger cambium produces less dense wood; in line with the engineering concepts of Woodcock and Shier's theory, there is less need for stronger wood near the tip of the tree, because the bending stresses exerted there by the wind are less than near the base of the tree. Often, there seems to be markedly less dense wood produced by the cambium for about 6–10 years and then density increases in outer layers of wood beyond that. This column of less dense wood up the centre of the tree stem is often termed juvenile wood. More recent work (Burdon et al. 2004) has considered how both the age of the cambium and the stage of maturity of the tree affect the development of wood, both across the stem cross-section and with height up the stem.

Not only does wood density tend to vary generally both across the stem cross-section and with height up the stem, it also varies at a very small scale across the stem cross-section because of seasonal growth of the trees (Bowyer et al. 2003). When seasonal conditions are most favourable for growth, when the weather is warmer or wetter, tree stems grow faster and less dense wood is produced; this is commonly called **earlywood** (or sometimes springwood). When conditions are less favourable, growth is slower and denser wood is produced; this is termed **latewood** (or summerwood). The alternation of earlywood and latewood in tree stems makes their growth rings.

In some species, in cooler climates especially, growth rings are produced very regularly as the seasons change, one ring per year. Their ring widths can be used to estimate changes in weather conditions from year to year to reconstruct past climates (a discipline known as dendrochronology). In other species, and especially in more tropical areas where seasonal weather changes are not so marked, growth rings are still produced, but not necessarily at clearly defined annual intervals.

The small-scale differences in basic density across growth rings can be very large. Evans et al. (2000) used X-ray equipment to measure basic density at less than 0.5-mm intervals across stem sections of 15-year-old, plantation grown shining gum (*E. nitens*) trees. They found that basic density differed by as much as 700 kg/m^3 between the least dense and densest part of some growth rings. Similarly, Bouriaud et al. (2004) found differences as large as 600 kg/m^3 across stem sections of 55-year-old trees of the hardwood beech (*Fagus sylvatica*), growing in native forests in France.

Wood density is considered an important characteristic to help define wood quality because it correlates reasonably well with a number of the characteristics important in various end uses of wood. Denser wood tends to be stronger and stiffer, properties important for its use in construction. There is an increasing tendency around the world to harvest fast-growing, commercial plantations at younger and younger ages. This means that the wood will tend to contain an increasing proportion of less dense, juvenile wood. Yang and Waugh (1996a, b) found that logs from 19–33 year old plantation-grown eucalypts could be sawn to produce high-quality structural timber, but the average strength of the sawn boards was lower in younger trees; hence, it was slightly less valuable commercially. Bowyer et al. (2003) have discussed the variety of problems that the presence of juvenile wood can lead to when processing timber.

The thicker cell walls of denser wood tend to contain larger amounts of water than the cell walls of less dense wood. This makes denser wood subject to more shrinkage when it is being dried in kilns for use in service (Desch and Dinwoodie 1996; Hillis 1984; Blakemore 2004). As well, the need to remove more water from it makes it more expensive to dry; however, because it is stronger, denser wood is less likely to suffer collapse of cells during drying and the development of small splits (commonly referred to as checks) in the dried wood (Ilic 1999; Oliver 2000).

The strength of particle-based panels and timber is determined more by the strength of the glue than by the strength of the wood; thus, less dense wood can be used to make panel products which are just as strong but are cheaper to press, cut and transport. Denser eucalypt wood has been found to be more difficult to peel when making veneer to produce plywood (Macmillan 1984).

Because it contains more wood per unit volume, denser wood tends to give a higher yield of pulp in paper-making. However, it is easier to collapse the thinner cells in less dense wood. This leads to better bonding between the cells when they are made into paper, which, in turn, leads to stronger paper (Higgins 1984; MacDonald and Hubert 2002).

3.3.4 Grain Angle

The orientation of fibres and tracheids (Sect. 3.3.1) within the stem is known as the grain direction. When they are aligned parallel to the axis of the stem, the wood is said to be straight-grained. Commonly, they tend to be at a slight angle and, more extremely, may even be spirally arranged around the stem. In some cases the direction of the spiral changes as a tree grows over several years, when the tree is said to have interlocked grain (Desch and Dinwoodie 1996; Bowyer et al. 2003). The grain direction may vary systematically across the stem cross-section in some species (Hillis 1984; MacDonald and Hubert 2002).

Localised variations in grain direction, together with the changes in colour which occur with annual growth rings (Sect. 3.33), can lead to highly attractive grain features. These can be very important for the manufacture of high-quality furniture or in other timber to be used as a decorative feature. Different sawing patterns can be used to emphasise these highly desirable features of some timbers.

For more prosaic uses of timber, such as in construction, timber strength tends to decrease as the grain angle increases and the degree of distortion during drying tends to increase (MacDonald and Hubert 2002). These effects can be exacerbated if the tree stem was bent during growth, because the angle of the grain in the sawn board will vary as the saw cuts straight through a bent stem; distortions during drying will then vary greatly at different distances along the board. For products where wood is reduced to small particles or is macerated (paper or particle-based panels and timber), grain angle is of little importance.

3.3.5 Sapwood and Heartwood

As the stem of a tree increases in diameter with age, it eventually contains more wood than is necessary to transport from the roots the water required by the leaves (Sect. 2.1.1). Older wood, closer to the stem centre, is then converted to what is known as heartwood, through which water is no longer transported. Sufficient sapwood is left in the outer part of the stem cross-section to provide the water-transport needs.

Heartwood is often darker in colour than sapwood. Its formation involves the deposition by the tree of a wide range of chemical substances known as polyphenols; because they can be extracted from heartwood by boiling in water, alcohol or other solvents, they are known also as extractives (Hillis 1984; Bowyer et al. 2003). These substances fill the empty cells in heartwood and are deposited in their walls. The blockage of the cells prevents any water transport through them. Any live parenchyma tissue (Sect. 3.3.1) dies.

Some of the extractives deposited in heartwood may be toxic to, or at least repellent to, fungi (Sect. 11.1) or insects (Sect. 10.2) (Bowyer et al. 2003). Fungi and insects can rot or decay timber in service, or even the stem wood in the tree when it is still alive. Because of its resistance to rot and decay, heartwood may be much more durable in service than sapwood (Hillis 1984); the durability of both varies widely between species. The blockage of the cells in heartwood may make it more difficult to dry than sapwood. It may also make heartwood more difficult to penetrate with liquids, such as the anti-fungal or insecticide chemicals with which round wood products, such as posts, are often treated to preserve them when in use (Bowyer et al 2003). Other extractives may affect pulping properties of wood, the adhesion of glues or the degree of shrinkage and collapse of cells during drying (Higgins 1984; Hillis 1984). However, the deposition of extractives does not affect the basic characteristics of cell walls or the basic density of the wood and so there is no difference in strength between timber sawn from heartwood or sapwood.

As fast-growing, commercial plantations are harvested at younger and younger ages, there will be an increasing need to deal with wood which contains a lower proportion of heartwood than has been usual in the past.

3.3.6 Reaction Wood

Trees which lean, or are bent by winds which blow consistently from one direction, develop what is known as reaction wood. This is positioned eccentrically around the stem, reaching its greatest development in the direction of the force which promotes its development. Reaction wood is present consistently in branches, which are bent normally under their own weight.

In softwoods, the reaction wood develops on the side of the stem or branch that is under compression (that is, on the side towards which the tree is leaning, on the side towards which it is being blown or on the under-side of branches) and is known as compression wood. In hardwoods, the reverse occurs and tension wood develops on the side of the stem or branch which is under tension (Bowyer et al. 2003). In some hardwood species, appreciable amounts of tension wood have been found in stems of vertical trees, which do not seem to have been subjected to any excessive bending forces; this is probably a response to normal swaying in the wind (Washusen 2002; Washusen et al. 2002; Washusen and Clark 2005).

Compression wood is denser and darker in colour than surrounding tissues. Its tracheids tend to be shorter in length than in normal wood and the cell walls are more heavily lignified and have a higher microfibril angle, giving it reduced strength. When dried, compression wood shrinks far more than normal wood. This gives particular problems if a sawn board contains some compression wood; when dried, the excessive shrinkage of the compression wood may cause the board to bow (Desch and Dinwoodie 1996).

Tension wood contains fewer and smaller-diameter vessels than normal wood. It has a lower proportion of lignin in the cell walls and so tends to be paler in colour. The inner cell walls of its fibres have a gelatinous layer, made largely of cellulose. As with compression wood, tension wood has abnormally high shrinkage (Hillis 1984; Washusen et al 2000a; Washusen and Evans 2001) and its cells may collapse when dried, causing further defect in timber (Bootle 1983). Its low lignin content means that fibres tend to be pulled out of the timber as it is sawn, rather than being cut cleanly; this leads to a more roughly cut surface (Desch and Dinwoodie 1996).

3.3.7 Growth Stresses

As a tree grows and wood cells form from the cambium (Sect. 2.1.1), the newly formed cells become stressed, a phenomenon in trees known as growth stress. The word ‘stress’ is being used here in an engineering sense, where an object is stressed when a force stretches, compresses or twists it; the stress is measured as the force applied per unit area of the object. When the stress on an object is relieved by removing the force causing it, the object returns to its original size and shape; the change in dimensions of an object caused by stress is known as **strain** (Raymond et al. 2004).

It is not known exactly how wood cells become stressed during their development (Wilkins 1986). Recent theories suggest that it has to do either with the way lignin is deposited as it binds microfibrils in the cell wall or as cellulose contracts as microfibrils are formed (Yang and Waugh 2001). Whatever their cause, the relief of growth stresses in tree stems when timber is sawn from them can cause such serious distortions in the sawn boards that they may be rendered useless. Even when the tree is first felled, relief of growth stresses can cause major cracks to appear right across the stem, extending for several metres up the tree (Page 1984; Kauman et al. 1995; Yang and Waugh 2001).

Different species vary greatly in the extent to which growth stresses occur within their stems. It has been thought generally that, as trees grow in diameter, the stresses within the stem are spread over larger areas; hence, the stresses tend to be less in the outer parts of the stems of large trees, which can then be sawn with less damage occurring (Hillis 1984). So severe are the growth stresses towards the centre of stems in some species that the structure of the wood can be damaged for as much as one third of the width of the stem. This wood is known as brittleheart and is much reduced in strength (Hillis 1984; Desch and Dinwoodie 1996; Yang and Waugh 2001). However, more recent work with eucalypts (Raymond et al. 2004) has found that the distribution of growth stresses across stems can be highly variable and very difficult to predict for any particular tree stem.

Growth stresses are generally much less in softwoods than in hardwoods. Some eucalypts (which are hardwoods) are particularly renowned for their high levels of growth stress. This causes considerable damage to the quality of boards sawn from them, particularly in younger, plantation grown trees (Yang and Waugh 2001; Yang et al. 2002; Chauhan and Walker 2004). However, the larger (that is, the faster-growing) trees in a plantation at a particular age may have a lower gradient of stress from the inner to the outer parts of the stem cross-section and, hence, show less damage in sawn boards (Wilkins and Kitahara 1991a, b).

Methods have been developed to assess and predict the level of growth stresses within stems of trees, even whilst they are still standing (Fourcaud and Lac 2003; Forcaud et al. 2003; Yang and Ilic 2003; Raymond et al. 2004; Yang et al. 2005). Various ways of handling stressed stems have been developed to minimise consequent damage during processing (Page 1984; Yang and Waugh 2001).

3.3.8 Knots

Where a branch has grown from the stem of the tree, a knot develops in the stem wood. Eventually, as the tree grows taller, branches on the lower part of its stem lose their leaves, die and are shed by the tree (Sect. 9.1). However, the knot remains in the stem wood, extending for a distance determined by the length of time the branch was held on the tree.

Knots can give a very attractive appearance to the surface of sawn timber and, in some species, can make the timber extremely valuable for the manufacture of the highest-quality furniture or panelling (Desch and Dinwoodie 1996). In general however, knots are the chief source of defect in sawn timber, either in its structural strength or the quality of its surface appearance (Kellomäki et al. 1989; Waugh and Rosza 1991; Yang and Waugh 1996a, b; Washusen et al. 2000b; MacDonald and Hubert 2002; Yang et al. 2002; Wang et al. 2003; Pérez et al. 2003; Washusen and Clark 2005). Other sources of defect are the presence of gum or resin, splits, stem pith, sloping grain (Sect. 3.3.4), holes left by agents such as wood-boring insects or excessive taper or crookedness of the stem (McKimm et al. 1988; Waugh and Rosza 1991). These are less common problems than knots and will not be discussed further.

The larger the knot, the more likely it is to cause a serious defect in sawn wood. It is common in plantation forestry to remove branches by **pruning** to minimise the occurrence and size of knots. Pruning is discussed in detail in Chap. 9.

4 Choosing the Species and Site

Chapters 1–3 have introduced plantation forestry in general and established the biological principles surrounding plantation growth and wood quality. The rest of this book will be concerned with more specific details of the silvicultural practices which are used today in growing plantation forests.

If a grower is setting out to develop a new plantation forestry programme, the crucial decisions to be made initially are:

- What species is to be grown?
- What specific pieces of land are to be used for the plantations?

Section 4.1 deals with species choice. Section 4.2 considers the constraints that may simply make some land unavailable for plantation use. The final section in this chapter considers the techniques used to assess how well a plantation will grow on any particular piece of land.

4.1 Selecting the Species

The principal factors which govern the choice of the species for a plantation programme are:

- The ability of the species to provide whatever products are desired ultimately from the plantation
- Its ability to grow at an adequate rate on the sites available.

The choice may be simple. If there is already a successful plantation programme in a region, with well-established markets for the products, chances of commercial success are probably highest if a new plantation enterprise uses the same species. However, if a plantation programme is to be started in a region where none exists already, the choice is far more problematic. Pragmatic considerations will limit the possibilities; only species with qualities appropriate to the desired products and for which seed is available would be considered. The final choice may then rest on the ability of the species to grow adequately in the environment of the region.

Where a species from some other part of the world is being considered, Booth (1985) has developed a tool to assess its growth capability in the region of interest. Booth's method is based on the assumption that a species may grow adequately anywhere that the climate is similar to that of the region where the species occurs naturally. His method involves collating information on long-term, average weather conditions across the region of natural occurrence of the species. A computer program is then available to draw maps showing where else in the world climatic conditions are similar (Booth et al. 2002). Booth and his co-workers (Booth 1985, 1991, 2005; Booth et al. 1989; Booth and Jones 1998; Jovanovic et al. 2000) have applied the system to consider the potential for various species in various parts of the world. As an example, Fig. 11.1 shows where Booth's system predicts that climatic conditions might be suitable in South Africa for the establishment of plantations of flooded gum (*Eucalyptus grandis*), a species native to the east coast of Australia.

Of course the present natural distribution of a species does not reflect necessarily the full range of climatic circumstances within which that species may grow adequately; thus, Booth's system may fail to identify some regions of the world where a species will in fact do well. It also takes no account of soil characteristics, which may be just as important as climate in determining whether or not a species will grow well. Hackett and Vanclay (1998) described a system which predicts plant growth in relation to both climatic and soil characteristics. To apply their system requires some knowledge of how well any particular plant species grows in relation to these characteristics; the system is available publicly[1], and can be used for a number of forest tree species. Booth (1998) applied a system, closely related to Hackett and Vanclay's, to assess where climatic and soil conditions in Africa might be suitable to establish plantations of river red gum (*E. camaldulensis*), an Australian species grown widely around the world, especially in areas where soils are saline. Thwaites (2002) used the system also in mapping locations most suitable to establish plantations in parts of tropical north Queensland, Australia.

Ultimately, tools such as these will give only guidelines as to which species might be appropriate for use in a new plantation enterprise. Once the choices have been limited to a few possibilities, it will be necessary to establish trials of the various species across the region of interest to compare their performance. These may show that different species perform better in different environmental circumstances across the region and,

[1] At the time of writing (December 2005), this information was available from the web site of the Center for International Forestry Research, Jakarta, Indonesia at http://www.cifor.cgiar.org/docs/_ref/research_tools/tropis/plantgro-infer.htm

hence, that more than one species should be used in the plantation programme; sophisticated techniques are available to judge which species is most suited to which circumstances (Caulfield et al. 1993). For major plantation programmes, it may be decades before final decisions are made as to which one or several species are most suitable for the programme.

4.2 Land Availability

Once decisions have been made about what species are to be used in a plantation programme, it is necessary to decide which areas within the region are both available and suitable for planting.

Some land will be unavailable simply because it is not for sale; its present owner may wish to use it for other purposes. Even if it is for sale, it might be so expensive that it would be unfeasible commercially for plantation forests.

Often there are government land-use planning constraints imposed on land availability. Other agricultural pursuits, urban use, water catchment or environmental protection might be deemed as more appropriate uses for particular land areas.

Societal constraints also limit land availability. Communities may feel threatened by the advent of a new plantation enterprise in their region; an interesting example of the threats felt by communities to an expansion of radiata pine (*Pinus radiata*) planting in Victoria, Australia is given by Spencer and Jellinek (1995). Various other works discuss the many social and policy issues to be faced in developing forestry programmes (e.g. Hall 1997; Race and Curtis 1997; Whiteman and Aglionby 1997; Walters et al. 1999, 2005; Herbohn et al. 2000; Gerrand et al. 2003; Lockie 2003; Schirmer and Tonts 2003; Spencer et al. 2003; Williams et al. 2003; Herbohn et al. 2005; Luckhert and Williamson 2005; Muhammed et al. 2005).

To rationalise these conflicting land-use and societal issues may require complex and protracted negotiations with government agencies and community groups to establish finally what land is potentially available for plantation establishment in any particular region. Sophisticated computer-based tools are available to assist in assessing the conflicting options during these negotiations (Ive and Cocks 1988; Baird and Ive 1989; Johnson et al. 1994; Kangas and Kangas 2005; Mendoza and Prabhu 2005; Sheppard 2005; Sheppard and Meitner 2005).

4.3 Site Productive Capacity

Once the land areas available in a region for plantation establishment have been identified, the grower will wish to determine which particular sites are most suited and will offer the best return on the investment. In commercial forestry, this usually means determining how fast a plantation will grow on any particular site under consideration, that is, determining its **site productive capacity**.

By site is meant 'an area of land which can be managed homogenously and will produce a more or less constant wood **yield** across it from a particular plantation species', a definition adapted loosely from Louw and Scholes (2002). In West (2004), I defined site productive capacity as 'the total stand biomass produced by a stand on a particular site, up to any particular stage of its development, when the stand has been using fully the resources necessary for tree growth which are available from the site'; this is a rather formal way of saying simply how fast a plantation will grow on a site.

We will discuss only briefly here what measures of site productive capacity are used in forestry. They are discussed in more detail in texts on forest measurement (e.g. West 2004). One measure which has been used for many years in forestry science is known as **site index**. It is defined as the **dominant height** (which is the average height of the tallest trees in a stand) at a particular age. Research over many years and forest types has shown that this measure correlates very closely with biomass production. A second measure, being used more commonly in recent times, is the maximum mean annual increment (Sect. 3.1) in stand stem wood volume observed on a site. It is usually one or other of these measures of site productive capacity which it is attempted to predict when assessing the suitability of a site for plantation establishment.

4.4 Predicting Site Productive Capacity

When a new plantation programme has been started recently in a region, there will be little specific information available to allow proper assessment of the productive capacity of any site. Only the experience available from other agricultural crops, or observation of the growth of native forests, will provide a guide as to which particular sites will grow plantations rapidly.

However, once some plantations have been established, it will be possible to gather information from them and develop systems to relate plantation productivity to the environmental circumstances of the sites on which they are growing. Such systems can then be used to predict plantation productivity on sites which have not been considered previously for planting.

From time to time and place to place around the world, many different systems of this nature have been developed. Louw and Scholes (2002) and Ryan et al. (2002) have reviewed a number of the approaches which have been used. They will be summarised in the next three subsections.

4.4.1 Site Classification Approach

The first approach involves classifying sites into those which are more productive and those which are less productive. These systems can be quite simple. They may be used initially in a region, before substantial amounts of information are available from detailed research studies of the plantations already established there.

A simple example of this approach is described by Kerr and Cahalan (2004), for plantations of common ash (*Fraxinus excelsior*) in Great Britain. Common ash is becoming an important plantation species in Europe and parts of Asia and Africa, because it grows rapidly and produces wood of high quality with a variety of uses. In Kerr and Cahalan's case, the productivity of existing ash plantations in Great Britain was assessed as high, medium or low in relation to each of four environmental factors. The factors were temperature (which was assessed as being in one of eight classes, from colder to warmer), rainfall (eight classes), soil fertility (six classes) and soil moisture availability (eight classes).

Their system is used by determining the class, for each of the four environmental factors, to which a particular site of interest belongs. If ash is planted on that site, its productivity is then predicted to be that which is the lowest for the corresponding classes of the four environmental factors. This assumes that ash growth is determined by whichever environmental factor most limits its growth on any particular site.

This concept of growth at a site being determined by the environmental factor which is in most limited supply derives from work with agricultural crops (Hackett 1991). Another forestry example, rather more sophisticated than Kerr and Cahalan's, is a system developed by Laffan (1994, 2000), where the same concept was used to predict productivity of eucalypt plantations in Tasmania, Australia; Osler et al. (1996a) tested Laffan's system and showed it was quite reliable.

There are various other examples of productivity assessment systems based on classification of sites (e.g. Gasana and Loewenstein 1984; Turvey et al. 1990; Ares 1993; Andersen et al. 2005; Bravo-Oviedo and Montero 2005). Some of these systems require far more detailed measurement of the environmental characteristics, both of the climate and soil, of a site than does Kerr and Cahalan's. Good examples of the application of such systems are given by Foster and Costantini (1991a, b) and Turner et al. (1996, 2001).

4.4.2 Regression Approach

Where more detailed information is available from a region, the regression approach is used commonly to develop a system to assess site productivity capacity. Later, it will become evident from where the term regression approach derives.

This approach requires the accumulation of substantial quantities of data on both the productivity of previously established plantations and the environmental characteristics of the sites on which they have been grown. The data are then used to develop an equation to predict productivity from the environmental characteristics. Different researchers have used different sets of environmental characteristics to do this, sets which they consider include the environmental characteristics which are most important in determining plantation growth in their region of interest. These characteristics may include climatic factors (such as rainfall or temperature), soil factors (such as soil depth or content of nutrient elements) and topographic factors (such as slope of the site or its aspect). Both Louw and Scholes (2002) and Ryan et al. (2002) have listed a variety of environmental characteristics which have been used by various researchers, from time to time and place to place, in developing regression systems.

An interesting example of the regression approach comes from Shoulders and Tiarks (1980), for pine plantations in Louisiana and Mississippi in the southern USA. Experimental plantations of four species, loblolly (*Pinus taeda*), slash (*Pinus elliottii*), longleaf (*Pinus palustris*) and shortleaf (*Pinus echinata*) pine, had been established at each of 87 sites over the region; these four species are native to the region and are important for plantation forestry there. The average height of the taller trees in the experimental plots was measured at 20 years of age; these measurements gave a site index measure (Sect. 4.3) of site productive capacity for each species at each site.

Previous experience suggested to Shoulders and Tiarks that the amount and seasonal distribution of water available from the soil was the most important environmental factor determining the rate of growth of pine species in the region; hence, for each experimental site Shoulders and Tiarks used long-term weather records to determine its average warm season (April–September) and cool season (October–March) rainfalls. They measured also the water holding capacity of the soil, which is determined principally by its depth and texture (Sect. 5.1), and the slope of the site, which influences how much rainfall runs off it. Using the data from all the experiments, they were then able to develop a separate equation for each of the four species, which showed how site index was related to these various environmental variables. For example, the equation they developed for slash pine was:

$$S = -3.8 + 0.1075R_W - 0.0612R_C - 0.0000314R_W^2 + 0.0000983R_C^2 - 0.0000929R_W R_C - 0.298/\theta + 0.591W_C - 0.0563W_C^2 + 0.00146W_C^3, \quad (4.1)$$

where S is site index (metres), R_W and R_C are the annual average rainfalls (millimetres) in the warm and cool seasons, respectively, θ is the slope of the site (percent) and W_C is the water holding capacity of the soil (percentage of the soil volume). The equations developed for the other three species were similar in form, although the values of the constants (known as the parameters of the equation) differed for each species.

It is not the place here to discuss exactly how scientists decide what form such an equation should take, nor how the values of its parameters are determined. Suffice to say that the mathematical statistical technique known as regression analysis is used; hence, we use the term regression approach. Regression analysis is used widely across all areas of science, wherever it is desired to determine how one particular variable (site index in this example) is related to others (the environmental characteristics in this case). Standard texts are available which describe regression analysis and its use (e.g. Draper and Smith 1988).

To use Shoulders and Tiark's equations, the seasonal rainfall, ground slope and soil water holding capacity must be determined for a site where consideration is being given to the establishment of a plantation. Equation 4.1 can then be used to predict how well slash pine would grow on that site. Their other equations would predict how well each of the other species would grow.

I chose Shoulders and Tiarks' example because it considered four species and so allows comparison between them. Figure 4.1 illustrates this. It was drawn using the equations for each of the four pine species. These equations indicate that slash pine generally grows faster than any of the other three species; wherever the cool season rainfall exceeded about 705 mm and the warm season rainfall exceeded about 660 mm it certainly does, as indicated by the region marked as slash pine in Fig. 4.1. However, where seasonal rainfall falls below those limits, one or other of the other three species may grow faster than slash pine; the regions marked in the figure for each of the other three species indicate where each would grow faster than slash pine. Hence, this site productivity prediction system will

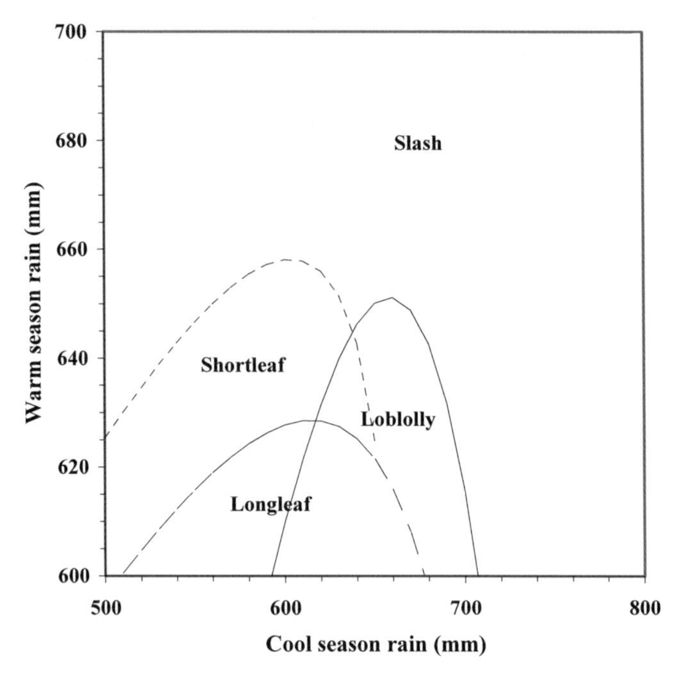

Fig. 4.1. Regions, delimited by cool season and warm season rainfall, where height growth to 20 years of age of loblolly (——), longleaf (— - —) or shortleaf (- - - -) pines will exceed that of slash pine. The results derived from experimental plantations grown in Louisiana and Mississippi in the USA. In drawing these diagrams, it was assumed the plantations were growing on a site with a slope of 3° and with a potentially available soil moisture content of 10% (derived from the site index prediction models of Shoulders and Tiark 1980)

assist growers to decide which species would be most productive on any-particular site.

There are many other examples of the regression approach to prediction of site productive capacity. Some recent examples are Corona et al. (1998), Uzoh (2001), Hökkä and Ojansuu (2004), Romanyà and Vallejo (2004) and Stape et al. (2004). Although similar in principle to the regression approach, some different statistical techniques are now being applied to establish the relationship between productivity and environmental variables (e.g. Wang et al. 2005)

4.4.3 Process-Based Model Approach

In more recent times, process-based mathematical models of forest growth are being developed for use as systems to estimate site productive capacity. By process-based is meant that the models are based on knowledge of the physiological processes which occur in plants and how those processes are affected by the environmental circumstances in which the plants are growing.

Process-based models are the most recent development in forest growth and yield modelling. The discipline of growth and yield modelling is very important in commercial forestry; mathematical models capable of predicting the amounts of wood which plantations will produce some time in the future are used by forestry organisations throughout the world as part of their planning.

In the past, growth and yield models have been empirical models. That is, they are developed using large amounts of data collected from plantations which have already been established. These data are then used to derive equations which relate forest growth behaviour to the environmental circumstances of the forest. Equation 4.1 is a good example of an empirically derived equation.

Process-based models attempt to predict directly the photosynthetic production by the plantation canopy, usually on a daily or monthly basis. Photosynthetic production is determined by the specific physiological characteristics of the species concerned. These include:

- The way its leaves are arranged in the canopy to intercept sunlight
- The chemical efficiency of its photosynthetic system
- The way its metabolic processes are affected by air temperature
- Its ability to keep its stomata open as soil water availability decreases seasonally
- The size and distribution of its fine roots in the soil and their efficiency in taking up water and nutrients.

With research by specialists in these fields, these characteristics can be determined and described mathematically. They can then be incorporated into a process-based model to predict growth of the plantation as sunlight, temperature, the amount of water available from the soil and the availability of nutrients from the soil all change seasonally and from year to year as the forest grows. Whilst the development of process-based models requires detailed study of the physiological characteristics of the species concerned, their advantages over empirical models are:

- They require less data from previously established forest for their development.
- They are more likely to make reliable predictions when novel environmental circumstances are encountered or novel plantation management practices are proposed. Empirical models are notoriously unreliable if they are used to make predictions for new circumstances,

outside the range of conditions included in the data used to develop them.

The development of forest growth and yield models is a highly specialised field of forestry science and will not be considered further here. Many works discuss both older and more recent approaches to forest modelling (e.g. Vanclay 1995; Battaglia and Sands 1998; Mäkelä et al. 2000; Peng 2000a, b; Davis et al. 2001; Le Roux et al. 2001; Avery and Burkhart 2002; Porté and Bartelink 2002; Mäkelä 2003; Matala et al. 2003; Almeida et al. 2004; Battaglia et al. 2004; Valentine and Mäkelä 2005).

A model named ProMod (Battaglia and Sands 1997; Sands et al. 2000; Booth 2005) is an example of a process-based system being used for prediction of site productive capacity. ProMod predicts the maximum mean annual increment in stem wood volume production as its measure of site productive capacity (Sect. 4.3). It has been used to draw a large-scale map showing where, across the whole of Tasmania, Australia, sites might be found where growth rates would be appropriate for establishment of plantations of Tasmanian blue gum (*E. globulus*) (Mummery and Battaglia 2001). The same model system has been used to produce small-scale maps of how productivity might vary across individual plantation areas, say of tens of hectares (Battaglia et al. 2002). Of course the scales at which the environmental characteristics (rainfall, temperature and soil information) of sites must be measured will vary from large scale to small scale correspondingly; obtaining suitable environmental data at both those scales can present difficulties in itself (Mummery et al. 1999; Thwaites and Slater 2000; Mummery and Battaglia 2001, 2002; Battaglia et al. 2002; Ryan et al. 2002; Thwaites 2002). Swenson et al. (2005) give another example of the use of a process-based model to map potential forest productivity over a large area.

As a concluding comment to Sect. 4.4, it is important to realise that none of the methods developed to predict site productive capacity will do so exactly. The predictions can differ quite substantially from the eventual productivity realised on a site (Verbyla and Fisher 1989; Osler et al. 1996a; Battaglia and Sands 1998). Inevitably, there will be environmental factors which influence growth at a site and which are not included as part of the prediction system. It may be that such factors have simply not been identified, or they are so difficult to measure it would be impractical to include them as part of the system. As well, the silvicultural practices actually applied to a plantation may differ from those that were used when the productivity prediction system was developed; this may lead to either faster or slower growth than predicted. Nevertheless, systems such as those described here are used very widely around the world, wherever it is necessary to assess the potential productivity of land for plantation growth.

5 Establishment

Once the land and the species have been chosen for a forest plantation, the real work of establishment can begin.

Initially, there will be various tasks to get the plantation area into a condition suitable for tree planting. Forest or other vegetation pre-existing on the area may have to be cleared and the debris stacked and burnt. The boundary may have to be fenced for security or to keep out larger animal **pests** (Sect. 10.4) or grazing livestock, both of which might damage newly planted seedlings. Roads and tracks will have to be installed to allow vehicles to move about the plantation area. Minor engineering works, such as small bridges or culverts, may have to be undertaken. Drainage works might be necessary to make parts of the area trafficable or to make waterlogged areas suitable for tree planting. Firebreaks may have to be established around the outer perimeter of the plantation.

Areas inappropriate for plantation establishment will have to be identified. These might be steep areas, where there are risks of erosion, or strips along stream edges (called riparian strips) which are left unplanted to avoid silting of streams or to act as refuges for wildlife. There are often government regulations that specify in detail these environmental-protection measures. They may even include retention of clumps of pre-existing native vegetation or large, old trees, both of which may provide suitable habitats or refuges for native animals. Large, old trees retained in a plantation can reduce appreciably the growth of the plantation trees in their vicinity (Bassett and White 2001). Foster and Costantini (1991c) have given a good example of the issues to be considered in planning the layout of a plantation area.

None of these preparations for plantation establishment will be considered in detail here. This chapter will deal principally with the issues involved in establishing the seedlings in a plantation and ensuring their survival and rapid growth over the first few years, the period when the seedlings are small and most vulnerable to damage (Sect. 2.2). The last section of the chapter will consider the important technique, known as **coppice**; after a plantation has been harvested, this technique can be used sometimes to re-establish the plantation without having to plant new seedlings.

5.1 Cultivation

Before seedlings can be planted out, the final stage in site preparation is cultivation. This involves disturbing and breaking up the soil to provide the optimum environment for root growth and development.

A wide range of equipment is available for different types of cultivation to suit different soils and crops, whether tree or agricultural crops (Hillel 1980; Brady and Weil 2001; Charman and Murphy 2001). Ploughs are used for primary cultivation, to lift, turn and break up the soil. Ripping (also called subsoiling) is also done commonly. It involves pulling a large, vertical tine through soils which are deeply compacted or have a hard, subsurface layer; these would otherwise prevent soil drainage and lead to waterlogging or restrict root penetration. Secondary cultivation refines the surface. It may involve harrowing, furrowing or the building of mounds (also called beds), depending on the requirements of the crop.

The type of cultivation used on a particular site depends on the properties of the soil and the requirements of the tree species to be planted. So varied are the practices used in different parts of the world, it is impossible to generalise as to which are the most appropriate for any particular site. However, it is reasonable to say that cultivation is often one of the largest costs involved in establishing a plantation. It requires large and expensive machinery which is costly to run. This means it is important to evaluate whether or not the gains in crop yields achieved ultimately as a result of cultivation are sufficient to justify the expense.

Two physical properties of soil, its air porosity and its strength, are particularly important to the ability of tree roots to grow and develop. Both may be modified by cultivation. Much research remains to be done to determine the optimum values of these characteristics for different plantation species growing in different parts of the world. However, it will at least be explained here what these characteristics are and what is known at present of their optimum values.

Soil is made up of mineral and **organic matter** (the rotting detritus of living organisms) particles, between which are spaces (called pores) containing water or air. Different names are given to soil particles of different sizes: clay particles are less than 0.002 mm in diameter, silt particles are 0.002–0.02 mm, sand particles are 0.02–2 mm and gravel is greater than 2 mm (in different parts of the world, scientists vary slightly in their definitions of these particle sizes). The texture of soil is determined by the proportion of its particles which are of clay size: sandy soils have less than 10% clay, loamy soils have 10–35% clay and clay soils have more than 35% clay.

Soil air porosity is the proportion of the soil volume which consists of pores filled with air. It is measured in units of cubic metres of air per cubic metre of soil volume. Obviously, the wetter the soil, the more of its pores will be water-filled and the fewer air-filled, so the lower will be its air porosity. To function and grow, plant roots need at least some air in soil, from which they obtain oxygen for their metabolism (they use oxygen when they consume the food manufactured in the leaves by photosynthesis). Waterlogged soils are those containing too much water and insufficient air for the needs of plant roots. Tree root growth and development has been found to slow when soil air porosity falls below about 0.15 m^3/m^3 (Zou et al. 2001).

Soil strength represents the resistance offered by soil to penetration by roots. It is measured as a pressure (many different units are used for pressure, but we will use mega-pascals, abbreviated as MPa[1]) and is determined often by measuring the resistance to penetration of the soil by a specially designed metal probe (called a penetrometer). Sandy soils have lower strength than clay soils. Strength declines also as the soil porosity (the proportion of the soil volume occupied by both water and air) increases and as the amount of water in the soil increases, hence, air porosity declines. Zou et al. (2001) and Sands et al. (1979) have found that growth rates of radiata pine (*Pinus radiata*) tree seedling roots decline as soil strength increases and become severely restricted when soil strength exceeds about 3 MPa.

Compaction by the passage of heavy machinery over the ground is one very important way in which soil air porosity can be reduced, or soil strength increased, beyond the values which allow adequate root growth (Powers 1999). This can result from the use of logging machinery, when a previous tree crop is harvested, or by using heavy machinery to clear a site of logging debris before the next tree crop is planted.

Ilstedt et al. (2004) gave a good example of the effects of heavy machinery in Sabah, Malaysia. On tracks over which logging machinery had passed, soil air porosity was sufficiently low to restrict root growth (0.1 m^3/m^3), whereas it was adequate for roots (0.16 m^3/m^3) where the machinery had not passed. Growth of seedlings of the hardwood *Acacia mangium*, a plantation species used extensively in tropical countries, was reduced appreciably when they were planted on the compacted tracks. In this case, simple cultivation with a power tiller appeared to be sufficient to remove the affects of the compaction and allow the seedlings to grow adequately. There are various other examples of the consequences of the use

[1] The conversion factors for other units of pressure you may encounter are 1MPa =10 bar=10.2 kg/cm^2=9.87 atmospheres=750 cm of mercury=145 lb/in^2.

of heavy machinery for logging or site clearing (e.g. Turnbull et al. 1997; Costantini and Doley 2001a, b, c; Blumfield et al. 2005).

It should not be assumed that soil compaction is always detrimental to subsequent plantation growth. Cases have been reported where it leads to seedling roots having improved contact with the soil and to reduced loss of water from the soil during periods of drought (Mósena and Dillenburg 2004).

Whatever the natural soil conditions and the past history of the site, cultivation prior to planting can lead to substantial gains in the subsequent growth of seedlings. The work of Lacey et al. (2001) illustrates this. They conducted a cultivation experiment on a site being prepared for a blackbutt (*Eucalyptus pilularis*) plantation, in New South Wales, Australia. Blackbutt is a tall forest tree, native to the coastal region of eastern Australia, from southern New South Wales to southern Queensland. It is an important timber species in native forests and is also being grown in plantations in subtropical eastern Australia.

In Lacey et al.'s case, the soil was not particularly deep (just less than 1 m) and its strength generally exceeded 2.5 MPa, sufficient to impede root growth appreciably. The experiment considered the effects of cultivation by mounding (to 0.2–0.4 m above ground), combined with ripping to various depths (0.8 m at most) with various types of ripping tine. Planting seedlings on top of mounds can aid in keeping their roots clear of seasonal waterlogging on a site. Mounding also accumulates nutrient rich topsoil from the adjacent soil into the mound; this can make more nutrients available to roots. Ripping reduces soil strength to depth and may aid in draining waterlogged soil. Both practices are used widely in plantation forestry around the world (Fig. 5.1).

Table 5.1 gives some results on tree growth and development from Lacey et al.'s experiment, 19 months after the seedlings were planted. Cultivation increased seedling survival substantially, from only 52% without cultivation to nearly 90% or more with cultivation; so poor was survival without cultivation that the plantation would normally be considered to have failed. The trees planted with cultivation were appreciably taller and larger in diameter than those planted without cultivation. These differences, at only 19 months of age, are sufficiently large that they would lead ultimately to the production of commercially useful quantities of timber a number of years earlier on cultivated than on uncultivated sites. However, the table shows also that the costs of cultivation were substantial. They were much greater with the deeper ripping, which required the use of machinery which was larger, more powerful and more expensive to operate. In this example, the gains with deeper ripping were little different from those with the much cheaper, shallower ripping. For any plantation enter-

Fig. 5.1. A recently cultivated mound, prepared for a eucalypt plantation in northern New South Wales, Australia. At the same time as the mounding was done, the site was ripped to a depth of 0.6 m, along the middle of the mound line. In the background, note that ripping and mounding followed the land countours, to minimise water runoff in storms and, hence, erosion losses of soil. A burnt remnant of a pile of debris left after clearing the site appears in the *top right-hand corner*. The site had been treated with weedicide to clear it of weed growth (Photo—P.W. West)

Table 5.1. Survival and growth to 19 months of age of blackbutt (*Eucalyptus pilularis*) seedlings, established in an experiment with various cultivation treatments in New South Wales, Australia. The costs of the cultivation treatments (in Australian dollars at 2001 prices) are shown also (Source—Tables 2, 5, 6 of Lacey et al. 2001, for none, mounding and ripping to 0.4-m depth with a winged main tine, VS, and mounding and ripping to 0.8-m depth with a winged main tine and winged leading tines, DWW, cultivation treatments)

Cultivation treatment	Tree survival (%)	Average tree height (m)	Average tree diameter at breast height (cm)	Cost of cultivation ($/ha)
None	52	1.4	0.8	–
Mounding and ripping to 0.4-m depth with a winged main tine	87	3.5	4.4	87
Mounding and ripping to 0.8-m depth with a winged main tine and winged leading tines	95	3.2	3.9	128

prise, it will always be necessary to evaluate economically whether or not cultivation leads ultimately to worthwhile gains.

Whilst Lacey et al.'s experiment illustrates a case where cultivation was worthwhile, it must be recognised that both the need for and the type of cultivation will vary greatly, depending on the circumstances of any particular site. There are many reported cases of substantial growth gains from cultivation and many others where there were no gains, either because the soil conditions were already appropriate for tree growth or because environmental factors other than soil characteristics were more limiting to growth (Flinn and Aeberli 1982; Francis et al. 1984; Attiwill et al. 1985; Mason and Cullen 1986; Turvey and Cameron 1986b; Mason et al. 1988; Varelides and Kritikos 1995; Lacey et al. 2001; Pennanen et al. 2005; Varelides et al. 2005). Only with detailed study of the physical properties of soils and their relation to growth of the species to be planted, will it be possible to determine if, and what type of, cultivation is appropriate (Foster and Costantini 1991b; Smith et al. 2001).

There are effects of cultivation other than simply growth effects owing to improved root development. In an experiment with Sitka spruce (*Picea sitchensis*) plantations in Sweden, Nordborg and Nilsson (2003) found that deep soil cultivation, by completely inverting the soil, led to increased survival and growth of seedlings. This was due to much reduced weed growth on cultivated sites (Sect. 5.4), a common advantage of cultivation (Nilsson and Örlander 1995). However, it was not just the reduced competition from weeds which encouraged seedling growth. Voles, mammal pests (Sect. 10.4) which destroy seedlings, enjoy the protection afforded by a dense weed cover; the lack of weed growth on cultivated sites kept them free of voles.

Another important benefit of cultivation can be increased stability of the trees. Where their root systems are poorly developed, trees may be uprooted by strong winds (Stokes et al. 1995). Ripping to encourage deep root penetration can be effective in preventing this. However, root systems can develop along the rip lines, so ripping should be aligned with the direction from which strong winds tend to blow (Somerville 1979; Mason et al. 1988; Balneaves and De La Mare 1989; Schaetzl et al. 1989; Papesch et al. 1997; Moore 2000).

5.2 Seedlings

Before they can be planted out, seedlings must be raised to a suitable condition. Once planted, their survival will depend on the ability of their roots to start taking up water and nutrients from the soil. Their rapid growth will then depend on the amount of photosynthesis their leaves can undertake.

Usually, planting is done shortly after rain and at times of year when rain is likely, so that as much water as possible is available to seedlings. Nevertheless, it is agreed generally that the greatest risk seedlings face immediately after planting is an inability of their root systems to take up sufficient water from the soil (Burdett 1990). Sands (1984) suggested that it is the initial lack of contact between the roots and the soil of the newly planted seedling which limits their ability to take up water, rather than any damage to the root systems suffered as the seedlings are raised or planted out (see also Örlander and Due 1986; Bernier 1993; Munson and Bernier 1993). Once seedling root systems have developed sufficiently to meet their water requirements, their ability to take up sufficient nutrients becomes the factor most likely to limit their growth.

Seedlings are raised in nurseries (Fig. 5.2). This section considers the issues faced in the nursery to ensure that seedlings reach a condition which maximises their chances of survival and rapid growth after planting out.

5.2.1 Nurseries

The management and operation of nurseries is a specialised task and will not be discussed in any detail in this book. Nurseries vary enormously in size and scale; the largest of them are highly automated seedling factories which may grow millions of seedlings per year, often to strict specifications of size and quality.

Seedlings are raised in two different ways, both in use widely throughout the world. Firstly, in open-rooted (also called bare-stock) nurseries, seedlings are grown in nursery beds, usually in the open air. Secondly, seedlings may be grown individually in containers, that is, in pots of various types. Container-grown seedlings may be raised in a greenhouse or in the open air (Fig. 5.2).

Generally, it is less expensive to raise open-rooted seedlings. However, it is more difficult to control their circumstances than when each seedling has is its own container and is grown under cover, where the environment can be controlled far more closely than in the open air. Furthermore, when they are to be planted out, open-rooted seedlings must be lifted from the nursery bed; substantial damage can occur to the root system as seedlings are removed from the soil. As well, they must be protected from drying out, both as they are transported to the planting site and when there are the inevitable operational delays that occur from time to in any major planting programme. On the other hand, container-grown plants can be transported

Fig. 5.2. A forest plantation nursery in subtropical eastern Australia. The seedlings are of the eucalypt species spotted gum (*Eucalyptus maculata*). They are being raised outdoors in root trainer containers (Fig. 5.3) supported on metal benches to keep them above ground. Sprinklers for irrigation can be seen scattered amongst the seedlings. This is quite a small nursery by commercial plantation forestry standards, but can raise several million seedlings per year (Photo—P.W. West)

to the planting site still in their containers and are then carried by the planter, so the seedling is removed from its container only as it is planted. Many examples can be found where container-grown seedlings have survived or grown better after planting out than open-rooted seedlings (Burdett et al. 1984; Nilsson and Örlander 1995; Thiffault et al. 2003; Mead 2005; South et al. 2005), although this is by no means always the case (Carlson et al. 1980; Neilsen and Ringrose 2001; McArthur and Appleton 2004; South et al. 2005).

5.2.2 Factors Determining Seedling Survival and Growth

Removing seedlings from the nursery and planting them out exposes them to a quite new environment in the field. It takes some time for the seedlings to acclimatise to this new environment; Close et al. (2005) have reviewed the physiological stresses to which seedlings are subjected during their acclimatisation. Generally, the characteristics which seem most important in determining seedling survival and rate of growth over the first year or so after planting are:

- Their size—this is measured usually as the seedling height (above the roots), its stem collar diameter (the diameter of the stem immediately above the point where the stem and roots join) or the oven-dry biomasses of its shoot (its aboveground parts) and its roots. Larger seedlings contain larger stores of food, so they can grow additional leaves more readily (Close et al. 2005), as well as having more leaves, so they are capable of more photosynthesis and faster growth.
- The relative size of their roots and shoots—this is called their root-to-shoot ratio and is measured as the oven-dry biomass of the roots of a seedling divided by the oven-dry biomass of its shoot. Seedlings with a higher root-to-shoot ratio tend to survive better, because they have a relatively large root system to supply the water requirements of their relatively small shoot.
- Their root growth capacity—this is measured by harvesting seedlings from the nursery and growing them under ideal conditions (with plenty of water, light and nutrients and at favourable temperatures) for a few weeks and measuring the increase in size over that time of their root systems (Burdett 1979; Ritchie and Dunlap 1980; Burdett 1983). The higher the root growth capacity, the more rapidly can the root system develop to supply the water and nutrient requirements of the growing seedling.
- The amount of the various nutrient elements they contain—this is measured by chemical analysis of the root and shoot biomass. The more nutrients a seedling contains, the more time is available for the root system to establish itself and become capable of taking up from the soil the nutrient requirements of the growing seedling.

There are many examples from around the world which illustrate how seedling growth and survival over the first few years after planting is related to these characteristics (e.g. Benson and Shepherd 1976; Burdett 1979; Albert et al. 1980; Ritchie and Dunlap 1980; Sutton 1980; van den Driessche 1982, 1984, 1992; Bernier 1993; Munson and Bernier 1993; Burdett et al. 1984; Margolis and Waring 1986; Balneaves 1988; Timmer et al. 1991; Egnell and Örlander 1993; South et al. 1993a, b, 2005; Krasowski 2003; Close and Beadle 2004; Rikala et al. 2004; VanderSchaaf and McNabb 2004; Davis and Jacobs 2005; Jacobs et al. 2005a; Weih and Nordh 2005).

Sometimes these characteristics interact in complex ways, so it is difficult to predict exactly how seedlings will perform after planting out. An example of this comes from Spain, where Villar-Salvador et al. (2004a) compared the survival and growth, 2 years after planting out, of seedlings of the hardwood Holm oak (*Quercus ilex*), which had received different amounts of nitrogen fertiliser in the nursery. The seedlings were 15 months old when they were taken from the nursery and planted. At that time, the fertilisation in the nursery had increased the size of the seedlings (16-cm tall on average for fertilised seedlings as against 13 cm for the unfertilised ones), increased their root growth capacity by 3 times and increased their nitrogen content (9,000 mg nitrogen/kg of oven dry biomass for fertilised seedlings against 7,600 mg nitrogen/kg of oven dry biomass for unfertilised ones). However, because fertiliser was available so readily to them, the fertilised seedlings had not developed their root systems to the same extent as unfertilised ones; hence, fertilised seedlings had a lower root-to-shoot ratio (1.6 as against 2.1).

The site where the seedlings were planted out was rather hot and dry (an average rainfall of only 490 mm/year, with a pronounced summer drought period 3–4-months long, and with an average daily maximum temperature in the hottest month of 31.5°C). Two years later, 19% of the seedlings which had been fertilised in the nursery had died, whilst 41% of the unfertilised seedlings were dead. As well, the fertilised seedlings had grown much more; the average volume of their stems was 4.7 cm^3, whilst that of the unfertilised trees was only 2.2 cm^3. So, the larger size and higher nitrogen contents of the fertilised seedlings led to considerable gains in their survival and growth after planting out. So large were these gains, they more than offset the disadvantage those seedlings might have suffered from their lower root-to-shoot ratio.

5.2.3 Achieving Seedling Specifications

Given the discussion in Sect. 5.2.2, it is obvious that the plantation grower will need to specify what characteristics are required of seedlings from the nursery to ensure they have the best chance of survival and rapid growth once they are planted out. These characteristics will vary considerably, depending on the tree species concerned and the characteristics of the site where they are to be planted. It will be the task of the nursery manager to grow and manipulate seedlings in the nursery to achieve the required specifications. Some of the ways in which this is done are discussed in this subsection.

For open-rooted seedlings, the root systems can suffer substantial damage as they are lifted from the ground prior to planting out (Maillard et al. 2004). To avoid this, it is common to prevent seedling roots growing too deeply into the nursery soil by undercutting the roots and by wrenching the seedlings from time to time. Undercutting involves dragging a horizontal blade or wire through the nursery bed, at 10–15 cm below ground, to cut off deeper roots. Wrenching involves dragging a blade below the seedlings to loosen the soil and roots. These practices may render seedlings more resistant to drought when planted out (Rook 1971; Bacon and Hawkins 1977; Benson and Shepherd 1977; Bacon and Bachelard 1978; Nambiar 1980; van den Driessche 1983), but sometimes have no effect on survival and subsequent growth (Andersen 2004).

For container-grown seedlings, the type of container can influence root development substantially. An example is given by Annapurna et al. (2004), who used different types of containers to grow 6-month-old seedlings of sandalwood (*Santalum album*) in India. They compared the use of polythene bags, varying in volume over the range 600–1,500 ml, with 600-ml plastic pots and with various sizes (150–600 ml) of root trainer containers. These last are in common use in forest nurseries (Fig. 5.3). They are tapered containers, with an open top and base. They have several small ridges running down their interior to train roots to go straight down and not coil around the inside of the container; this prevents unacceptable distor-

Fig. 5.3. A view from the top of a tray of root trainer containers. Note the five small ridges running vertically down the inside of each container to train roots to go straight down (Fig. 5.4). The open bottom allows air pruning of the base of the root system. The top opening of these containers is 4 cm in diameter and the containers are 8-cm long. Tree planters carry these trays into the field and remove the seedlings as they are being planted out (Photo—P.W. West)

Fig. 5.4. The root system of a eucalypt seedling which has been grown in a root trainer container (Fig. 5.3). Note how the larger roots have been trained to grow vertically down by the ridges inside the container. This seedling was raised in a commercial potting mix (Photo—P.W. West)

tion of the root system (Fig. 5.4). As well, these containers prevent root systems becoming too large; as roots grow through the open end, they simply dry out and die (commonly referred to as air pruning). These containers are sufficiently small and convenient that they can be carried into the field by planters, so that the seedling is not removed from its container until it is actually planted.

At 6 months of age, Annapurna et al. assessed the quality of the seedlings on the basis of their biomass (the larger the better), their sturdiness (those with a higher root collar diameter relative to their height were assessed as better) and their root-to-shoot ratio (the higher the better). They found that seedlings raised in 600-ml root trainer containers were of appreciably higher quality than those raised in the other types of containers. Polythene bags produced seedlings of particularly poor quality; problems with this type of completely closed container included coiling of the roots, poor aeration of the potting soil and the relatively large size of the containers, which used excessive potting mixture and made them heavy to transport.

In later work, Annapurna et al. (2005) found also that the type of potting medium used could affect the seedling quality. They tested various combinations of sand, soil and organic matter (peat, compost, burnt rice husks and charcoal) and found that a combination of sand, soil and organic matter produced seedlings of the highest quality. Their best combination of these gave very favourable aeration of the potting medium, held water well and was of relatively low density, reducing the weight to be transported to the planting site. Interestingly, there are reports that whilst peat as the potting medium can produce desirable seedlings in the nursery, it may dry out rapidly after the seedlings have been planted out and subject them to high water stress (Bernier 1993; Thiffault et al. 2003).

Another important nursery practice is fertilisation of seedlings. The timing and amounts of fertiliser applied can affect substantially the size of seedlings and their subsequent survival and growth after planting out (Burdett et al. 1984; Margolis and Waring 1986; Timmer and Armstrong 1987; Burgess 1991; Timmer et al. 1991; van den Driessche 1992; Hawkins et al. 2005); of course, if nutrients are available readily from the soil at the planting site, seedling fertilisation in the nursery may have little value (van den Driessche 1985). Advantage may be gained also by applying fertiliser to seedlings as they are planted out, to increase the amount of nutrients available immediately from the soil (Close et al. 2005; Jacobs et al. 2005b). However, the fertiliser can induce changes in soil chemical properties, which may be deleterious to seedling growth (Jacobs and Timmer 2005). Benefits for seedling nutrition after planting have been gained also by inoculating seedlings in the nursery with mycorrhizas (Sect. 2.1.4), although often the mycorrhizas already present in the soil colonise the seedlings adequately after planting (Wilson et al. 1987; Bogeat-Triboulot et al. 2004; Parladé et al. 2004; Teste et al. 2004; Brundett et al. 2005; Turjaman et al. 2005).

Various other treatments can be used in nurseries to pre-condition seedlings to field conditions before they are planted out. For example, droughting the seedlings in the nursery has proved helpful in ensuring they can tolerate a relatively low availability of water after planting out (van den Driessche 1992; Villar-Salvador et al. 1999, 2004b; Stape et al. 2001; Helenius et al. 2005). In cold climates, seedlings grown in containers in greenhouses may be moved outdoors to acclimatise them to the cold before being planted out. Seedlings grown in nurseries at lower altitudes may have to be moved to holding areas at higher altitudes, to acclimatise them to the cold to avoid frost damage when they are planted out on higher-altitude sites (Close et al. 2005). Depriving seedlings of the nutrient element nitrogen in the nursery can render them less attractive to animal pests

(Sect. 10.4.2) which may browse them after planting out (McArthur et al. 2003). Low nitrogen levels may also render seedlings less liable to damage to their leaf chemical systems, which can occur in sunny but frosty conditions after planting out (Close et al. 1999, 2000). Of course, plants deprived of nitrogen may not grow as well as those well supplied, but at least they will have survived under these particular circumstances. Estes et al. (2004) showed that inoculation of loblolly pine (*Pinus taeda*) in the nursery with certain **bacteria** could render them less liable to damage after planting out from industrial ozone pollution in the atmosphere.

There are no particular rules about how long seedlings should be raised in the nursery, nor how large they should be before they are planted out. In warmer climates, seedlings are often only 3–9 months old before they are ready for planting in the field. Where growth is slow in cold climates, they may be raised outdoors in the nursery for 2–3 years. Practical constraints limit the size to which seedlings are grown, so they can be handled easily by planters in the field. Mason et al. (1998) found that seedlings of radiata pine (*Pinus radiata*) averaging 39-cm tall were too large to plant conveniently; as well, they suffered root distortions after planting and tended to topple over in the first few years after planting. Seedlings which averaged 23-cm tall suffered these problems to a much lesser extent. Burdett (1990) suggested that root systems are usually restricted to no more than 10–25 cm in length; if the roots are any longer than that, the planting hole (Sect. 5.3) would have to be unreasonably deep to avoid any distortion of the root system. The specifications for seedlings of *Cupressus macrocarpa* in New Zealand were that they should be 30–35-cm tall, with a root collar diameter of 0.6 cm or more and with roots no more than 10-cm long (Balneaves 1988). In Brazil, eucalypt seedlings are considered ready for planting out when aged 90–100 days, 25–35-cm tall and with a root collar diameter of 0.25–0.5 cm (Stape et al. 2001).

An interesting example which illustrates the need for careful research to determine appropriate specifications for seedlings comes from Tasmanian blue gum (*E. globulus*) in southern Australia (Close et al. 2003). Some growers there specified that seedlings from the nursery should appear to be hardy, as they believed these would survive better after planting. Hardy seedlings had relatively small and thick leaves, were red to purple in colour and had relatively short internodes (the distance between the leaves on the stem). Soft seedlings had relatively large and thin leaves, were green in colour and had long internodes. Six months after planting out, Close et al. found that growth was greater in seedlings which had higher concentrations of nitrogen in the biomass of their leaves at the time of planting out. Most interestingly, they found that hardy seedlings had low nitrogen levels (which led to the reddish-purple coloration of their leaves), whilst soft seedlings had higher nitrogen levels (and hence green leaves). Thus, growers who believed that hardy seedlings would do well were specifying the very characteristic which led to poor growth.

It has not been possible in this section to discuss in detail all the manipulations that are possible in nurseries to raise seedlings best suited for planting out. The most appropriate seedling conditions will vary considerably from species to species and from site to site on which they are to be planted. However, it should be evident that considerable research work will be necessary to identify what specifications are most appropriate for any particular circumstances. There are various reviews available of nursery practices used top raise seedlings in various parts of the world (e.g. Gadgil and Harris 1980; Burdett 1990; Knight and Nicholas 1996; Prado and Toro 1996; Stape et al. 2001; Zwolinski and Bayley 2001; Krasowski 2003).

5.3 Planting

When cultivation has been completed and suitable seedlings have been raised, planting out can proceed. Choices about the number of seedlings to be planted and the spacing between them have important consequences for the longer-term growth and development of the trees; this will be discussed in Sect. 7.2.

The first practical problem encountered during planting is to ensure that seedlings are available when required. One of the advantages of seedlings grown in containers is that this problem is largely avoided; the seedlings are transported in their containers and may be stored safely at the planting site if any delays occur in their actual planting. However, open-rooted seedlings must be lifted from the nursery bed and then perhaps stored for some time to accommodate the logistics of the planting operation.

Mena-Petite et al. (2004) have provided an example of these storage problems. They considered the effects of storage after lifting on 9-month-old open-rooted seedlings of radiata pine (*Pinus radiata*) in Spain. They studied the effects of keeping the roots either free of soil or surrounded by soil, of the length of storage (up to 15 days after lifting) and of the temperature at which seedlings were stored (in temperature-controlled chambers at 4 or 10°C). The seedlings were then planted out in a glasshouse and a detailed study was made of their physiological condition over the next 2 weeks. Lack of soil around roots, higher temperature and longer storage all increased the water stress suffered by the seedlings after planting out, reduced their root growth capacity and increased their chance of dying. Mena-Petite et al concluded that to maximise the chance of survival of seedlings they could be stored after lifting from the nursery bed for 1 week at most. Even then, storage would need to be in a cool room and with soil surrounding their roots.

In parts of the world with cold climates, it is common to lift open-rooted seedlings from the nursery just before winter and store them at near freezing temperatures before they are planted out in spring (Webb and von Althen 1980). This helps large nurseries ease their workload in spring and meet their shipping schedules. It can prevent also losses of seedlings in the nursery over winter, from animal browsing or extremely low temperatures. Jiang et al. (1994) examined the effects of this practice on 3-year-old seedlings of white spruce (*Picea glauca*) in Canada. After planting out in spring, seedlings which had been under storage over winter did not start to grow quite so early as seedlings which were lifted immediately before planting. However, after 1 year of growth, there was no difference in growth or survival of either group of seedlings. This showed that winter storage was quite appropriate for this cold-climate species which, unlike warmer-climate species, goes into a dormant phase over winter. Container-grown seedlings may also be stored over winter in freezing conditions (Helenius 2005).

Once seedlings are delivered to the planting site, there is nothing particularly complicated about the act of planting (Nyland 1996). With use of a suitable tool, a hole is dug in the ground, deep and wide enough to contain the seedling roots without distorting them. The seedling is inserted carefully into the hole, without bunching or coiling the roots, and the soil firmed around it, usually with the heel. This ensures that the roots make firm contact with the soil and that no air gaps are left within the hole. Distortions of the root system during planting can lead to long-term distortions, which may persist even when the tree is some years old; this can reduce the ability of the roots to take up water and nutrients from the site and render seedlings more likely to be blown over by the wind (Carlson et al. 1980; Nambiar 1980; Mason 1985; Harrington et al. 1989; Burdett 1990), but is not always deleterious to growth (VanderSchaaf and South 2003). The depth at which the seedling is planted is often not of great importance, although commonly they are planted with root collars a few centimetres below ground level (Stape et al. 2001; VanderSchaaf and South 2003). Planting as deep as 15 cm below ground level may still allow seedlings to develop adequately (VanderSchaaf and South 2003).

Commonly, planting is done by hand. However, there are planting machines. They are towed behind a tractor and make a furrow in the ground with a blade. Seedlings are dropped by hand into the furrow and wheels on the machine then firm the soil around the seedling. Planting machines have the advantage that, in effect, they cultivate the soil as planting is done and allow other activities, such as fertilisation, to be done at the same time as planting. However, their use is restricted to sites with even topography, with few surface rocks and without any debris on the ground.

Following planting, the two most important silvicultural treatments applied in commercial plantation forestry are control of weed growth (Sect. 5.4) and fertilisation (Chap. 6). Under difficult conditions, practices such as the use of mulches to prevent water loss, of irrigation in very dry climates or of tree guards to restrict animal browsing of seedlings (Sect. 10.4.1) or termite damage (Evans and Turnbull 2004) can be used. Plastic tubes, about 1-m-long, placed around seedlings in a tropical environment were found to boost their growth rate, probably by maintaining a humid environment around each seedling and so minimising its water use (Applegate and Bragg 1989). However, these various methods are usually too expensive to apply to large plantation areas. Rather, it is preferred that seedlings be produced in a condition which affords them the best opportunity to survive and grow without the need for other aids.

5.4 Weed Control

One of the most commonly used silvicultural practices in newly established plantations is to keep the site free of weeds for the first few years following planting. A weed is defined simply as any plant that is growing where it is not wanted. Weeds may be herbaceous (that is, non-woody) plants, like grasses or bracken fern, which grow near to the ground or perhaps to only 1–2-m high. Other weeds are woody plants, which may be bushes growing to several metres tall or may be other tree species.

5.4.1 Loss of Growth due to Weeds

The long-term benefits of weed control to growth of plantation trees can be very substantial, as several examples will illustrate. The first comes from an experimental plantation of radiata pine (*Pinus radiata*) in New South Wales, Australia (Snowdon 2002), where weed growth was either controlled or not. In this experiment, the weeds were herbaceous and were controlled by application of weedicides (chemicals which kill weeds—they are a herbicide, which is a general term for chemicals which kill plants). Weedicides were applied at the time of planting and annually for the next 3 years. By 3 years of age, the trees, both with and without weed control, would have grown large enough for their canopies to shade the ground sufficiently to prevent weeds growing, so further weed control was unnecessary. After the weed control treatments had finished, the **stand basal area** of each experimental plot was measured annually until the plantation was 16 years old. Stand basal area is the cross-sectional area at breast height of all the trees on an experimental plot, expressed per unit area of the plot (usually as square metres per hectare). It is a measure of tree growth used very commonly in forestry science, both because it is easy to measure and because it correlates quite closely with stand stem wood volume; further details about stand basal area measurement and its usefulness can be found in texts on forest measurement (e.g. West 2004).

The results of the experiment are shown in Fig. 5.5. Quite clearly, the weed control over the first 3 years led to increased growth of the trees. Their initial advantage over the trees without weed control continued to be maintained for the next 13 years.

The second example comes from eight separate experiments with plantations of loblolly pine (*Pinus taeda*) in southern states of the USA (Lauer et al. 1993). Loblolly pine is a very important species commercially in that region. It occurs naturally there and is planted extensively also (Allen et al. 1990). The experiments were conducted very similarly to Snowdon's radiata pine experiment, except that herbaceous weed control continued for

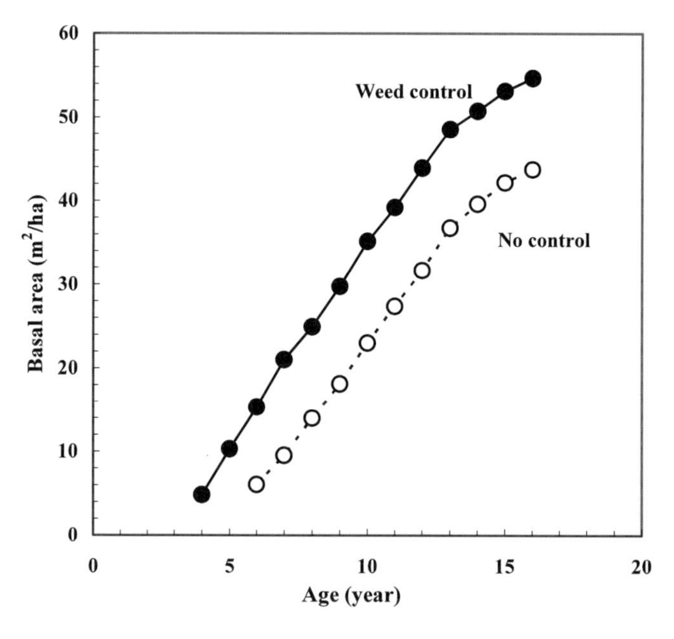

Fig. 5.5. Growth in stand basal area in an experimental plantation of radiata pine (*Pinus radiata*) in New South Wales, Australia. Weeds were either controlled for the first 3 years after plantation establishment (●——●) or were not controlled (O- - - -O) (Source—Fig. 5 of Snowdon 2002)

only 2 years, rather than 3 years, after planting. The stand stem wood volumes at 9 years of age, of plots with and without weed control, are shown in Table 5.2 for each of the eight experiments. Just as in Snowdon's case, there were very substantial gains in growth as a result of the 2 years of weed control. Gains in some of the experiments were considerably larger than in others, reflecting differences between the environmental circumstances of the different experimental sites.

The third example differs from the first two in that it concerns a plantation where the weed problem was with woody, rather than herbaceous, weeds (Glover and Zutter 1993). As in Lauer et al.'s case, the plantation species was loblolly pine (*Pinus taeda*) being grown in the southern states of the USA. The plantation was established on a site previously forested with a mixture of pine and hardwood species. The large trees in this forest had been felled for timber, leaving behind many woody bushes and seedlings and saplings of various hardwood tree species, which had formed the **understorey** of the original forest.

The experiment involved removal of the understorey remnants before establishing the plantation, either by clearing the site thoroughly with a bulldozer or by girdling (also called ringbarking) the larger remnant saplings (that is, cutting away a narrow strip of bark through the phloem down to the wood right around the stem of the tree, so its roots can no longer receive any food from its leaves and it dies—Sect. 2.1.1) and simply cut-

Table 5.2. Stand stem wood volume (cubic metres per hectare) at 9 years of age in eight experimental plantations of loblolly pine (*Pinus taeda*) in the southern USA. Weeds were either controlled for the first 2 years after plantation establishment or were not controlled (Source—Tables 1, 5 of Lauer et al. 1993, using data for check, i.e. control, plots and for those with 2-year broadcast application of weedicide)

Experiment location	Without weed control	With weed control
Chapman, Alabama	66	108
Fort Davis, Alabama	89	123
Dixie, Alabama	100	147
Cedar Bluff, Alabama	40	79
Whitesburg, Georgia	52	117
West Point, Virginia	48	81
Iuka, Mississippi	50	76
Auburn, Alabama	69	127

ting off the smaller saplings and bushes. The loblolly pine plantation was then established.

The difference between this experiment and the two others just described is that the remnant hardwood tree species are capable of resprouting after they have been girdled or cut off (a phenomenon known as coppicing—Sect. 5.5). Furthermore, they are capable of continuing to grow even when they are shaded by the canopy of the developing pine plantation. In native forests, these hardwood species normally grow slowly in the shade of taller, faster-growing species, a phenomenon known as succession in vegetation **ecology** (the study of the relationships between plants and their environment). This is quite unlike the herbaceous weeds in the two previous examples; such weeds are unable to grow in shade and disappear from the plantation site after the plantation trees have grown sufficiently large to shade the ground below.

Figure 5.6 shows the growth in stand basal area, measured from time to time until 27 years of age, of both the loblolly pine trees and the hardwood trees. In the experimental plots cleared with a bulldozer, some hardwoods were able to resprout (either from root remnants or from seeds), but their total basal area remained quite small, whilst the plantation pines grew well.

By contrast, where hardwoods were cleared by girdling, many resprouted and grew substantially, leading to much reduced growth of the pines. Furthermore, the difference in pine growth between cleared and girdled plots continued to increase with age because of the continuing growth of the hardwoods in the girdled plots. That is to say, where hardwood weeds were allowed to develop vigorously, there was a large reduction in growth of the plantation pines; this continued as long as the hardwoods continued to develop. This problem of the development of hardwood weeds occurs quite commonly in plantations in the southern USA (Allen et al. 1990).

In commercial forestry terms in these three examples, the loss in growth of plantations in which weeds were not controlled would probably have been sufficient to justify the cost of undertaking the weed control. As discussed later, that is not always the case and for any particular site it will be necessary to determine whether or not there are financial benefits from weed control. There are many other examples of the magnitude of the effects of weed control on plantation growth and/or the survival of seedlings (Nambiar and Zed 1980; Sands and Nambiar 1984; Oppenheimer et al. 1989; Allen et al. 1990; Allen and Wentworth 1993; Clason 1993; Fredericksen et al. 1993; Lowery et al. 1993; McDonald and Fiddler 1993; Perry et al. 1993; South et al. 1993a, b; Richardson 1993; Bi and Turvey 1994; Boomsma et al. 1997; Harrington and Edwards 1999; Hunt et al. 1999; Tahvanainen and Rytkönen 1999; Neilsen and Ringrose 2001; Stone and Birk 2001; Little et al. 2003a, b; Nordborg and Nilsson 2003; Thiffault et

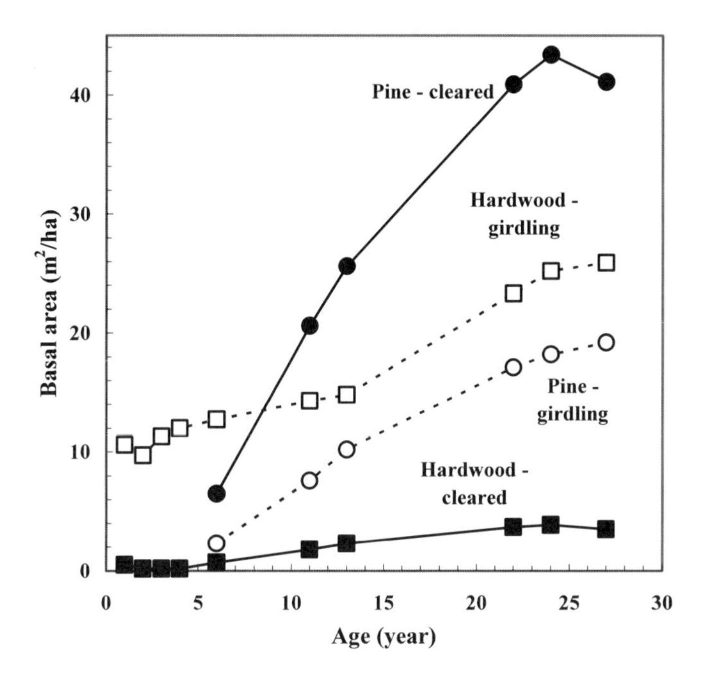

Fig. 5.6. Growth in stand basal area of loblolly pine (*P. taeda*) in an experimental plantation in the southern USA, where the site was cleared of hardwood weed species before planting (●——●) or where hardwood species were girdled only before planting (O- - -O). Also shown is the basal area growth in the same stands of the hardwood weed species, where they were cleared from the site before pine planting (■——■) or where they were girdled only before pine planting (□- - - -□) (derived using data from Table 1 of Glover and Zutter 1993)

al. 2003; Watt et al. 2004; Westfall et al. 2004; Haywood 2005; Mercuri et al. 2005; Rose and Rosner 2005). Watt et al. (2005a) found there were substantial differences between the properties of the wood of young radiata pine (*Pinus radiata*) in New Zealand, which had been grown with or without weed control. However, the effects appeared to result from the increased growth rate of the trees where weeds were controlled, rather than as a result of the weed control itself.

5.4.2 Causes of Growth Losses

When growth of plantation trees is reduced by the presence of weeds, it is believed to be a result principally of competition between the weeds and the trees for water and/or nutrient elements from the soil. An example will illustrate these effects. It comes from an experimental plantation of radiata pine (*Pinus radiata*) in the southeast region of South Australia (Nambiar and Zed 1980). This region has hot, dry summers, during which plantations can suffer a shortage of water from the soil. In this experiment, seedlings were planted in mid-winter and a detailed study was made of them the following summer. Herbaceous weeds develop vigorously on the planting sites in this region and experimental plots were established with varying degrees of weed control, using both weedicides and hand weeding.

Figure 5.7 shows the results of measurements made of the water potential of the leaves of the seedlings when they were 6 months old. Water potential is a measure of the water stress to which the leaves are subject; in effect, it is a measure of the drop in water pressure between the roots of the seedling and their leaves. The less water available from the soil and the hotter and drier the air which is evaporating water from the leaves through their stomata, the greater will be the pressure drop and the more water stress will the leaves experience. Water potential is measured in units of pressure. Conventionally, it is recorded as a negative value; the lower the value, the greater the water stress. To limit water loss by the plant, stomata respond to leaf water potential and start to close when it falls below a cer-

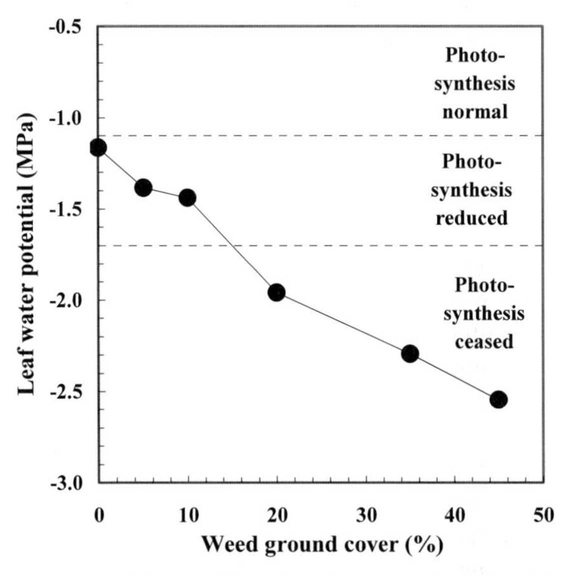

Fig. 5.7. Leaf water potential, at midday in mid-summer, of 6-month-old radiata pine (*P. radiata*) seedlings in relation to the proportion of the ground covered by herbaceous weeds, in an experimental plantation in southeastern South Australia. The *dashed lines* indicate water potential ranges where photosynthesis of leaves can continue normally, is reduced or ceases altogether because of complete closure of the stomata (Source—Fig. 1 and text of Nambiar and Zed 1980)

tain level. In radiata pine, stomata start to close when water potential falls below about −1.1 MPa and close fully when it reaches less than about −1.7 MPa. Once the stomata are closed completely, no photosynthesis can occur because carbon dioxide from the air can no longer enter the leaves.

The measurements shown in Fig. 5.7 were made at midday on one day in mid-summer, in experimental plots which had different amounts of the ground covered by the foliage of weeds. It is obvious that the more weed cover there was, the more water stress were the seedlings suffering. This shows that the weeds were competing with the tree seedlings for water from the soil. Of course, the degree of water stress in the tree seedlings would have varied during the day as the air temperature changed. Also, stress would have been reduced on days following rain when more water became available from the soil. Hence, the amount of photosynthesis which the seedlings could undertake would have varied from day to day and from time to time during the day. Nevertheless, the higher water stress suffered from time to time by seedlings in plots with more weed cover would have led to an overall reduction in their photosynthesis and, hence, their growth; by 9 months of age, seedlings in the experiment with thorough weed control averaged 52-cm tall, whilst those in plots without weed control averaged only 23-cm tall. Deaths were more common also in the plots without weed control; at 1 year of age none of the trees had died in the plots with thorough weed control, whilst 40% were dead in the plots without.

The concentrations of nutrient elements in the leaves of these seedlings varied also with the level of weed control. At 11 months of age, seedlings in plots with thorough weed control had higher concentrations of both nitrogen and potassium (22,000 and 9,900 mg/kg respectively) than plots without weed control (15,000 and 4,500 mg/kg). Phosphorous concentrations were similar with and without weed control (1,400 mg/kg). The lower nitrogen and potassium concentrations where weeds were present may have been a result of the weeds competing with the trees for the nutrients; later work (Smethurst and Nambiar 1989) confirmed that weeds indeed compete with the trees for nutrients in these plantations. Many other research papers have reported on the water stress and nutrient uptake by tree seedlings in plantations with and without weed control (e.g. Sands and Nambiar 1984; Elliott and Vose 1993; Fredericksen et al. 1993; Richardson 1993; Hunt and Beadle 1998; Harrington and Edwards 1999; Adams et al. 2003; Nordborg and Nilsson 2003; Thiffault et al. 2003, 2005; Watt et al. 2003, 2004).

Less commonly than competition for water and nutrients, it has been suggested (Lowery et al. 1993; Richardson 1993; Sands et al. 2000; Nordborg and Nilsson 2003) that growth losses due to weeds can be caused also by:

- Weeds reducing air temperatures near ground level and so increasing the risk of frost damage to seedlings
- Parasitism of the trees by some weed species
- The exudation from roots of some weeds of substances which inhibit the growth of trees (allelopathy as this is called)
- Increased concentrations of carbon dioxide in the soil resulting from the presence of weeds
- Increased damage to the trees by animal pests sheltering amongst weeds.

5.4.3 Controlling Weeds

Where it is necessary, weed control is one of the major expenses incurred over the first few years of the life of plantations, until the trees have grown sufficiently large to restrict or eliminate the growth of weeds altogether.

Cultivation can be effective in restricting weed growth initially. However, pre-emergent weedicide sprays are applied commonly following cultivation; these restrict germination of the seeds of weeds. Over the next few years, post-emergent weedicides are often used; these are capable of killing weeds which have already grown. There is a wide range of chemicals available for weed control and they differ both in their capabilities of killing different weed species and in the extent to which they damage the trees if they are sprayed accidentally on them. Weedicides are commonly applied using boom-sprays carried on tractors, which are often mounted with protective guards to prevent the spray drifting onto the tree seedlings. Some weedicides perform adequately only if they are applied at certain stages of development of the weeds, so care is required to spray the weeds during the appropriate season.

Of course, there must be concern also that the broad-scale spraying of chemicals such as weedicides may have deleterious environmental effects. The fate of the weedicides in the soil and the possibility that they might be carried eventually into watercourses must be considered (Wilkins et al. 1993; Bubb et al. 2003). Because of the threats these chemicals may pose, governments in some parts of the world have restricted the use of weedicides in forest plantations.

Methods of weed control which are potentially less threatening to the environment include the use of grazing animals (provided they are allowed into the plantation only when the trees have grown sufficiently that the animals will not damage them), mulches, weedicides developed from natural chemicals, soil cultivation or biological agents such as insects or diseases (Markin and Gardner 1993; McDonald and Fiddler 1993; Green 2003; Clay et al. 2005; Thiffault et al. 2005). Various authors have reviewed the techniques available for and use of weed control in plantations (Margolis and Brand 1990; Campbell and Howard 1993; Lowery et al. 1993; Markin and Gardner 1993; McDonald and Fiddler 1993; Richardson 1993; Teeter et al. 1993; Wagner 1993; Green 2003; Prado and Toro 1996; Weih 2004; Mead 2005)

5.5 Coppice

The establishment of a new plantation following harvesting of the previous plantation crop usually involves a sequence of events similar to those described earlier in this chapter. However, for many species and plantation circumstances, it is possible to simply allow the trees from the previous plantation to resprout from the cut stumps left after harvesting. This resprouting is known as coppice. The stem form and wood quality of the trees which grow eventually from coppice may be just as good as trees grown from seedlings. Re-establishment of a plantation by coppice may be repeated a number of times, so that several coppice **rotations** may follow a rotation established initially with seedlings (Sect. 5.5.2).

Where it is possible to re-establish a plantation using coppice, a great deal of expense can be avoided (Rosenqvist and Dawson 2005b; Whittock et al. 2004; Bailey and Harjanto 2005). Coppice has been used quite extensively, from time to time and place to place, around the world. It may become increasingly important if plantations being grown for bioenergy production become more common. Because these are grown on short, perhaps 3–5-year, rotations and are planted with high stocking densities of seedlings, re-establishment will be a frequent event requiring large numbers of seedlings. Under these circumstances, it may be economically feasible to grow these plantations only if they can be re-established using coppice.

Coppice is a form of **epicormic shoot** development. These are shoots which develop from buds in the phloem. The buds may remain dormant, until the tree is damaged in some way (often through causes such as fire, severe drought or heavy browsing by leaf-eating insects) or may develop in the cambium at cut surfaces (when they are known as callus). After the damage, the buds develop to form new leaves, branches or stems.

Not all species produce coppice readily. Generally, softwood species do not, although there are exceptions. Many hardwood species produce coppice readily, although again there are some species which do not (Evans and Turnbull 2004). Variation in the environmental circumstances of different sites may affect coppice development of different species in different ways (Little and Gardner 2003). The coppice from different genetic strains of the same species can develop rather differently (Robinson et al. 2004).

5.5.1 Wood Production by Coppice

Apart from the cost savings in re-establishing a plantation using coppice, another important advantage is that the growth rate of coppice is often greater than that of seedlings. This productive advantage seems to occur particularly during the first few years after coppicing.

It is not understood fully why coppice grows faster than seedlings. Some research has shown that plants must have available a food reserve (usually stored as starch) in their roots, if coppice is to develop (Bamber and Humphreys 1965; Taylor et al. 1982; Tschaplinski and Blake 1994, 1995; Luostarinen and Kauppi 2005). This can be drawn upon by the newly developing shoots until they produce new leaves and can undertake photosynthesis themselves. As well, coppice has available the root system of the cut tree and, unlike a seedling, does not have to develop a new root system. Other research has found that coppice behaves as if the plant has been rejuvenated or reinvigorated (Bachelard 1969; Taylor et al. 1982; Blake 1983; Tschaplinski and Blake 1989a, b, 1995; Florence 1996, p. 155). None of this research is definitive and much remains to be done to explain why coppice often shows very high growth rates.

In practical terms, Evans and Turnbull (2004) suggest that over longer rotations (say, 10–15 years or more), production by coppice often differs little from seedling rotations, even though coppice grows faster in its early years (see also Bailey and Harjanto 2005). However, after several coppice rotations its production eventually falls below that of a seedling rotation; perhaps the trees eventually lose their capability to rejuvenate or be reinvigorated. Evans and Turnbull (2004) suggest that usually three or four coppice rotations can be grown before it is necessary to replant with seedlings. However, this is not always so; Pawlick (1989) refers to *Markhamia lutea* in Kenya, a species which has been found to coppice readily and repeatedly, without loss of production, over periods as long as 100 years.

There is a growing body of practical studies which suggest that the early, rapid growth of coppice may be used to advantage in the short (3–5 years, say) rotations of bioenergy plantations. In their review of hardwood bioenergy plantations in Europe and North America, Cannell and Smith (1980) concluded that coppice produced 10–30% more biomass than seedlings over 4–5-year-long rotations. However, after several rotations the advantage seems to be lost and a new seedling rotation would need to be established. Similar effects have been observed with numerous hardwood species in various parts of the world (Carter 1974; Kaumi 1983; Oliver 1991; Willebrand et al. 1993; Kopp et al. 1993, 1996; McKenzie and Hay 1996; Bergkvist and Ledin 1998; Pontailler et al. 1999; Liesebach et al. 1999).

An interesting example of the potential for the use of coppice in bioenergy plantations comes from 15 years of experimental work, with 19 different eucalypt species, in the north island of New Zealand (Sims et al. 1999). Following a single seedling rotation of 3 years, four successive 3-year coppice rotations were established. In terms of stem wood volume production, the best performing species was swamp gum (*E. ovata*), a species native to southeastern Australia. It had mean annual increments in stem wood volume to 3 years of age of about 24, 31, 54, 39 and 100 m^3/ha/year over each of the seedling rotations and four successive coppice rotations, respectively.

Comparison with the information in Fig. 3.2 will show that the growth rate of 100 m^3/ha/year over 3 years, for the fourth coppice rotation, is well above the highest growth rates which have been encountered usually in plantations of hardwood species; it is one of the highest growth rates for a plantation forest that I have come across ever in the scientific literature. Certainly part of the reason for it was that many coppice shoots had developed on each stump by the fourth coppice rotation, so there was a total of about 10,000 trees per hectare in the plantation (which had been planted initially with about 2,200 seedlings per hectare); higher stocking densities produce higher wood yields (Sects. 7.1.3, 7.2.1). The differences in growth rates of coppice from rotation to rotation in Sims et al.'s results may reflect weather differences during each rotation period. However, there was certainly no evidence that coppice production was declining after several rotations.

5.5.2 Silviculture of Coppice

Given that coppice arises from buds within the bark of tree stumps, it is obviously important that whatever harvesting system is used to fell the trees of the previous rotation it does not strip the bark from the stump. Generally, the use of harvesting implements which leave a clean, although not necessarily smooth, cut surface of the stump seems to be recommended for most coppice systems. Pawlick (1989) suggested that a ragged cut, or one which leaves bark hanging loose, may allow moisture to be trapped and encourage insect or fungal attack (Chaps. 10, 11). It has been suggested that a rougher, sawn surface may be better than a smooth surface (such as that from an axe cut) because callus development to heal the stump may occur more readily on a rougher surface (Evans and Turnbull

2004). However, Oliver (1991) used pruning shears, which would be expected to leave a smooth surface, with apparent success to establish coppice on stumps of shining gum (*E. nitens*) in New Zealand. Both chainsaws and bow saws have been used to establish coppice in eucalypt plantations (Kaumi 1983; Oliver 1991; Brandenburg 1993; Evans and Turnbull 2004).

A variety of observations have suggested that the success and vigour of coppice vary with the diameter of the cut stem (Luoga et al. 2004) or the height at which the stump is cut (Blake 1983; Evans and Turnbull 2004). There seems to have been little research done to determine the reasons for this, although Sakai et al. (1997) suggested that food reserves in taller stumps may be necessary to allow buds to develop in some species. For eucalypts, Evans and Turnbull (2004) suggested that stump height was not critical for coppice development. However, successful coppice has resulted in various species when stumps were cut 'as near to ground level as possible' (Kaumi 1983), whilst others have used or recommended heights of 10–12 cm (Nicholas 1993; Luoga et al. 2004) or 20–50 cm (Cremer et al. 1984; Oliver 1991; Brandenburg 1993; Hummel 2000; Luoga et al. 2004). Evans and Turnbull (2004) suggested that if the stump height is too high, the coppice may be more susceptible to being blown over by strong winds.

On the basis of observations of coppice from various species in New Zealand, it has been suggested also that the stump should have its cut face sloping, to allow rain to run off (Nicholas 1993), at 20–45° below the horizontal and with its cut surface not facing the sun (Brandenburg 1993). Experience in New Zealand has found also that application of a fungicide solution to the cut surface may improve stump survival (Nicholas 1993; Brandenburg 1993) and, presumably, the sloping cut assists in keeping the surface dry and reducing fungal development (or insect attack; Pawlick 1989).

Because the availability of food reserves is essential to coppice development (Sect. 5.5.1), it should be best if trees are cut at the beginning of the season of active growth when food reserves have accumulated during the season of little growth (Cremer 1973; Clarke 1975; Blake 1983; Webley et al. 1986). An example of the importance of the season of cutting for coppice development was given by Clarke (1975), for 6-year-old flooded gum (*E. grandis*) plantations in northern New South Wales, Australia. When stumps were cut in spring (September or October), immediately after the period of slower growth in autumn and winter (March–August), about 90% of the stumps developed coppice. This proportion declined steadily as cutting was done in later months, until less than 50% of stumps produced coppice if they were cut during autumn and winter.

Normally more than one coppice shoot will develop from a cut stump. As many as 20 have been observed in eucalypts (Florence 1996, p. 154).

Where a plantation is being grown for timber products or paper pulp, it is usually necessary to thin multiple coppice stems to a single stem on each stump, or perhaps two stems (Little and du Toit 2003). If this is not done, the higher stem stocking density of the unthinned coppice plantation will produce final-crop trees with an average stem diameter lower than desired (Sect. 7.1). This is less of a problem for bioenergy plantations, where small tree sizes are of little concern. As well, the large number of shoots on each coppice stump may reduce the efficiency of the final harvesting operation. Perhaps even more importantly, if more than a few shoots arise, they tend to lean out from the stump and become liable to breakage (Florence 1996, p. 154).

6 Nutrient Management

It is of considerable importance to ensure that sufficient of the nutrient elements essential to the growth of trees (Sect. 2.1.4) is available as and when they are required, or tree health and growth rates will suffer. Furthermore, the nutrients available on a site are such a valuable resource that it is important that any losses from the site be minimised, both in the short term during any one plantation rotation and in the long term over many rotations.

Any shortages that occur in nutrient supply can be made up by fertilisation; however, this is expensive and should be used only if the resulting gains in production can be justified economically. Furthermore, if fertiliser is to be used, it is important that it is of the right type and is applied at just the right time and in just the right amounts to match the requirements of the growing trees. If insufficient is added, growth may suffer. If too much is used, the nutrients may simply be lost after dissolving in the water in the soil and being washed through the soil with rainfall (a process known as **leaching**) (Bruijnzeel 1997).

For most forest plantations, nutrient availability is the only one of the resources essential for the growth of plants which can be managed realistically by the plantation grower. Irrigation can be used to make more water available, but it is an extremely expensive practice, generally used in plantations only for very special purposes. The availability of light and carbon dioxide cannot be controlled in any broad plantation programme.

It is very difficult to generalise about how nutrients should be managed on any particular plantation site. Soils differ greatly, both in their ability to supply the nutrient needs of a plantation and in the extent to which they may lose nutrients through leaching. Tree species differ in their nutritional requirements. This means that the need for fertilisation must be assessed individually for any particular site and plantation species. At present we have no simple and straightforward method of going to any site and assessing easily whether or not fertilisation will be necessary; it often requires complex, long-term experiments to do so. The later chapters in Attiwill and Adams (1996) illustrate how varied is the need for fertilisation in different parts of the world.

This chapter describes how the demand for nutrients changes during the life of a plantation and how fertiliser regimes can be developed to match that demand. As well, it considers the issues surrounding the long-term maintenance of site fertility.

6.1 Early Growth in Relation to Nutrient Supply

The first few years of growth of a plantation are particularly crucial for the supply of nutrients. As the trees grow and their canopies and root systems expand, their nutrient requirements increase correspondingly until the canopy closes and fine roots and leaves reach a more or less constant stand biomass (Sect. 2.2).

During this period, it is possible to control the plantation growth rate by ensuring that the right amounts of nutrients are available to the trees at the right times. Our understanding of how to do this derives from some basic research, carried out in the late 1970s and the 1980s by a Swedish scientist, Torsten Ingestad, and his colleagues. The theory that Ingestad developed is discussed in Sect. 6.1.1. The practical consequences of the theory for fertilisation practice is discussed in Sect. 6.1.2.

6.1.1 Ingestad's Theory

Ingestad studied how the growth rate of very young tree seedlings was affected by altering the rate of supply of nutrient elements to their roots. He carried out his experiments using the apparatus shown diagrammatically in Fig. 6.1. Small tree seedlings were supported in holes through the lid of a bowl about the size of a domestic washing machine. Water containing dissolved nutrient elements was sprayed into the bowl, up and around the roots of the seedlings. Meters measured the concentration of nutrients in the water as it entered and left the bowl. A computer system controlled additions of nutrients to the solution at frequent intervals (hourly, say), so that the amounts of nutrients being supplied in the water to the roots was controlled very closely. The entire apparatus was kept in a greenhouse or controlled environment room. More details of the apparatus and its use can be found in Ingestad and Ågren (1992).

Ingestad conducted experiments in which he supplied all but one nutrient element to the seedlings at rates well in excess of the maximum that the seedlings could take up. The one nutrient, he supplied at a very carefully controlled rate, known as the **relative addition rate**, chosen so that the seedlings would have less of the nutrient element available to them

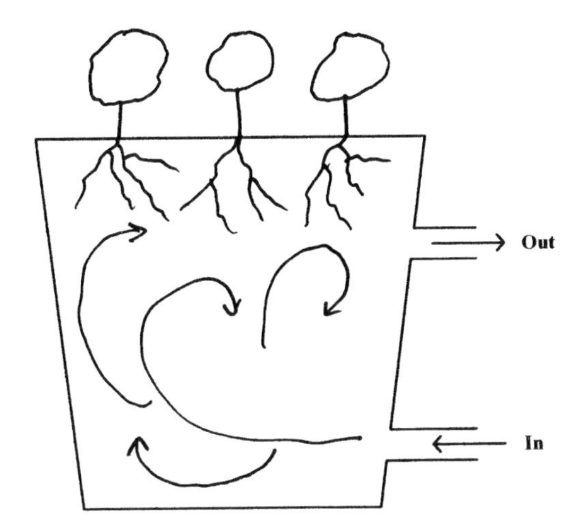

Fig. 6.1. Representation of the apparatus used by Torsten Ingestad for experiments relating tree seedling growth rate to nutrient supply rate. A water solution of nutrient elements entered and left the apparatus, as indicated by the *arrows*, and was sprayed directly onto the roots of the seedlings (adapted from Ingestad and Ågren 1992)

than they were actually capable of taking up. Relative addition rate is the percentage increase, over any day of the experiment, in the amount of a nutrient element in the seedlings. From the measurements he made of seedling biomasses during an experiment, which would run usually for several weeks, Ingestad could determine their **relative growth rate**, that is, the percentage increase over any day of the experiment of the biomass of the seedlings.

Ingestad conducted such experiments with many different tree species and many different nutrient elements. He found that if the relative addition rate of a nutrient was kept constant over the duration of the experiment, the relative growth rate of the seedlings would remain constant also and be *equal* to the relative addition rate. Put simply, this means that the rate at which nutrients are supplied to seedlings will determine exactly the rate at which they grow (provided of course that the availability of no other factor essential to plant growth limits the seedling growth rate).

Results from two of Ingestad's experiments, one with silver birch (*Betula pendula*) and one with Scots pine (*Pinus sylvestris*), are shown in Fig. 6.2. In both cases, nitrogen was the growth-limiting nutrient.

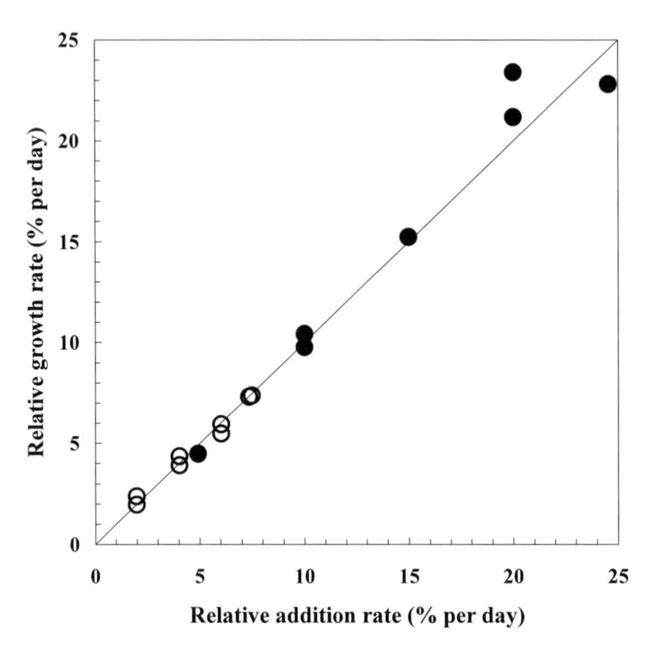

Fig. 6.2. Results from two experiments conducted by Ingestad, where seedlings of the hardwood silver birch (*Betula pendula*) (●) or the softwood Scots pine (*Pinus sylvestris*) (O) were grown at different relative addition rates of the nutrient element nitrogen. Seedling relative growth rates are shown plotted against relative addition rates. The *line* shows where the points would lie if relative growth rates were exactly equal to relative addition rates (adapted from Ingestad and Kähr 1985 and Ingestad and McDonald 1989, using data for seedlings lit with a photon flux density of 22 mol/m^2/day)

It is obvious from the data points plotted in Fig. 6.2 that at any relative addition rate of nitrogen for either species, the corresponding relative growth rate was nearly the same. If they had been *exactly* the same, all the plotted points would have fallen on the solid line which has been drawn diagonally across the figure. However, the plotted points fall sufficiently close to the line that Ingestad could propose confidently his theory that relative growth rate does equal relative addition rate. Ingestad and other researchers have conducted similar experiments with many other tree species and with other nutrient elements, always with results similar to those of Fig. 6.2 (e.g. Ingestad 1982; Ericsson and Ingestad 1988; Cromer and Jarvis 1990; Ericsson and Kähr 1993; Kriedemann and Cromer 1996; McDonald et al. 1996).

6.1.2
Conclusions for Fertilisation Practice

From Ingestad's theory, there are important ramifications for fertilisation practice in the early stages of forest plantation growth as follow:

- The higher the relative addition rate at which nutrients are supplied to seedling roots, the faster will seedlings grow. In effect, applying fertiliser to a plantation forest is an attempt to increase the relative addition rate of nutrients from the soil and, hence, the seedling growth rate.
- To maintain those seedling growth rates, the actual amount of nutrient which must be supplied from the soil daily to the seedling roots will increase progressively as the seedlings grow larger (this follows directly from the mathematics of Ingestad's theory[1]). That is to say, an ever increasing amount of nutrients must be available from the soil daily to supply the needs of the seedlings.

Of course, the growth rates of seedlings in a plantation can never be controlled with the precision possible in an experiment using Ingestad's apparatus. There are various reasons for this as follow:

- The relative addition rate of nutrients to plant roots from soil is much more difficult to control than spraying a water solution of nutrients directly on the roots as in Ingestad's experiments; experiments attempting to emulate Ingestad's results, with seedlings growing in pots or

[1] For the mathematically minded, Ingestad's theory, that relative growth rate equals relative addition rate, can be written as

$$(1/W)\mathrm{d}W/\mathrm{d}t=(1/N)\mathrm{d}N/\mathrm{d}t=r \ , \qquad (6.1)$$

where r (proportion per day) is both the relative growth rate and the relative addition rate (the latter being maintained constant during an experiment), and, at time t (days), W is the oven-dry biomass (grams) of seedlings and N is the weight (grams) of the limiting nutrient element they contain.

By integrating Eq. 6.1, it follows that, at any time t, the oven-dry biomass of the seedlings will be given by

$$W=W_0\exp(rt) \ , \qquad (6.2)$$

and the content of the seedlings of the limiting nutrient will be given by

$$N=N_0\exp(rt) \ , \qquad (6.3)$$

where W_0 and N_0 are, respectively, the oven-dry biomass (grams) and nutrient content (grams) of the seedlings at t=0, which occurs some time after seedlings have been exposed to and become acclimatised to that particular nutrient relative addition rate

It follows from this theory that, for any given r, the concentration of the limiting nutrient element in the seedlings will remain constant throughout the duration of an experiment at a value of N_0/W_0 (grams per gram). Experiments have found that this concentration is directly proportional to r, although the constant of proportionality varies from species to species (Ingestad and Kähr 1985; Ingestad and McDonald 1989).

trays filled with soil, have had varying success (Timmer and Armstrong 1987; Burgess 1991; Timmer et al. 1991; Hawkins 1992).

In soil, nutrients are released into water held in the spaces between the soil particles; the nutrients are either dissolved from the soil particles themselves or are released from organic matter in the soil as it is broken down by microorganisms (mainly bacteria and fungi). The soil chemical and physical properties will determine how much of and how quickly the nutrients are released into soil water and, hence, what their relative addition rates will be.

Further, as plant fine roots take up the nutrients from the soil water, they deplete the amount of nutrients in the water immediately adjacent to the root. More nutrients must then diffuse from the surrounding water to the depleted region. There is no 'stirring' of the water in soil, equivalent to the spraying of nutrients in Ingestad's apparatus. This means that the availability of nutrients at root surfaces will depend on the rate at which nutrients can diffuse through the soil water to nutrient-depleted zones immediately adjacent to the roots; the availability of water in the soil is very important itself in determining how rapidly this can occur (Gonçalves et al. 1997).

- Other environmental factors, such as seasonal changes in temperature or periods of drought over a dry summer, may limit the growth rates of seedlings from time to time during a year. Over those periods, nutrient availability will not be the factor limiting seedling growth rates; the extent to which low availability of water will often override the importance of nutrient availability in controlling plantation growth has been discussed by many authors (e.g. Kimmins et al. 1990; Nambiar 1990, 1990/91; Nambiar and Sands 1993; Gonçalves et al. 1997).
- Application of fertiliser to plantations is a costly practice and it will be possible to apply fertiliser only on two or three occasions at most during the first few years of the life of a plantation. Thus, it will never be possible in practice to control the relative addition rates of nutrients from soil with the precision possible in Ingestad's apparatus.

Given these various considerations, it is now accepted that over the early years of plantation growth, until the canopy closes, fertiliser should be applied as often as is practicable and the amounts should increase progressively with time to keep pace with the ever-increasing amounts the trees require. If more fertiliser is applied than the trees require when they are very small, the excess nutrients will simply be leached from the soil and wasted. If insufficient is applied when the trees are larger, they will be unable to maintain their growth rate. Hence, the amounts applied at each fertiliser application should be carefully judged to match the nutrient requirements of the trees.

6.2
A Fertiliser Regime for Sweetgum in North America

This section will give an example of the modern approach to the development of a fertiliser regime for the early years of plantation growth (that is, when and how much fertiliser should be applied), drawing on the principles of Ingestad's theory.

6.2.1
Example Details

The example comes from experimental work with a plantation of sweetgum (*Liquidamber styraciflua*) in South Carolina, USA (Scott et al. 2004a, b). There is increasing interest in North America and Europe in growing plantations of hardwoods, on 3–20-year-long rotations, to produce wood either for bioenergy or paper pulp production (Rosenqvist et al. 1997; Scholes 1998; Toivonen and Tahvanainen 1998; Updegraff et al. 2004; Andersen et al. 2005; Rosenqvist and Dawson 2005a). Sweetgum is one such species.

On the sites they were considering, Scott et al. believed it was the availability from the soil of nitrogen which limited the growth rate of sweetgum. To explore this, they established an experiment on a site which had been used previously to grow a plantation of loblolly pine (*P. taeda*). This had been harvested shortly before establishing the sweetgum plantation.

Both nitrogen and phosphorus fertilisers were applied to the sweetgum plantation shortly after planting. The experiment started when the plantation was 2 years of age. Experimental plots received either 0, 56 or 112 kg/ha of nitrogen in nitrogen fertiliser. The plots were fertilised again with the same amounts of nitrogen at 4 years of age.

Scott et al. took detailed measurements, year by year as the experiment continued. They measured the growth of the trees and the amounts of nitrogen both in the trees and available to them from the soil.

6.2.2
Growth Response to Fertilisation

Some results from the experiment, at 6 years of age, are given in Table 6.1. They show that trees in plots which had been fertilised were appreciably larger in average height and diameter and in the stand biomass of both stems and foliage. However, there was little advantage gained by applying 112 kg/ha of nitrogen in each fertiliser treatment over the gain with 56 kg/ha of nitrogen. That is to say, applying 56 kg/ha of nitrogen biannually apparently led to the maximum growth on the site; adding more fertiliser

Table 6.1. Average tree and stand characteristics, at 6 years of age, in an experimental plantation of sweetgum (*Liquidamber styraciflua*) in South Carolina, USA. Experimental plots received different amounts (as indicated in the first column) of nitrogen in fertiliser at 2 and 4 years of age (Source—Tables 2–4 of Scott et al. 2004a)

Nitrogen applied at each fertiliser treatment (kg/ha)	Tree average diameter at breast height (cm)	Tree average height (m)	Stand stem oven-dry biomass (including bark) (t/ha)	Stand foliage oven-dry biomass (t/ha)
0	6.3	7.0	8.4	1.8
56	7.0	8.6	13.6	2.7
112	7.0	8.5	13.7	3.5

would simply be wasteful. By 6 years of age, the fertilised plots had grown to the stage where the tree canopies had closed and the stand foliage biomass had reached its maximum and could be expected to remain more or less constant thereafter.

To this point, Scott et al.'s results are typical of experiments done by many forest scientists, with many different tree species in many parts of the world. These experiments have shown varying degrees of growth advantage from application of fertiliser at various stages of development of plantations on particular sites. Typical also of such experiments is the rather obvious conclusion which can be drawn from Scott et al.'s results; applying nitrogen fertiliser to sweetgum plantations on their site led to a tree growth advantage. However, these results do not answer certain questions about what would be the most appropriate fertiliser regime for sweetgum, particularly:

- If fertiliser had been applied more frequently, would there have been an even greater growth advantage?
- Should the amount of fertiliser applied have increased in successive applications, to keep pace with the increasing demand of the trees as they grew larger, consistent with Ingestad's theory?

Because Scott et al. had measured much more in their experiment than just the tree sizes at 6 years of age, they were able to develop answers to these questions. This involved developing an annual nitrogen budget for their experimental plantation. This is described in Sect. 6.2.3.

6.2.3 Nutrient Budget

Scott et al.'s first step in developing a nitrogen budget was to determine, year by year, how much nitrogen was present in the foliage of the trees.

From time to time during the experiment, they took samples of foliage from the trees in their experimental plots. These samples were analysed chemically in the laboratory to determine the concentration of nitrogen in the foliage.

When Scott et al. combined all the growth and foliage nitrogen concentration data they had collected over 6 years, they concluded that trees with a concentration of 17,500 mg/kg or more nitrogen in their foliage would be growing as fast as was possible on that site. This is somewhat less than the 25,000 mg/kg, shown in Table 2.1 as being the nitrogen concentration in foliage of normal, vigorously growing plants. This emphasises that the information in Table 2.1 is simply an average for plants in general and specific values need to be determined for any particular species of interest.

Given these results and other information Scott et al. obtained from the trees and soils in their experiment, they could determine what the nitrogen budget should be annually, if the plantation was to grow as fast as possible on the site. An example of this budget, for growth between 4 and 5 years of age, is shown diagrammatically in Fig. 6.3.

Sweetgum is a deciduous species, so loses all of its foliage annually in autumn and replaces it in spring. Their observations of the fastest-growing plantations on their site suggested to Scott et al. that, between 4 and 5 years of age, the plantation would grow an oven-dry biomass of foliage of 3.3 t/ha. If the concentration of nitrogen in that foliage was 17,500 mg/kg, the total amount of nitrogen required by the leaves would be 58 kg/ha (the leaf oven-dry biomass multiplied by nitrogen concentration in leaves). This amount of nitrogen is shown in the 'foliage' box in Fig. 6.3.

Scott et al. were then able to determine the sources from which that 58 kg/ha of nitrogen would be obtained. When leaves fall from trees, the trees are able to withdraw at least some of the nutrients from the leaves before they die and store them in other tissues for reuse in new leaves as they develop. This process is known as nutrient retranslocation; different species differ greatly in the extent to which they are able to do this and some nutrient elements can be retranslocated much more readily than others (Ericsson et al. 1992; Miller 1995; Negi and Sharma 1996). In the sweetgum case, Scott et al. found that 23 kg/ha of nitrogen would have been retranslocated when the trees lost the leaves which had grown the previous year; this amount would be made available to the new leaves which developed between 4 and 5 years of age, as indicated by the 'retranslocation' amount in the figure.

They determined also that a total of 69 kg/ha of nitrogen was available from the soil, as shown in the 'soil' box in the figure. Some of that would not be available to be taken up by the tree roots, simply because roots do not reach every point in the soil. Also, some would have been taken up for

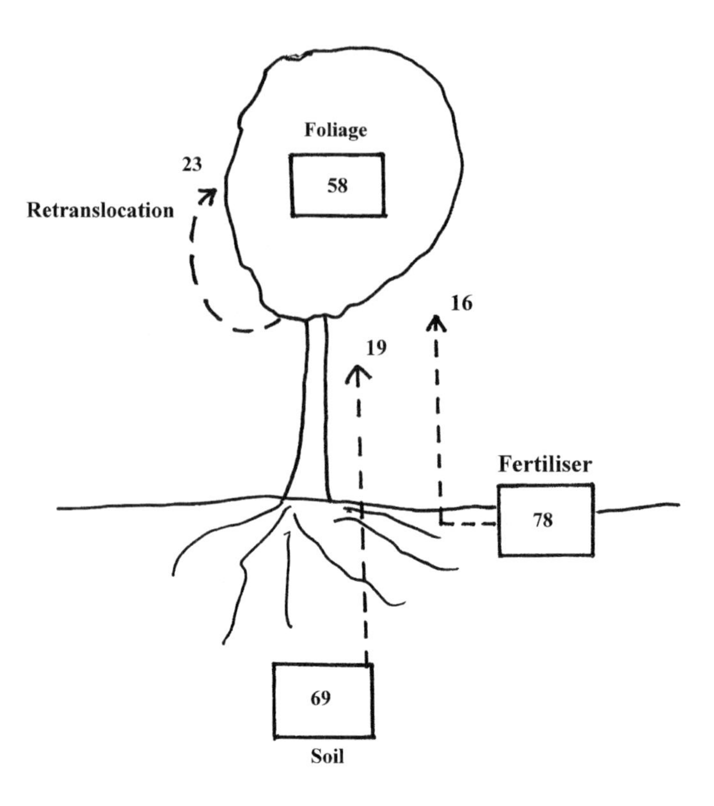

Fig. 6.3. The nitrogen budget, between 4 and 5 years of age, to achieve maximum growth of a sweetgum (*Liquidamber styraciflua*) plantation in South Carolina, USA. The values shown are amounts of nitrogen in a stand in kilograms per hectare. The values in the *annotated boxes* are the amounts required by the foliage, the amount available for uptake by the trees from the soil and the amount to be added at the start of the year in fertiliser. The *dashed arrows* indicate the transport of nitrogen to the foliage from the various sources. The foliage oven-dry biomass was 3.3 t/ha, of which 17,500 mg/kg was nitrogen (derived using Eq. 10 and Fig. 3 of Scott et al. 2004b)

use by newly developing fine roots. Scott et al. found that 19 kg/ha of it would be taken up and made available to the leaves.

Thus, Scott et al. identified that 42 kg/ha (23+19) of the foliage nitrogen requirement would have come from retranslocation and uptake from the soil. This meant that a further 16 kg/ha of nitrogen would have to be made available to the foliage by adding it in fertiliser, so that the leaves would have the total 58 kg/ha of nitrogen they required. They found that 78 kg/ha of nitrogen would have to be added to the soil as fertiliser to supply this, the amount shown in the 'fertiliser' box in the figure; as for the nitrogen in the soil, not all of the nitrogen applied in fertiliser would be available to the foliage.

Other researchers have developed far more detailed nutrient budgets of forests than Scott et al., for nitrogen and for other nutrient elements (e.g. Miller et al. 1979; Mahendrappa et al. 1986; Beets and Pollock 1987; Eckersten and Slapokas 1990; Bargali et al. 1992; Thomas and Mead 1992a, b; Nilsson et al. 1995; Grove et al. 1996; Negi and Sharma 1996; Fölster and Khanna 1997; Neilsen and Lynch 1998; Parrotta 1999; Laclau et al. 2003, 2005). However, Scott et al. felt they had obtained sufficient information from their budget to make recommendations for the fertiliser needs of their sweetgum plantation (Sect. 6.2.4). In general, the derivation of a nutrient budget for any particular type of site and tree species is an essential part of developing an appropriate fertiliser regime for a plantation.

6.2.4 Fertiliser Regime

From their annual nutrient budgets, Scott et al. determined that maximum growth of their sweetgum plantation would be achieved if 15, 70, 78 and 82 kg/ha of nitrogen in fertiliser was applied at 3, 4, 5 and 6 years of age, respectively. These fertiliser amounts increase annually, to keep pace with the increasing demand for nitrogen as the stand foliage biomass of the plantation increases year by year. By 6 years of age, the canopy of the plantation would have closed and the stand foliage biomass should remain more or less constant thereafter.

The development of this fertiliser regime by Scott et al. is completely consistent with the principles established from Ingestad's theory (Sect. 6.1). It aims to maintain nutrient concentrations in tissues at a level appropriate to ensure maximum growth. As well, fertiliser is added from time to time in ever-increasing amounts to try to keep the relative addition rate of nutrients from the soil constant as the demand of the trees for nutrients increases year by year as they grow.

Of course Scott et al.'s regime is based on the biological requirements of the trees. In practice, it would be necessary to determine if the gains in income achieved from the higher wood yields from a fertilised plantation more than offset the costs of fertilisation. If they did not, fertilisation would not be worthwhile economically. The effects of fertilisation on wood properties will also affect economic returns. Much work remains to be done on the effects of fertilisation on wood quality. Lower wood density and increased branch size and number have been observed in trees from fertilised plantations; however, these often seem to be the effects of faster growth from the fertilisation, rather than from any effects of the nutrients themselves. There are various studies on these issues (e.g. Wilkins

1990; Clearwater and Meinzer 2001; Mäkinen et al. 2001, 2002b, c; Anttonen 2002; Amponsah et al. 2004).

6.3 Long-Term and Later Age Fertilisation

Whether or not close attention has been paid to the nutrient requirements of seedlings during their first few years of growth, the situation becomes rather different once the canopy of the plantation has closed and its stand foliage (and fine-root) biomass has reached a more or less constant level (Sect. 2.2). A process of nutrient cycling through the plantation system will then start. Before older leaves and fine roots die and are lost as surface and soil litter, some of the nutrients they contain will be retranslocated from them to their replacements. Then, the litter will be broken down by microorganisms and fungi and the nutrients remaining in them released back to the soil. Those nutrients will then continue to cycle through the plantation system as roots take them up to supply newly developing leaves and fine roots.

Nambiar (1990) and O'Connell and Sankaran (1997) have reviewed how this cycling operates. There will always be some retention of nutrients in non-living tissues of the trees (not all nutrients are retranslocated from living cells as they die to form wood cells); however, the amounts retained are insufficient generally, although not always, to prejudice the requirements of newly developing living tissues which are to be met through the nutrient cycling (Miller 1981; Laclau et al. 2003). Provided nothing occurs to disrupt this cycling, the nutrients in the plantation should remain more or less in a steady equilibrium (Miller 1995). If all the nutrient demands of the seedlings have been met, up to the point where the plantation canopy closed, there should be no requirement for additional nutrients to be supplied to the plantation for the remainder of its life.

However, if a fertiliser regime has not supplied fully the demands of the plantation during its early years, there may be benefits from applying fertiliser at later ages, after canopy closure. The additional nutrients should then allow the foliage and fine-root biomasses to increase, leading to increased growth of the plantation. This will occur only if there is sufficient water available from the site to allow these biomass increases to occur; if there is not, fertiliser additions at this stage would be wasted. Commercially speaking, it is usually found to be most profitable to ensure that growth increases from fertilisation are achieved as early as possible in the life of the plantation, rather than leaving them to a later stage of plantation development.

There are many reported cases where the nutrient cycling system at later ages of a plantation has been disrupted. One particularly important way this can happen is that nutrients are retained in the litter and not released to the soil to be taken up again by the trees (Miller et al. 1979; Gholz et al. 1985a, b; Dighton and Harrison 1990; Bargali et al. 1992; Miller 1995; Bubb et al. 1998; Carlyle and Nambiar 2001; McMurtrie et al. 2001). An example where this appears to be the case, is the loblolly pine (*P. taeda*) plantations in the southeastern states of the USA. Experiments on many sites have shown that growth responses continue to occur with applications of fertiliser long after the plantation canopy has closed and reached its constant state. This occurs even when the plantations had a more than adequate supply of nutrients during their early years of growth (Martin and Jokela 2004; Jokela et al. 2004). Whilst research remains to be done to confirm the reasons, it appears that cycling of nutrients through litter does not occur very efficiently in these plantations, so the trees constantly need additional supplies of nutrients from fertiliser. Blazier et al. (2005) have shown that absence of weeds in these plantations can reduce the activity of microorganisms in the soil, which can affect the recycling of nutrients; this suggests that some careful balance between weed control and fertiliser application will be necessary in these plantations. However, it appears that to achieve the maximum growth potential of these plantations will require a commitment to long-term fertilisation, continuing from time to time throughout their life. The extent to which growers in the region will be willing to do this will depend on the balance between the gains in wood yields achieved and the costs of fertilisation.

When there does appear to be a need for fertilisation at later stages of plantation growth, Miller (1981) suggested that plantation growth rate would increase *only if* the plantation was thinned at the same time as the fertiliser was applied. Thinning is discussed in Chap. 8; it involves harvesting some of the trees from a plantation part way through its life. Over 2–3 years following a thinning, the remaining trees expand their canopies and increase their fine-root biomasses, until the foliage and fine-root biomasses of the thinned stand reach the same levels as those of the unthinned stand.

Of course, the new foliage and fine roots developing on the trees remaining after a thinning will require nutrients from the soil. Much of that nutrient will be obtained from breakdown of the leaves and fine roots of the trees which were removed at thinning (as long as the leaves of the thinned trees were left behind on the site). However, if the nutrient supply from the soil had become somewhat inadequate anyway, because of disruptions to the nutrient cycling system, there might be an additional advantage in growth gained by adding fertiliser at the same time. So firm is the belief that later-age fertilisation will not boost growth without thinning at the same time that Carlyle (1995) would say 'thinning, which returns the canopy to an aggrading [that is, increasing its biomass] phase, may be a prerequisite for a fertilizer response'.

However, clear evidence of the need for thinning in conjunction with later-age fertilisation seems limited. Often, a growth response to fertilisation has been observed in a thinned stand in an experiment, but the experiment has failed to consider whether or not there would have been a growth response to fertilisation anyway, even if the stand had been unthinned (Crane 1981; Carlyle 1995, 1998; Snowdon 2002; McGrath et al. 2003). In fact, experiments which have considered fertilisation in both thinned and unthinned stands have found there was a growth response to fertilisation in both (Woollons 1985; Beets and Madgwick 1988; Jozsa and Brix 1989; Stegemoeller and Chappell 1990; Messina 1992; Knight and Nicholas 1996). The question as to whether or not thinning is essential in conjunction with fertilisation at later ages in plantations remains open. Whatever is the case, research studies will be necessary to determine if the gains from later-age fertiliser applications are sufficient economically to offset the costs involved (Donald 1987; Donald et al. 1987; Turner et al. 1996).

6.4 Assessing the Need for Fertilisation

As soil properties and other site conditions vary widely from place to place, even over distances of tens or hundreds of metres, the need for fertilisation can vary similarly. At present, our ability is rather limited in assessing the need for fertilisation from simply measured characteristics of a particular site. Much of this difficulty arises because the chemical and physical properties of soils are highly complex and vary widely from soil type to soil type. For many soils, our understanding is limited, both of those properties and of how they affect the availability of nutrients to trees (Fölster and Khanna 1997; Gonçalves et al. 1997; Lal 1997).

Because of these problems, it is usually essential to conduct fertilisation experiments to determine reliably if fertiliser is necessary and what fertiliser regime is appropriate for any particular site. Scott et al.'s example in Sect. 6.2 is typical of the quite complex, time-consuming and expensive experiments which need to be carried out for this purpose. Through extensive research of this nature, guidelines can be established as to what fertiliser regimes are appropriate from site to site over large plantation areas (Carter and Klinka 1992; Turner et al. 2001).

Sometimes, more immediate information about the need for fertilisation can be obtained by measuring the concentration of nutrient elements in the leaves of trees growing already on a site. As discussed in Sect. 6.2.3, Scott et al. (2004a) determined that sweetgum trees needed a nitrogen concentration in their leaves of at least 17,500 mg/kg of their oven-dry biomass for their growth to be maximised. Long before Ingestad showed that nutrient concentration in living tissues increases as plant relative growth rate increases (see footnote to p. 89), it had been recognised that measurement of the concentration of nutrient elements in tree foliage might be a suitable diagnostic technique to determine if the supply of nutrients to them from the site was sufficient to maximise their growth rate.

When using nutrient concentrations in foliage as a diagnostic technique, it has been found important to use leaves at a particular stage of development and even harvested at a particular time of year (Richards and Bevege 1972; Mead and Will 1976; Evans 1979; Ward et al. 1985; Knight 1988; Lambert and Turner 1988). Often, newly developed leaves, near the top of tree canopies and which have just reached their full size are most appropriate for the purpose; such leaves are likely to be at their peak metabolic activity. Nutrient concentrations in older leaves, or in leaves in more shaded parts of the canopy, often do not reflect the nutrient requirements of the tree.

Foliage diagnosis is used also to identify when there is a gross lack of supply of nutrients from a site. Plants have a physiological requirement that nutrients should be present in their living tissues at a concentration above some minimum level, or else their tissues are simply unable to perform properly their chemical functions. Foliage analysis can identify these gross deficiencies, but, at the same time plants usually display other quite obvious symptoms of the nutrient deficiency. These symptoms are often quite specific to deficiencies of particular nutrient elements in particular species (Dell 1996). Yellowing of leaves, blotchy marks on leaves and deformities of leaves, twigs or the stem are all typical symptoms of gross nutrient deficiencies (Fig. 6.4).

Monitoring of nutrient element concentrations in foliage has been, and continues to be, used to diagnose a need for fertilisation in many plantations (Mahendrappa et al. 1986; Cromer 1996; Herbert 1996; Knight and Nicholas 1996; Negi and Sharma 1996; Gonçalves et al. 1997; Lambert and Turner 1998; Carter and Klinka 1992; Jones and Dighton 1993; Yang 1998; Turner et al. 2001; Merino et al. 2003); however, it has been found unreliable sometimes (Dighton and Harrison 1990). An alternative is to measure nutrient availability from the soil and, particularly, to assess the nutrient availability to roots in the soil water (Smethurst 2000); in effect, this is an attempt to assess the relative addition rate of nutrients from the soil (Sect. 6.1.1). Considerable advances have been made in finding methods to do this, although the results are often rather specific to a particular site and species (Mahendrappa et al. 1986; Dighton and Harrison 1990; Tahvanainen and Rytkönen 1999; Prasolova et al. 2000; Smethurst 2000;

Fig. 6.4. View through the canopy of a New Zealand plantation of radiata pine (*P. radiata*) suffering a severe deficiency of the nutrient element phosphorus. The foliage was sparse and clumped, with rather short, yellowing needles, all symptoms typical of phosphorus deficiency in this species (Photo—P.W. West)

Turner et al. 2001; Mendham et al. 2002; Paul et al. 2002; Moroni et al. 2004; Romanyà and Vallejo 2004; Smethurst et al. 2004). Other work has shown that laboratory measurement of the uptake rates of nutrients by fine roots excavated from a plantation can be a reliable method of assessing the nutrient needs of trees (Dighton and Harrison 1990; McDonald et al. 1991a, b; Harrison et al. 1992; Dighton and Jones 1992; Dighton et al. 1993; Jones and Dighton 1993).

6.5 Sustaining Nutrients on the Site

Over the life of a plantation there will be some losses of nutrients from the site, either through leaching or through chemical reactions in the soil by which nutrients become bound permanently to soil particles. At the same time, there may be some additions, through weathering of the rocks under the soil to form new soil and by deposition of nutrients (usually in small amounts) from the atmosphere.

The biggest threat of nutrient losses occurs when the plantation is harvested (Kubin 1995; Fölster and Khanna 1997; Gonçalves et al. 1997). Some nutrients will be lost in the timber which is removed, although wood contains relatively small amounts of nutrients. If there happen to be severe storms after harvesting, there may be erosion losses; these may be exacerbated if harvesting machinery has disturbed the soil. Large amounts of debris (leaves, branches and roots) will be left on the site and may be broken down quite rapidly, releasing their nutrients to the soil; some of those nutrients may be leached from the site before replacement seedlings are planted. This can be exacerbated if the site is cleared of debris by burning before replanting; large amounts of nutrients can be volatilised by the heat of the fire, lost as particles in smoke or rendered highly soluble and leached from the site.

Two of the biggest threats to nutrient loss after harvesting can be managed. Firstly, bark contains quite large amounts of nutrients. Stripping it from logs and leaving it on the site will reduce losses. Secondly, there are alternatives to burning debris. An example will illustrate the importance of this. It comes from Birk (1993), for a 21-year-old plantation of radiata pine (*Pinus radiata*), planted in New South Wales, Australia. At the time of harvesting, Birk measured the amounts of various macro-nutrient (Sect. 2.1.4) elements (nitrogen, phosphorus, calcium, magnesium and potassium) in the aboveground parts of the trees (their leaves, branches, stem wood and stem bark), as well as the amounts in the soil.

Birk's information can be used to determine what the nutrient losses would be from the site, with or without bark removal from the harvested logs and with or without subsequent burning of the harvesting debris. The results are shown in Fig. 6.5. It is quite clear that failing to remove the bark and burning the debris both increase appreciably the proportion of the various nutrient elements which will be lost.

It will vary greatly from site to site and tree species to tree species as to how great nutrient losses will be as a consequence of harvesting (Augusto et al. 2000; Johnson and Curtis 2001). However, Birk's results are not unusual. They suggest there could be appreciable nutrient losses from a site over several plantation rotations, losses which might eventually prejudice the growth of plantations on that site. Other studies have shown that similar losses can occur with various plantation species in various parts of the world (e.g. Flinn et al. 1979; Smethurst and Nambiar 1990; Herbert 1996; Judd 1996; Fölster and Khanna 1997; Gonçalves et al. 1997; Bubb et al. 2000; Mendham et al. 2003; Chen and Xu 2005; Merino et al. 2005).

Only by appropriate management of the site can the need for fertilisation in successive rotations be avoided. The importance particularly of removing bark from harvested logs and leaving it on the site is now being recognised (Ericsson et al. 1992; Fölster and Khanna 1997; Martin et al. 1998; Ruark and Bockheim 1988; Ganjegunte et al. 2004; Adler et al. 2005; Merino et al. 2005). As well, large chopper-roller machinery is used in some places, to physically mulch the debris left after harvesting. This avoids any need to clear the site by burning the debris, to allow people and

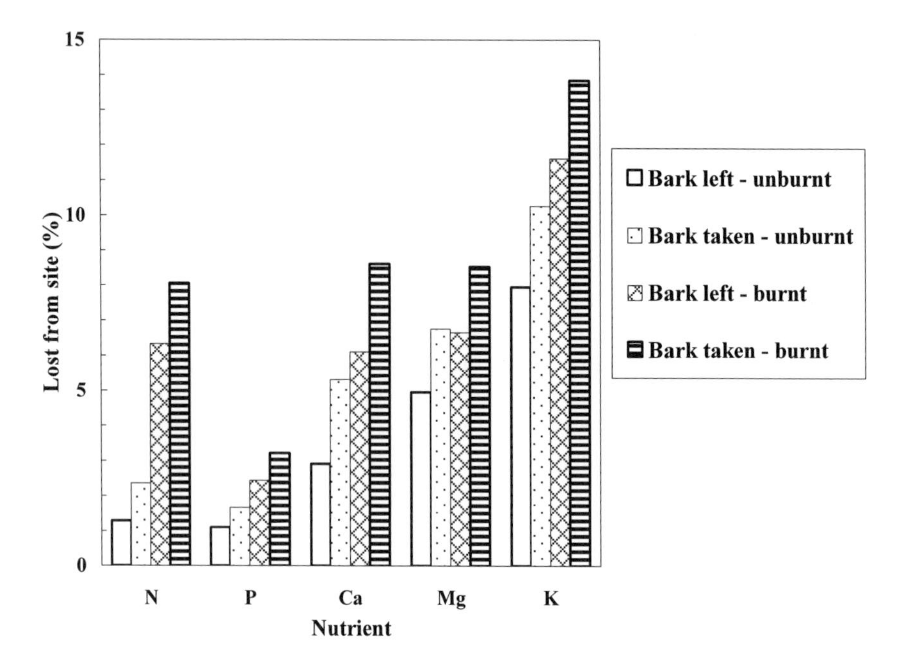

Fig. 6.5. The proportion of nitrogen (*N*), phosphorus (*P*), calcium (*Ca*), magnesium (*Mg*) and potassium (*K*) which would be lost from the site if a 21-year-old plantation of radiata pine (*P. radiata*) in New South Wales, Australia, was clear-felled. Results are shown if the debris left on the site after harvesting was or was not burnt before planting the next tree crop and if bark was or was not removed from harvested logs, before they were transported away from the site. The total amounts of the nutrients in the plantation before clear-felling (to a depth of 38 cm in the soil, plus leaf and branch litter on the ground, plus the living trees) were 8.4 t/ha of N, 1.3 t/ha of P, 4.0 t/ha of Ca, 1.0 t/ha of Mg and 1.8 t/ha of K (derived from information in Birk 1993)

machinery to move about easily as the next rotation is being established (Turvey and Cameron 1986a; Bekunda et al. 1990; Smethurst and Nambiar 1990; Nyland 1996; Costantini et al. 1997b; Carlyle et al. 1998; Little et al. 2000; Johnson and Curtis 2001).

7 Stand Density and Initial Spacing

Together with Chap. 8, this chapter is concerned with the way in which the **stand density** of forest plantations is managed during their lifetime. Stand density can be defined as the 'degree of crowding of trees in a plantation' and it has very important effects on how the trees grow and develop.

Section 7.1 discusses in general how density affects the growth and development of trees in a plantation. Given this, Sect. 7.2 considers the issues to be considered when choosing the initial spacing and arrangement of the trees when a forest plantation is established.

7.1 Stand Density

From the point of view of commercial forestry, stand density is important for several reasons. The higher the density of a plantation:

- The smaller will be the average stem diameter and stem wood volume of the individual trees in the plantation at any age. This leads to a reduction in the availability of larger-diameter, more commercially valuable logs from which timber might be sawn (Sect. 3.3).
- The smaller will be the average size of the crowns of the individual trees in the plantation. This means their branches will be smaller and less likely to lead to defects in sawn timber from knots (Sect. 3.3.8).
- The more stem wood volume will the plantation produce in total to any age; hence the more wood will be available for sale.
- The earlier in the life of the plantation will tree deaths from competition between the trees occur and the greater will be the number of deaths (Sect. 2.4); loss of wood in dying trees represents a commercial loss to the plantation grower.

It will depend on the purpose for which a plantation is being grown as to which of these four effects of stand density will be important to the plantation owner. For example, in bioenergy plantations, the amount of wood produced is the main concern and the sizes of the individual trees are of little importance. However, in plantations being grown for high-quality sawn timber, large individual tree sizes are required to maximise the availability of large logs.

The ways of managing stand density during the life of a plantation are:

- By selecting an appropriate number and spatial arrangement of trees at the time of planting.
- By judicious removal of some trees from time to time during the life of the plantation, a practice known as thinning. This will be discussed in Chap. 8.

7.1.1 Maximum Density

Any site has a limited amount of space which the trees may occupy. Once the trees in a plantation stand have grown large enough to use all the available space, the stand will have reached its maximum degree of crowding, hence its maximum density. As the trees continue to grow, they will compete with each other (Sect. 2.4) and some must die to prevent excessive crowding.

In 1963, Japanese scientists (Yoda et al. 1963) proposed a theory to describe the condition of stands at their maximum density. Their theory applies to plant stands in general, not just trees. It proposes that if a stand is self-thinning (that is, plants are steadily dying in the stand owing to competition), it will be at its maximum density and, furthermore, the average biomass of the plants in the stand will be directly proportional to their **stocking density** (that is, the number of plants per unit area of a stand) raised to the power of −3/2 (that is, −1.5). Mathematically, this theory can be expressed by the equation,

$$b=\kappa S^{-3/2} , \tag{7.1}$$

where b is the average biomass of the plants in the stand, S is their stocking density and κ is the constant of proportionality. Without describing fully here Yoda et al.'s reasoning, they argued that plants have a three-dimensional space available to them in which they may grow their biomass (the ground surface and the air above it), but only a two-dimensional space (the ground surface) over which they may distribute the many plants growing in a stand; this argument led them to the values of 3 and 2 in the power −3/2 in Eq. 7.1.

Yoda et al. tested their theory with data from several types of plants, but Gorham (1979) provided an even more remarkable test of it. From the scientific literature, he collated data of average aboveground oven-dry biomasses and stocking densities of self-thinning plant stands, for plants which ranged from large trees, through tree saplings, to much smaller plants (different types of sedges, rushes and ferns) and right through to very small plants, mosses. The stocking densities varied enormously in his data set, from tree stands with just over 700 stems per hectare to mosses with over 164 million stems per hectare. Likewise, the average aboveground biomasses of the plants varied enormously, from just over 1 t per plant in the trees to about 20 mg per plant in the mosses (1 mg is one billionth of a tonne). Gorham's data are reproduced in Fig. 7.1; so great is the variation in the values in the data set that they have been graphed as their logarithms, or else the diagram would not fit reasonably on the page.

Gorham determined a best-fitting trend line through his data (which was done using regression analysis—Sect. 4.4.2) and it had the equation

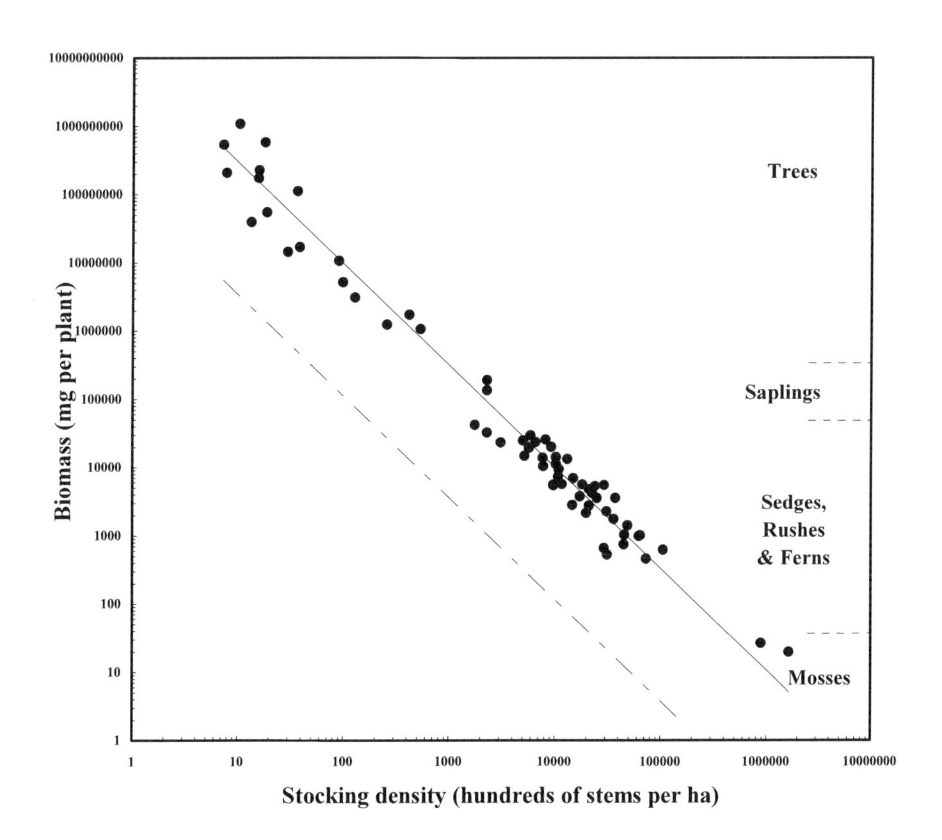

Fig. 7.1. A set of data points (•) of average aboveground plant biomass against stand stocking density, for self-thinning stands of a very wide range of plant types, from trees to mosses (as indicated). The *continuous line* represents a best-fitting trend line to the data and may be considered a maximum density line for plant stands. The *broken line* defines a stand relative density of 5% (Sect. 7.1.2) (adapted from Gorham 1979)

$$b=9{,}235{,}000{,}000S^{-1.49} \quad , \tag{7.2}$$

where b is the average biomass of plants in a stand (milligrams) and S is the stand stocking density (hundreds of stems per hectare). The power to which S was raised in Gorham's trend line (−1.49) was so close to −1.5 that Gorham's results can be considered as confirmation of Yoda et al.'s theory. The trend line represents a maximum density line for plant stands; stands will not be found in nature with a biomass and stocking density which would position them appreciably above the line.

Yoda et al.'s theory has become known as the '−3/2 power law of self-thinning'. Since Gorham's work, the theory has been tested more exhaustively, by many scientists working with a wide range of plant species. For some species under some circumstances, some departures from the law have been found; Jack and Long (1996) have summarised many of these findings. However, by and large the law seems to hold well for most plant species under most circumstances. Different values of the constant of proportionality (κ in Eq. 7.1) have been found to apply to different species and, for any one species, its value has been found to increase somewhat as site productive capacity increases (Schulze et al. 2005).

The practical importance of Yoda et al.'s theory to plantation forestry is that it provides an upper limit to the density of plantation stands, although the maximum density line needs to be determined specifically for any particular plantation species. As will be discussed in succeeding sections, this provides a framework for the management of plantation stand density.

7.1.2 Measuring Density

Given a maximum density line as in Fig. 7.1, it can be used to define a useful measure of stand density, known in forest science as relative density. In effect, relative density is a measure of how closely any particular stand is approaching its maximum density.

Formally, relative density is defined as the actual stocking density of a stand divided by the stocking density it would have *if* it was at maximum density and had the same average tree size. We will not consider here exactly how relative density is calculated; forestry texts such as Davis et al. (2001) or Avery and Burkhart (2002) discuss these details. Any value of relative density can be represented by a line drawn parallel to the maximum density line. As an example, one such line has been drawn in Fig. 7.1, representing a relative density of 5%. Any plant stand which had an average plant biomass and stocking density which positioned it on that line would have that relative density. Similar lines could be drawn for any other value of relative density. The higher the relative density, the closer would the line be to the maximum density line; of course the maximum density line itself represents stands with a relative density of 100%.

This measure of stand density is not the only one which has been developed for use in forest science; however, the others are all based on the principle of determining the degree of approach of a stand to maximum density. West (1983) discussed many such measures and showed how each relates to the others.

7.1.3 Stand Development in Relation to Density

For much of the life of a plantation stand, it will not have grown to the point of maximum density. Furthermore, if a stand was allowed to ever reach its maximum density, trees would then be lost from it as deaths due to competition (Sect. 2.4); this would represent a loss to the plantation grower of wood which could otherwise be harvested and sold. For both these reasons, it is necessary to know how plantations grow and develop before they reach maximum density.

Figure 7.2 illustrates this for an example plantation. Its form differs from that of Fig. 7.1, by showing values of stand stem wood volume on the vertical axis, rather than corresponding values of average tree biomass; this in no way alters the principles established by Gorham's theory. As will be seen in Sect. 8.4, this is the form of the diagram used often for forestry purposes. Super-imposed on the diagram is the growth trajectory with time, from 2 years of age, of a plantation stand of blackbutt (*Eucalyptus pilularis*), planted with 1,000 trees per hectare and growing in subtropical northern New South Wales, Australia. The solid line in the figure represents a maximum density line, which the experience of the author suggests may be appropriate for this species.

Initially, the blackbutt trees were small and not using fully the resources for growth available from the site (Sect. 2.2). With time, the stand stem wood volume increased until, at about 6 years of age, the trees had grown large enough to utilise fully the site resources; the lower relative density line drawn in the figure represents this point. As growth continued, competition between the trees started (Sect. 2.4). After 18 years of age, substantial mortality started to occur owing to the competition, at the point marked by the upper relative density line in the figure; stocking density of the stand then started to decline substantially. By about 40 years of age, the stand reached maximum density and thereafter its growth trajectory would follow the maximum density line.

Research (reviewed by Jack and Long 1996) suggests that stands reach the point at which they are utilising fully the resources of the site at a particular relative density, no matter what their stocking density is at the time.

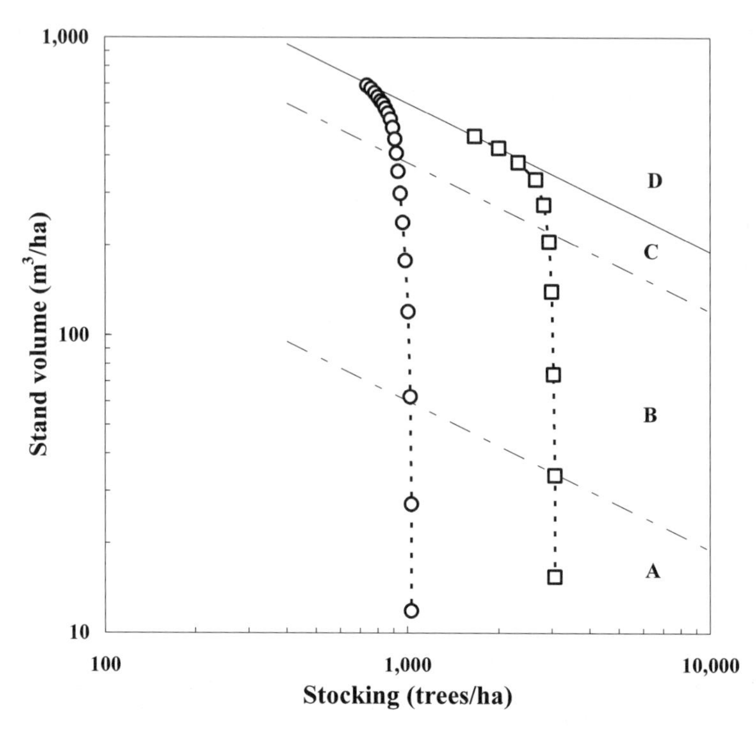

Fig. 7.2. The trajectory (o- - -o) of growth in stand stem wood volume, in relation to stand stocking density, of a plantation of blackbutt (*Eucalyptus pilularis*) growing in subtropical northern New South Wales, Australia, and planted with 1,000 trees per hectare. Each plotted data point (o) is for a specific age of the plantation, starting at 2 years of age for the lowest point and progressing at 2-year intervals. Similar data are shown (□- - -□) for a plantation where 3,000 trees per hectare were planted. The *continuous line* is a maximum density line for blackbutt. It and the relative density lines (*broken lines*) delimit the sections of the diagram marked *A*, *B*, *C* and *D*, which are discussed in the text. Diagrams of this nature have become known in forestry as density management diagrams (Sect. 8.4) (Source—author's unpublished information)

The same is true for the point at which they start to suffer substantial mortality. With this in mind, also shown in Fig. 7.2 is the growth trajectory for the same blackbutt plantation, if it had been planted at the much higher stocking density of 3,000 trees per hectare. Because of the many more trees present, this stand reached the point where it was using fully the resources of the site when it was about 4 years of age, rather earlier than the 6 years of age for the stand planted with 1,000 trees per hectare; however, the relative density at which this occurred was the same in both cases. Similarly, substantial mortality from competition set in at a younger age in the stand with the higher planting density, but at the same relative density as in the stand with the lower planting density.

Given this, a diagram such as Fig 7.2 can be subdivided into four sections, A–D, as marked. They allow assessment of the growth stage that *any* stand of interest has reached. If a stand is found to have a stand volume and stocking density which positions it in section A, it will not yet have reached the point where it is using fully the resources for growth available from the site. A stand in section B would be using the resources fully, but would not yet be undergoing substantial tree mortality as a result of competition between the trees. A stand in section C would be using the resources fully and also undergoing substantial mortality. Stands would not normally be found in section D, where their density would be above the maximum possible. The value of being able to identify the condition of stands in this way will become evident in Chap. 8.

There are several other important points of interest about growth trajectories of plantation stands in relation to their density:

- Any plantation stand will follow a growth trajectory similar to those of the examples in Fig. 7.2; however, the rate at which it moves along the trajectory will be determined by the productive capacity of the site (Sect. 4.3) on which the stand is growing. The higher the productive capacity, the more rapidly will a stand move along the trajectory and the earlier will it reach the various sections of the diagram.
- Consider any two stands with different initial stocking densities, but growing on sites of the same productive capacity (as in Fig. 7.2). Before each is using fully the resources for growth from the site (whilst each is in section A of the diagram), the stand with a higher stocking density will have a higher stand leaf biomass at any particular age (simply because it has more trees); thus, it will be able to carry out more photosynthesis and have a higher stand stem wood volume than the stand with the lower stocking density; this is apparent in the results in Fig. 7.2.
- Once the same two stands are using fully the resources available from the site, that is, both are moving through section B of the diagram, both would have the same stand leaf biomass in their canopies (determined by the level of site resources available—Sect. 2.2); hence, at any particular age, both would be able to carry out the same amount of photosynthesis and therefore would have the same biomass growth rates.

  However, in the stand with the higher stocking density, its biomass growth would be being distributed amongst more individual trees than in the stand with the lower stocking density. Inevitably then, the average size of the trees in the stand with higher stocking density would be smaller than in the stand with lower stocking density. This is important for plantation forestry; trees of larger sizes will be produced in stands with lower stocking densities than in stands with higher stocking densities.

- Whilst both these stands would have the same *stand biomass* growth rates at any particular age when both are in section B of the diagram, the *stand stem wood volume* growth rate would be higher in the stand with the higher initial stocking density.

  There are two reasons for this. Firstly, there would be more space between each tree in the stand with lower stocking density; hence, their crowns would be more widespread, with longer branches, than in the stand with higher stocking density. This means they will need a disproportionately greater branch biomass to support the weight of the branches and the leaves they carry (Sect. 2.3.2). As well, increased wind speeds through less dense stands lead to increases in branch sizes so they are stronger and better able to resist breakage (Watt et al. 2005b); hence, trees in stands of lower stocking density will need to use a higher proportion of their photosynthetic production in growing their branch biomass, leaving less for stem wood production (Kuuluvainen 1988; Jack and Long 1991, 1996; Kuuluvainen and Kanninen 1992; Sterba and Amateis 1998; Smith and Long 1989). Similarly, the wider spread of the woody root systems of trees in less dense stands would lead to use of a higher proportion of their photosynthetic production in growing their woody root biomass.

  Secondly, in the stand with higher stocking density, its trees would hold a higher proportion of the biomass of their leaves near their tops, because the stand has less space available to position its leaves to best advantage to intercept sunlight than in a stand with lower stocking density. Dean (2004) has shown that by distributing more of their crown weight towards the top of the tree, trees in denser stands will need to distribute a higher proportion of their photosynthetic production to their stems to ensure the stems maintain sufficient strength to continue to stand upright.

  It is only in recent years that this pattern has been proposed for stand stem wood volume growth rate in relation to stand stocking density (Zeide 2001; Long et al. 2004). Previously, a long standing theory in forest science (proposed in 1941 by A. Laengsaeter) held that stand stem wood volume growth rate was the same at any stocking density, when stands were in section B of the diagram (Möller 1954; Sjolte-Jorgensen 1967; Zeide 2001); I mention this old theory here, because it has become almost axiomatic in forest science and many foresters will find it difficult to accept that it no longer holds.

These considerations lead to the consequences of stand density, stated at the start of Sect. 7.1. At any particular age, stands established at higher stocking densities will produce more stem wood volume than stands established at lower stocking densities, but the individual trees will be of smaller average size and will have smaller crowns and branches. This may not continue after the stands have reached ages where mortality is well established (when they have entered section C of the diagram). Stands with higher stocking density would suffer greater mortality at earlier ages than stands of lower stocking density; the wood volume they lose may then reduce their stand stem wood volumes below those of stands of lower stocking density.

7.2 Initial Spacing

Having discussed in Sect. 7.1 how plantation stands develop in relation to stand density, this section now consider the choice of the spacing of the trees at planting. This is the first of the two ways in which stand density can be managed in practice.

There are two issues involved in this choice. Firstly, a decision must be made as to how many trees will be planted, that is, what the initial stocking density of the plantation will be. Consistent with the discussion of Sect. 7.1, this will be important in determining:

- The amount of wood available from the plantation when it is harvested ultimately
- The average sizes of the trees in the plantation
- The sizes of the branches which develop on the tree stems.

These issues are discussed in Sects. 7.2.1 and 7.2.2.

Secondly, a decision must be made as to how the trees will be arranged spatially in the plantation. Usually, plantations are established with the seedlings planted in parallel rows and then arranged in a square or rectangular pattern. Biologically speaking, there seems no particular reason why some other pattern of planting should not be used; Shao and Shugart (1997) have suggested that there may be advantages in an equilateral triangular arrangement. However, it is easiest practically to plant in rows and it also allows machinery to move easily between the rows, both at the time of planting and later in the life of the plantation.

Choice of the spacing between trees both within and between the rows determines the rectangularity of the initial spacing. Rectangularity is defined simply as the ratio of the between-row to within-row spacing. A high level of rectangularity (trees planted closely in rows with wide distances between rows) can affect stand density and subsequent tree development. This is discussed in Sect. 7.2.3.

Having discussed the effects on stand development of both initial stocking density and rectangularity of spacing, Sect. 7.2.4 will discuss the choices for initial spacing which are made generally in plantations around the world.

7.2.1 Effects on Stand Wood Yields and Tree Sizes

An example will be used to illustrate the effects of varying stocking density at planting on stand wood yields and tree sizes as the plantation develops. The example comes from McKenzie and Hay (1996), who reported results from an experimental plantation in New Zealand of Sydney blue gum (*E. saligna*), in which stands were planted at a very wide range of initial stocking densities.

The experiment used one of the rather novel designs of Nelder (1962). His designs allow plants to be grown at an extremely wide range of stocking densities, but they can be established on quite conveniently sized areas of land. The Nelder design used by McKenzie and Hay involves planting trees along many, equally spaced radii of a set of concentric circles. Figure 7.3 shows a typical example of the positioning of trees in such a design. In effect, trees nearer the centre of the experiment are planted at the highest stocking densities and the stocking density then declines progressively towards the outermost circle of trees. Any set of trees around the circumfer-

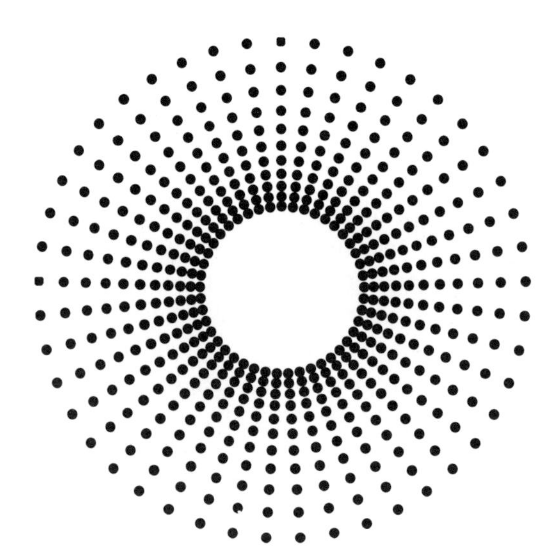

Fig. 7.3. Tree planting positions in a typical example of the circular experimental design of Nelder (1962)

ence of any one of the concentric circles are all planted with the same stocking density and can be considered as a stand with that stocking density. Other examples of the use of this experimental design can be found in Cole and Newton (1987), Aphalo et al. (1999) and Kerr (2003).

For McKenzie and Hay's example, results will be shown for stands in the experiment which were planted with stocking densities which varied from 158 to 5,827 trees per hectare and for measurements of the experiment made at each of 2.3, 3, 4, 5 and 6 years of age. The results for stand stem wood volume are shown in Fig. 7.4a. An attempt has been made to split the diagram into the sections A, B, C and D, as in Fig. 7.2, to denote the various stages of stand development. In this case, those divisions are rather speculative, because no detailed research has yet been done with Sydney blue gum to allow them to be determined with any certainty; they are based on some limited research with another eucalypt species (West 1991) and from trends apparent in the data from the experiment. The divisions suggest that, by 6 years of age, stands of any initial stocking density had reached the stage where they were using fully the resources for growth available from the site. Stands with the highest initial stocking densities were suffering appreciable tree mortality owing to competition by that age. However, the results show quite clearly that, at any age, stand stem wood volume increased with stand stocking density, consistent with the discussion in Sect. 7.1.3. Schönau and Coetzee (1989) have summarised many other studies which have shown also that stand stem wood volume at any age increases appreciably with stand stocking density and this continues to ages well beyond those shown in Fig. 7.4a.

Figure 7.4b shows the variation with stocking density of stand **quadratic mean diameter**. This is defined as the equivalent diameter of the tree of average stem cross-sectional area at breast height in a stand; it is a measure used quite commonly in forest science to describe the average size of trees in stands. The results show that once stands had reached ages where they were using fully the resources for growth available from the site, average tree size was reduced markedly at higher stocking densities, consistent with the arguments in Sect. 7.1.3. There are myriad reports from all parts of the world of average tree size declining with increasing stocking density (e.g. Opie et al. 1984; Schönau and Coetzee 1989; Baldwin et al. 2000; Gerrand and Neilsen 2000; Henskens et al. 2001; DeBell and Harrington 2002; Ares et al. 2003; Bishaw et al. 2003; Cucchi and Bert 2003; Plauborg 2004; Koch and Ward 2005).

The important issues that follow from McKenzie and Hay's example are that the higher the stand stocking density chosen at planting, the higher will be the stand wood yields at harvest. However, the higher the initial stocking density, the smaller will be the average stem diameter and stem wood volume of the trees and so the smaller will be the wood volumes

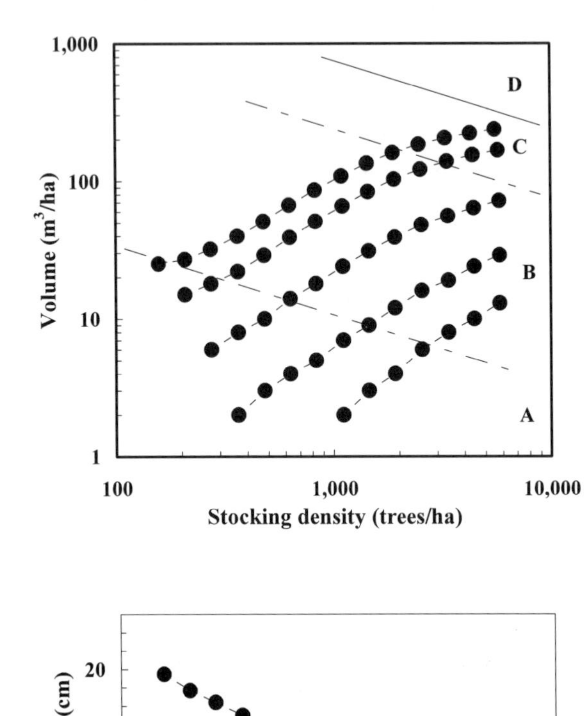

Fig. 7.4. Change with stand stocking density of **a** stand stem wood volume and **b** stand quadratic mean diameter in an experiment with Sydney blue gum (*E. saligna*) in New Zealand. Each line drawn (●- - -●) shows results measured at a different age of the experiment. The ages were 2.3, 3, 4, 5 and 6 years, for the *lines* progressively from the *bottom* to the *top*. In **a**, sections of the diagram *A*, *B*, *C* and *D* have been delimited, to indicate the stages of stand development as in Fig. 7.2. Stocking densities shown are the actual stocking densities at each age of measurement; these will sometimes be less than the stocking density at planting, since tree deaths occurred as the experiment continued (Source—data reported by McKenzie and Hay 1996)

available of logs of larger sizes. These results confirm the conclusions drawn in Sect. 7.1.

Recently, research is starting to suggest that the higher diameter growth rates of individual trees in stands with lower stocking densities will lead to a decline in their wood strength (Lassere et al. 2005); this may become an important issue in the future in deciding what is the optimum stocking density at plantation establishment.

7.2.2 Effects on Branch Size

As discussed in Sect. 3.3.8, knots are the chief source of defect in sawn timber. The stocking density of a plantation is the most important factor affecting branch size and, hence, the size of knots that will occur ultimately in sawn wood.

Figure 7.5 shows research results from experimental plantations of several different species in various parts of the world. All these plantations had been established with stocking densities in the range used normally in plantations from which sawn timber might ultimately be produced (Sect. 7.2.4). Each experiment consisted of a set of plots, which had been established with different stocking densities. In each case, measurements had been made of the diameter, at the base of the branch, of the largest branch on the lower part of the stems of the trees. The results are the averages of those diameters for the trees established at any particular stocking density.

These results show a clear tendency for larger branches to develop on stems of trees planted at successively lower stocking densities, consistent with the discussion in Sect. 7.1.3. The consequences of these effects of stocking density on branch size will be discussed in more detail in Sect. 9.3.

7.2.3 Effects of Rectangularity of Spacing on Trees and Stands

In a rectangular planting pattern, trees will be planted closer together within than between the rows. An extreme example of the effects of rectangularity on tree development is given by an experiment with slash pine (*Pinus elliottii*) in southeastern USA (Sequeira and Gholz 1991). The rectangularity of spacing in their experiment varied over the range 1.5–24. In the most extreme treatment, trees were planted 0.6 m apart within each of pairs of rows which were 2.4 m apart; the paired rows were then spaced 26.6 m apart. Sequeira and Gholz measured the experiment at 20 years of age.

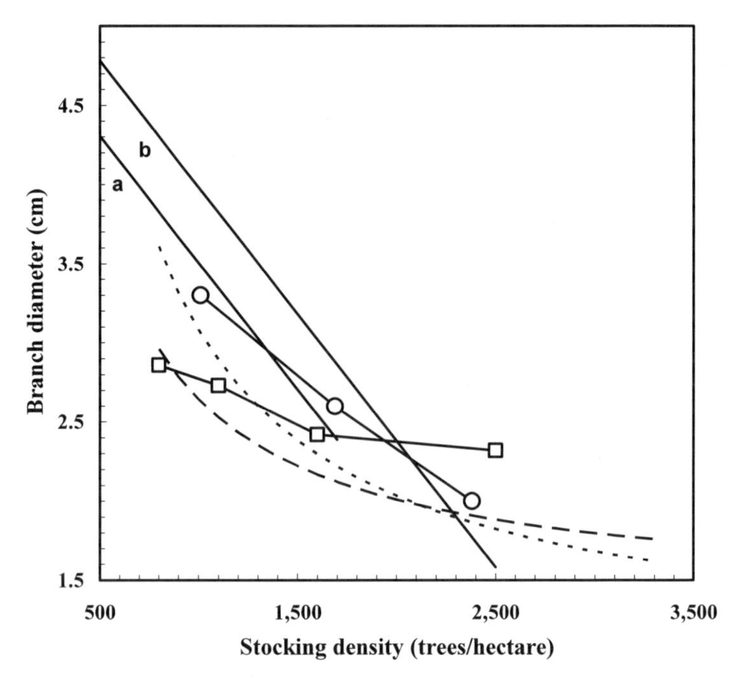

Fig. 7.5. Experimental results showing the trend in average largest branch diameter at branch base with changes in stocking density at planting. Results are for the lowest 6 m of the stem of 5-year-old shining gum (*E. nitens*) in Tasmania (——) in two different experiments (**a**) Neilsen and Gerrand (1999) and (**b**) Gerrand and Neilsen (2000), for the lowest 5.3 m of the stem of 5-year-old flooded gum (*E. grandis*) (– – –) or blackbutt (*E. pilularis*) (- - -) in northern New South Wales (Kearney 1999), for the lowest 3 m of the stem of 19-year-old *Pinus patula* in Tanzania (O—O) (Malimbwi et al. 1992) and for total tree height (which did not exceed about 7 m) of 14–17-year-old Scots pine (*P. sylvestris*) in Finland (□—□) (Salminen and Varmola 1993)

With such large, open spaces between the paired rows of trees in the most extreme treatment of this experiment, it might be expected that the trees would have developed rather eccentrically. Perhaps their root systems and crowns would have extended further into the more open spaces between the rows than along the rows. However, Sequeira and Gholz found little evidence of this occurring. Instead, tree development seemed to be constrained by the distance to the nearest neighbouring tree, no matter what the rectangularity of the spacing. The closer its neighbouring tree, the smaller was the crown and stem diameter of any tree; height growth was unaffected by the spatial arrangement of the trees.

A number of other experiments, with both hardwood and softwood species in various parts of the world, have confirmed generally Sequeira and Gholz's results (Hawke 1991; Salminen and Varmola 1993; Niemistö

1995; Bergkvist and Ledin 1998; Gerrand and Neilsen 2000; DeBell and Harrington 2002; Sharma et al. 2002). With rectangularities greater than about 2, competitive processes between the closely spaced trees in the rows can set in noticeably earlier than would otherwise occur. Tree crowns and stem diameters are then smaller and some tree deaths, owing to competition, can occur earlier than might otherwise be expected; in one example (DeBell and Harrington 2002), the early mortality in markedly rectangularly spaced, 12-year-old loblolly pine (*P. taeda*) stands was so substantial that stand stocking density was reduced sufficiently that the average diameter of the stems of the trees was appreciably greater than in stands established with square spacing.

Because of the wide spacing between trees across the rows in markedly rectangularly spaced stands, a delay can occur in trees growing large enough to utilise fully the resources for growth from the site; this can lead to some reduction in stand stem wood production at any age. Some eccentricity of development of crowns has been observed, with the crowns extending further into the wider spaces between the rows, and so with some increase in branch sizes in that direction; this probably reflects the higher availability of sunlight in the more open spaces between the rows.

However, research results in general suggest that rectangularity of spacing up to about 2–3 has only limited effects on tree and stand development. So, for example, the development of a stand established at 1,000 trees per hectare, with square spacing close to 3.2 × 3.2 m, would be expected to develop little differently from a stand with the same stocking density but with a rectangular spacing of about 2 × 5 m.

7.2.4 Initial Spacing in Practice

The preceding discussion has suggested that the initial spacing chosen at planting, hence, the stand stocking density, will have appreciable effects eventually on stand wood yields, the average diameter and average wood volume of trees in the stand and their branch sizes. Very marked rectangularity of planting can affect all of these also.

In bioenergy plantations, high initial stocking densities are preferred. This maximises wood production and the small average diameter of the harvested stems is of no importance to use of the product. As well, such plantations are grown usually for only 3–5 years, insufficient time for substantial losses to have occurred from tree deaths.

Around the world, quite a wide range of initial stocking densities have been used for bioenergy plantations. Reviews of the economic value of hardwood bioenergy plantations in Scandinavia suggested that 10,000–20,000 trees per hectare was an appropriate initial stocking density to maximise profitability (Willebrand et al. 1993; Toivonen and Tahvanainen 1998). In North America, hardwood bioenergy plantations have been established with densities varying over the range 1,000–35,000 trees per hectare, although more commonly in the range 2,500–4,000 trees per hectare (Kopp et al. 1993, 1996). However, to exploit the increase in wood production, even higher planting densities (40,000–440,000 trees per hectare) have been considered. These very dense systems were termed woodgrass and were initially harvested annually, with regeneration by coppice (Sect. 5.5). Later work found that harvests every 3 years were more productive than annual harvests and that planting densities in excess of about 110,000 trees per hectare led to excessive losses of production through mortality (Kopp et al. 1997).

Bioenergy plantations around the world have been planted generally with rectangularities in the range 1–1.5. Common practice in Scandinavia has been to use rectangularities as high as 2 in a double-row system, where pairs of rows are planted at about 0.7 m apart and with a spacing of 1.3 m between the row pairs and 0.5 m between trees within rows (Kopp et al. 1997; Tahvanainen and Rytkönen 1999); this has been found convenient for machinery access.

In plantations being grown for fuelwood, pulpwood for paper production or for larger log sizes for sawn timber production (Sect. 3.3), stocking densities at planting in the range 800–2,500 trees per hectare have been used throughout the world. The data collated to draw Figs. 3.2 and 3.4 all came from plantations established at stocking densities within this range. For any reasonable age of harvesting, planting densities higher than this will lead to the trees having diameters which are too small for the requirements of the products.

Choice of initial stocking density within the range 800–2,500 trees per hectare will vary with the species concerned and the products being grown. If high-quality sawn timber was to be produced, lower stocking densities might be favoured to maximise tree sizes. However, if the species concerned produces large branches which might lead to damaging knots, stocking densities towards the higher end of the range might be considered: branch pruning might then be undertaken to avoid this problem (Chap. 9). If thinning is to be undertaken (Chap. 8), higher stocking densities will minimise branch sizes and allow adequate choice of the trees to be retained at thinning.

If trees are to be harvested for posts or poles, higher stocking densities will favour development of straight stems with small branches; stocking densities as high as 10,000 trees per hectare have been used for plantations being grown for this purpose (Leslie 1991; Evans and Turnbull 2004).

To maximise production, stocking densities as high as 10,000 trees per hectare might be appropriate also for plantations being grown for pulpwood (Opie et al. 1984); however, such plantations around the world seem generally to be planted at much lower densities, say 1,500–2,000 trees per hectare, probably because harvesting costs, with the logging machinery used presently, rise very rapidly if individual tree sizes are very small (Roberts and McCormack 1991).

In general, planting densities within the range 1,000–1,500 trees per hectare seem to be used most commonly around the world in commercial plantations being grown for pulpwood or sawn wood products. Spacing between rows is often within the range 3.5–4 m, sufficient to allow machinery access. Spacing of trees within rows is often within the range 2–3 m, so rectangularity of spacing is usually less than about 2.

Very low stocking densities at planting are usually restricted to **agroforestry** plantations. The low stocking density allows either pasture for animal browsing or some other agricultural crop to be grown between the trees. The danger with very low stocking densities is that the trees may develop with poor stem form or very large branches, which will render them unsuitable for timber production.

One approach to avoid these problems is to establish the plantation at a conventional stocking density of, say, 1,500–2,000 trees per hectare and thin it heavily after a few years to the stocking density required at tree harvest. Pasture can be sown after the thinning and the trees can be pruned to avoid excessive branch development. This approach has been adopted with radiata pine (*P. radiata*) in agroforestry plantations in New Zealand (Hawke 1991; Knowles 1991).

Balandier and Dupraz (1999) have reviewed practices in agroforestry plantations in France, where hardwood species with highly valuable timber are grown at very wide spacings and animals are allowed to graze pasture in between them. Often these plantations have been established with about 2,500 trees per hectare and thinned early to the final stocking density required, as done in New Zealand. However, more recent experience has found it is feasible to plant at stocking densities as low as 100 trees per hectare, with a view to harvesting 50–80 trees per hectare. At these low planting densities the trees must be tended carefully because they are at risk of developing poor form, especially if they are browsed by animals or have forked stems already as seedlings. They are particularly vulnerable to damage on cold and frosty sites with strong winds and initial stocking densities of 200 trees per hectare are recommended for those sites.

Byington (1990) described similar practices in the USA with the hardwood black walnut (*Juglans nigra*), which is highly prized for both its timber and its nuts. Plantations are established with just over 250 trees per hectare and may be intercropped with wheat or soybeans, or pasture may be grown for animal grazing. The trees grow relatively slowly and may take 60 years to produce valuable timber logs, but produce nuts from 20 years of age. The trees may be thinned progressively during their lifetime until their stocking density is reduced to as low as 60–70 trees per hectare by the time of harvest for timber.

So varied are the practices of agroforestry (Sect. 1.2), it is impossible to generalise further about what initial spacings are appropriate for them. Many examples are provided by Jarvis (1991).

8 Thinning

Thinning involves changing the stand density (Chap. 7) of a plantation by the deliberate removal of some trees from it, from time to time during its life. As growth of the plantation continues, the stem wood produced by it will be distributed amongst the fewer trees which remain after the thinning. As a result, their stem diameter growth rates will accelerate and the trees will produce larger volumes of large-diameter logs at earlier ages than would have occurred if the plantation had not been thinned.

Thus, the principal reason for thinning plantations is to speed the production of larger, more valuable logs to be used for sawn wood production. This means it is usually appropriate to thin only in plantations being grown to produce large logs. In plantations being grown for bioenergy production, firewood or pulpwood, where log size is of much less importance, thinning is usually of no value. In agroforestry plantations, very heavy thinning (that is, removal of a very large number of trees) may be done to allow sunlight to reach the ground and encourage the growth of pasture or other crops.

Apart from accelerating the diameter growth rate of the remaining trees, it might be expected that the sudden removal of a large number of trees from a plantation could affect its subsequent growth and overall wood production in the long term. This is discussed in Sect. 8.1. There are also various hazards involved with such radical treatment of a plantation; these are discussed in Sect. 8.2. To achieve the results desired from thinning, it is necessary to design an appropriate thinning regime for any particular plantation. This involves deciding:

- How many times during its life should it be thinned
- At what ages the thinnings should be done,
- How many trees should be removed at each thinning
- Which particular trees should be removed at thinning.

These issues are discussed in Sects. 8.3 and 8.4. Finally, Sect. 8.5 gives some examples of thinning regimes that are used in practice around the world.

8.1 Growth Following Thinning

Thinning removes trees from a stand and with them their leaves, so a much reduced stand leaf biomass and leaf area index (Sects. 2.2, 2.3.1) will remain immediately after the thinning. Because of this, it might be expected that much less photosynthesis would occur in a thinned stand immediately following the thinning, so that its total growth would be appreciably less than if it had not been thinned.

However, experiments with many different plantation species in many parts of the world have found that the growth of stands immediately following thinning is often just the same as the growth that would have occurred if the stand had not been thinned (Aussenac et al. 1982; Cregg et al. 1988, 1990; West and Osler 1995; Cañellas et al. 2004; Connell et al. 2004; Hennessey et al. 2004; Mäkinen and Isomäki 2004c). Following thinning, the remaining trees seem able to adjust their overall photosynthetic production to just balance the loss of photosynthesis by the trees which have been removed (Whitehead et al. 1984; West and Osler 1995). It seems the principal ways in which the remaining trees do this are:

- Before the thinning, leaves in the lower parts of tree crowns would have been partially shaded by surrounding trees. Removal of some trees opens the canopy and exposes those leaves on the remaining trees to more sunlight. They are then able to undertake more photosynthesis than before the thinning (Donner and Running 1986; Ginn et al. 1991; Wang et al. 1995; Peterson et al. 1997; Tang et al. 1999; Medhurst et al. 2002; Whitehead and Beadle 2004; Medhurst and Beadle 2005).
- Water from the soil being used previously by the trees which have been thinned becomes available to the remaining trees. With the extra water, they are able to keep the stomata on their leaves open for more of each day and hence to undertake more photosynthesis (Whitehead et al. 1984; Brix and Mitchell 1986; Donner and Running 1986; Morikawa et al. 1986; Aussenac and Granier 1988; Stogsdill et al. 1992; Lieffers et al. 1993; Wang et al. 1995; Carlyle 1998).

There are some cases where growth of a thinned stand immediately following thinning has been less than the growth it would have had if it had not been thinned. This has been observed where:

- Rainfall is sufficiently high that, even in unthinned stands, none of the trees suffers any water shortage (Wang et al. 1991)
- Thinning has been so heavy that the remaining trees are no longer able to utilise fully the resources for growth available from the site (Mead

et al. 1984; Schönau and Coetzee 1989; Whyte and Woollons 1990; Pape 1999b; Medhurst et al. 2001)

- Nutrient availability from the soil is so low that it limits development of the trees retained after thinning (Valinger 1992; Valinger et al. 2000)
- Opening the canopy has encouraged vigorous weed growth, which competes with the retained trees for water and nutrients from the soil (Black et al. 1980; Cregg et al. 1990; West and Osler 1995).

In a few cases, growth of a thinned stand immediately following thinning has been greater than if the stand had not been thinned (e.g. Velaquez-Martinez et al. 1992; Carlyle 1998). These cases have occurred on sites where thinning alleviated a deficiency of nutrients from the soil. It did this either by making available to the retained trees the nutrients from the leaves and branches of the felled trees, or by increasing the rate at which nitrogen was released from organic matter, owing to increases in water availability in the soil or increases in temperatures at ground level resulting from opening the canopy.

Apart from these few specific cases where stand growth has increased or decreased, it appears that total growth of thinned stands, immediately following thinning, will generally be similar to the growth they would have had if they had remained unthinned.

Over several years following thinning, the remaining trees add additional leaves to their crowns, to restore eventually the leaf area index of the stand to its value before the thinning, a value determined by the availability from the site of the resources required for growth (Sect. 2.3). As they do so, the crowns increase in width to fill the space left by the trees which have been removed. As the retained trees grow taller, their crowns also become deeper, because the opening of the canopy allows light to penetrate to a greater depth within it (Barbour et al. 1992; West and Osler 1995; Peterson et al. 1997; Carlyle 1998; Baldwin et al. 2000; Valinger et al. 2000; McJannett and Vertessy 2001; Medhurst et al. 2001; Medhurst and Beadle 2001; Mäkinen and Isomäki 2004a, d). Because of the increase in crown width and leaf biomass, each tree requires longer branches, of much greater biomass, to carry the additional leaves (Medhurst and Beadle 2001). As well, the opening of the canopy by thinning allows more wind turbulence within the stand, increasing the threat that trees might be uprooted (Slodičák 1995; Mitchell 1995a); this means that trees in thinned stands must develop a larger woody root system to maintain their anchorage securely in the ground (Urban et al. 1994; Ruel et al. 2003).

The increases in size of branches and woody roots means that a higher proportion of the growth in biomass of the retained trees will be apportioned to their branches and woody roots and less to their stems than would have occurred if the stand had remained unthinned. In the long run, this means that the amount of stem wood produced by a thinned stand will be less than that from an unthinned stand, even though its total biomass growth is unaffected by the thinning (cf. Sect. 7.1.3). An example of this was given by Hennessey et al. (2004) for an experimental plantation of loblolly pine (*Pinus taeda*) in southern USA. Thinning was carried out at 9 years of age and again at 12 years of age. By 23 years of age, the thinned stands had produced a total of 10% less stem wood than the unthinned stands (a total which included all stem wood in trees which had been removed at thinnings or through tree deaths). However, because tree deaths were more frequent in the unthinned stand, it was unlikely that there would have been much difference between the total amounts of wood which could have been harvested from the thinned and unthinned stands. Long-term experiments to illustrate this trend are rare, but some results of a similar nature have been reported (e.g. Schönau and Coetzee 1989; Amateis et al. 1996; Curtis et al 1997; Baldwin et al. 2000; Medhurst et al. 2001; Mäkinen and Isomäki 2004b, c; Rytter and Stener 2005).

Following thinning, height growth of the retained trees changes little (Mead et al. 1984; Schönau and Coetzee 1989; Whyte and Woollons 1990; Valinger 1992; Peterson et al. 1997; Yang 1998; Pape 1999b; Baldwin et al. 2000; Mitchell 2000; Medhurst et al. 2001; Mäkinen et al. 2002a; Cañellas et al. 2004; Hennessey et al. 2004; Kanninen at al. 2004; Mäkinen and Isomäki 2004b, c; Varmola and Salminen 2004); however, there are other important effects on the subsequent growth of the retained trees as follow:

- Because their photosynthesis increases, their growth rates accelerate and they become larger, in both stem diameter and wood volume than they would have without the thinning (Hennessey et al. 2004; Cañellas et al 2004; Connell et al. 2004; Kanninen at al. 2004; Mäkinen and Isomäki 2004a, b, c, d; Varmola and Salminen 2004). This is consistent with the discussion of Sect. 7.1, that lowering stand density leads to an increase in individual tree sizes; it is the principal growth response desired of thinning in commercial forestry.
- They develop more sharply tapered stems (Newnham 1965; Barbour et al. 1992; Valinger 1992; Tasissa and Burkhart 1998; Pape 1999b; Baldwin et al. 2000; Karlsson 2000; Mäkinen and Isomäki 2004a, d). The opening of the canopy allows wind speeds to increase within the thinned stand and the weights of leaves and branches in the crowns of individual trees increase (see before). Both these effects alter the levels and distribution of stresses (in an engineering sense—Sect. 2.1.1) along individual tree stems. It is believed that tree stems are shaped so that the stresses to which they are subjected are constant along the

whole stem (see Dean et al. 2002 for a discussion of this theory); in thinned stands, the individual stems must become more tapered to maintain this constant stress along their length. The long-term result of this is to alter the set of logs of different size classes which can be cut from individual stems. In commercial plantation forestry, the economic advantages gained from thinning through increased stem sizes seem generally to outweigh any losses which result from increased stem taper.

8.2 Hazards of Thinning

Thinning exposes plantation stands to various hazards. Perhaps the greatest is posed by wind, which can damage trees either by snapping them off or by uprooting them. Uprooting can result either from breakage of large, woody roots, which anchor the trees in the ground, or from simply pulling the whole root system out of the ground (Schaetzl et al. 1989).

Wind storms can cause disastrous damage to plantations, whether thinned or not (Cremer et al. 1977; Somerville 1981; Liegel 1984; Neil and Barrance 1987; Bunce and McLean 1990; Everham 1995; Foster and Boose 1995; Studholme 1995). However, opening the canopy by thinning can render plantations more liable to damage by increasing wind turbulence within the plantation and allowing the trees to sway far more than when the canopy is closed (Moore and Maguire 2004, 2005). In some parts of the world where forests grow in windy regions, such as Great Britain, northern Europe and Canada, the potential for wind damage is of considerable concern and can discourage thinning (Quine 1995; Wollenweber and Wollenweber 1995; Quine and Bell 1998; Huggard et al. 1999; Gardiner and Quine 2000; Cameron 2002; Hale et al. 2004). Some research has even attempted to assess the suitability of sites for plantation forestry based on the risk of wind damage (Mitchell 1995b, 1998; Quine 2000).

The taller and thinner a tree, the further it can sway, rendering it more likely to snap off. As well, increased swaying increases the force exerted at ground level, increasing the risk of uprooting (Cameron 2002). Based on this, several researchers (Cremer et al. 1982; references in Wilson and Oliver 2000) have developed a simple index to assess the susceptibility of plantations to wind damage, at least for some softwood species in various parts of the world. The index is calculated as the average height of the 200 or 250 largest-diameter trees per hectare in a plantation before thinning, divided by their average diameter at breast height over bark (both measured in the same units, metres or feet say). They found that the risk of damage following thinning increased from a low level, when the index value was less than 60–70, to a high level, when the index value exceeded 90–100. However, the risk was generally far less before the average tree height reached 10–20 m, suggesting a first thinning should be done before a plantation reaches those heights. More detailed research has confirmed that the ratio of tree height to diameter is a reliable predictor of the risk of damage either by wind or by the weight of snow held in the crown (Moore 2000; Wilson and Oliver 2000; Wonn and O'Hara 2001). Considerably more sophisticated modelling systems are now being developed to assess in much greater detail the susceptibility of forests to wind damage (Gardiner et al. 2000; Ancelin et al. 2004; Zeng et al. 2004; Achim et al. 2005; Cucchi et al. 2005).

Whilst the risk of wind damage is probably the greatest hazard plantations face after thinning, there are other hazards also. If vigorous weed growth occurs in the plantation following the opening of the canopy, the weeds may compete successfully with the remaining trees for water or nutrients from the site. This can lead to a reduction in tree growth, with consequent loss of the volume of wood which can be harvested ultimately. If this problem occurs, it can be avoided by weed control; in thinned stands it may be economically practical to do this only by using grazing animals.

There may be effects of thinning on the quality of the wood produced ultimately in the stems of the retained trees. Larger branches or epicormic shoots may develop along the stems following thinning, increasing the size and number of knots in the timber (Sects. 3.3.8, 9.2, 9.4). Given that thinning accelerates the growth rate of retained trees, the greatest concern is that faster-grown wood might be less dense and, hence, less suited for various purposes (Sect. 3.3.3). However, whilst some changes in wood density have been observed in thinned stands, they have been sufficiently small generally to be of little practical concern (Gregg et al. 1988; Barbour et al 1992; Malan and Hoon 1992; Pape 1999a, b; MacDonald and Hubert 2002; Mörling 2002; Jaakkola et al. 2005). Some recent research with Tasmanian blue gum (*Eucalyptus globulus*) plantations in Australia (R Washusen, CSIRO, Australia, unpublished data) has found that the additional swaying of trees in thinned plantations can increase substantially the amount of tension wood that develops in the stems of the retained trees. As discussed in Sect. 3.3.6, this can reduce the quality of the wood produced ultimately from the trees. However, often the advantages in wood quality gained by selecting trees with better form to be retained at thinning (Sect. 8.3) may outweigh any deleterious effects on wood quality (Cameron 2002).

Damage to retained trees by being struck as other trees are felled is another hazard of thinning. The retained tree may simply be snapped off. However, if it is simply struck, the wood formed to heal abrasions on its stem may reduce the quality of the wood sawn ultimately from it (Terlesk and McConchie 1988). Abrasions may allow also the entry into the stem of wood decay fungi, which can lead to substantial loss of solid wood. Even slight abrasions, sufficient only to loosen the bark without stripping it from the stem, can lead to losses from wood decay (White and Kile 1991). Wood decay fungi and their effects are discussed in more detail in Sects. 9.4, 9.6 and 9.7.1. These problems of stem damage can only be avoided if the harvesting machinery operators or tree fellers are sufficiently experienced to avoid retained trees being struck by other trees as they are felled.

It is quite important that harvested wood to be sold is collected promptly from the forest after felling. Delays can allow fungi which stain wood to enter the cut end of logs, leading to discolouration of the wood and loss of value (Stone and Simpson 1987; Huber and Peredo 1988; Manion 1991). This is equally a problem when clear-felling is done (Keirle et al. 1983).

8.3 Tree Selection

In conducting a thinning, the aim is usually to select trees to be retained which:

- Have the best form to produce **sawlogs** of the highest quality, that is, have straight stems, are without multiple leaders (where a stem has split to produce more than one main stem) and have relatively small and well-formed branches
- Will grow most vigorously following thinning to produce the largest sawlogs at the earliest possible time.

As mentioned in Sect. 3.3.3, the species used for major plantation forestry developments around the world are commonly those which grow rapidly and well in full sunlight. In such species, trees of all sizes respond readily following thinning, both to the availability of extra sunlight from the opening of the canopy and to the additional resources for growth which become available by removing other trees (West and Osler 1995; Pukkala et al. 1998; Pape 1999a; Medhurst et al. 2001; Mäkinen and Isomäki 2004c, d). However, the larger trees in the stand have the larger canopies and root systems and so are able to produce larger amounts of new stem wood as their growth rates accelerate following thinning (Mäkinen and Isomäki 2004a). As well, since they are already the largest trees, they will inevitably be the individuals which will be the earliest to reach sizes to produce sawlogs. This means that thinning will usually favour the retention of the largest trees of the best form.

In the past, the types of thinning which have been used to favour retention of larger trees are:

- Thinning from above (also called high thinning), which involves retention of the largest, best-formed trees, which are felt most likely to be the best final-crop trees (that is, the trees to be retained until clearfelling). Trees around them are removed, some of which may be relatively large but which might interfere with the subsequent crown development of the retained trees. Smaller trees are retained, as long as they are unlikely to interfere with the development of the large retained trees.
- Thinning from below, which involves removing only smaller trees from the stand. The only very large trees removed in this type of thinning are those with bad form or, occasionally, one of two larger trees which are growing close together and would interfere with each other in their subsequent development.

One problem with thinning from above is that it limits the choice of which trees are to be retained at later thinnings (Arthaud and Klemperer 1988). By and large, thinning from below is preferred today in commercial plantation forestry, although there are cases where thinning from above has been found to be more successful commercially (e.g. Hytiäinen et al. 2004).

Whichever of these types of thinning is used, it is necessary usually to remove some rows of trees entirely to allow access to the stand by harvesting machinery. The distance between each of these outrows, as they are called, will be determined by how far the tree-felling arm of the machine can reach from the outrow into the surrounding stand.

A third type of thinning is called systematic or row thinning. It involves the removal of whole rows of trees only, perhaps every second or third row. It is a quick, easy and cheap way to conduct a thinning. However, unlike the other types of thinning, it does not favour retention of larger trees. It has been used successfully in various plantations around the world (Cremer and Meredith 1976; Bredenkamp 1984; Baldwin et al. 1989).

8.4 Developing Thinning Regimes

Drew and Flewelling (1977) brought to the western literature a tool which has proved very useful in making decisions about when and to what extent plantations should be thinned. Known as density management diagrams, they have the form of Fig. 7.2. The great value of these diagrams is that they allow the user to determine readily the stage of development of a stand, that is, in which section (A, B or C) of the diagram it is positioned. To do so requires measuring the stand stem wood volume and its stocking density; there are many ways of taking these measurements, as discussed in texts on forest measurement (e.g. West 2004).

Density management diagrams have been developed for a number of commercially important species, both in plantations and native forests (Drew and Flewelling 1977, 1979; Kikuzawa 1982; McCarter and Long 1986; Hibbs 1987; Smith 1989; Dean and Jokela 1992; Sterba and Monserud 1993; Newton and Weetman 1993, 1994; Kumar et al. 1995; Jack and Long 1996; Newton 1997a, b; Anta and González 2005). The detail and approach of these different diagrams differ sometimes from Drew and Flewelling's original concept, but their principle remains the same. Diagrams for many other species remain to be developed, perhaps notably for the eucalypts, given their importance in forest plantations around the world (Sect. 1.1).

On the basis of these diagrams, the following general principles have been established for thinning practice:

- Stands should be thinned before they enter section C of the diagram. This ensures no wood volume is lost owing to deaths of suppressed trees.
- Stands should remain unthinned until they leave section A of the diagram and enter section B. Thereafter, they should not be thinned sufficiently heavily that they re-enter section A. Any thinning that positions a stand in section A will lead to a loss of stand stem wood volume growth, because the stand will not be using fully the resources available for growth from the site.
- If a stand is thinned to near the bottom of section B of the diagram, it will have the fewest possible trees retained, whilst still using fully the growth resources available from the site. Individual growth rates of retained trees will then be at their maximum and larger, commercially more valuable trees will be available ultimately for harvest at the earliest possible time.
- Stand stem wood volume production will be maximised if stands are thinned so they remain near the top of section B of the diagram. As discussed in Sect. 7.1.3, stand stem wood volume growth rates tend to be higher in stands of higher density.

In essence, these principles suggest that thinning should be undertaken to maintain stands always in section B of the diagram. If the objective of thinning is to produce the largest trees possible in the shortest time, stands should be thinned so they remain always near the bottom of section B. If the objective is to maximise the volume of stem wood to be harvested, stands should be thinned so they remain always near the top of section B. For most commercial plantations around the world, the objective is to maximise tree sizes; hence, most thinning regimes aim to maintain stands near the bottom of section B.

The results from two thinning experiments will illustrate how a density management diagram can be used to assess thinning practice. The experiments were conducted in plantations of Douglas fir (*Pseudotsuga menziesii*), one in Washington, USA, and the other in Kaingaroa, New Zealand (Drew and Flewelling 1979). Douglas fir is one of the best known forest species of North America. A tall, softwood species, it occurs in magnificent native forests in western parts of the continent, forests which have been logged for timber for many years. In some places the Douglas fir forests are allowed to regenerate naturally and in other places plantations of the species have been established. Some plantations of Douglas fir have been established in New Zealand, although it is not the most common plantation species there.

In both experiments, some plots were left unthinned, whilst others were thinned several times, either more or less heavily. Their growth was measured for many years. In Fig. 8.1, the growth trajectories of the unthinned, lightly thinned and more heavily thinned plots of both experiments have been superimposed on the density management diagram for Douglas fir developed by Drew and Flewelling (1979). The sudden declines in stocking density and stand stem wood volume in the thinned plots denote the removal of trees at the thinnings; the Washington experiment was thinned three times, firstly at 16 years of age, and the Kaingaroa experiment twice. Table 8.1 records some data measured at the end of each experiment when the plantations were clear-felled, at 27 years of age in the Washington experiment and at an age unspecified by Drew and Flewelling in the Kaingaroa experiment, but probably considerably older.

Given the thinning principles discussed before, the results suggest thinning started far too late in the life of the Kaingaroa plantation if its aim was to maximise the sizes of individual trees by the time of clear-felling. The plantation had reached section C of the diagram before thinning started. Whilst the result of both the light and heavy thinnings was to increase the average size of the retained trees by the end of the experiment (Table 8.1), the heavier thinning more so than the lighter, none of the thinnings was ever heavy enough to shift the plantation stands near the bottom of section B of the diagram; if they had done so, average tree sizes would have been much greater at the time of clear-felling.

By contrast, the heavy thinnings in Washington appear to be too heavy,

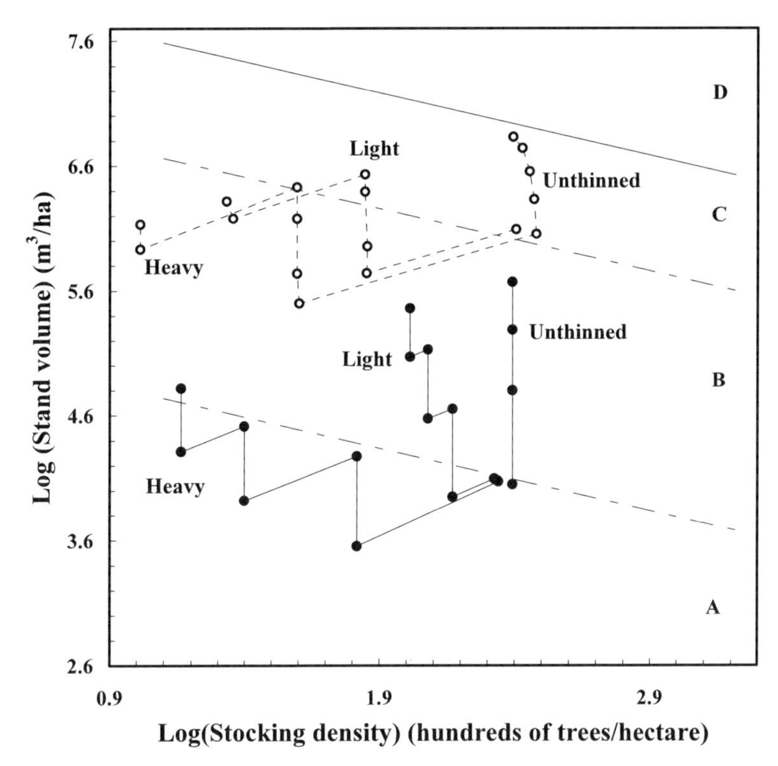

Fig. 8.1. A density management diagram for Douglas fir (*Pseudotsuga menziesii*). Superimposed are the growth trajectories of three experimental plantation plots in each of Washington, USA (●——●), and Kaingaroa, New Zealand (o- - -o), which were unthinned, lightly thinned or heavily thinned as indicated. All data are plotted as natural logarithms (adapted from Drew and Flewelling 1979)

Table 8.1. The average stem wood volume per tree at the age of clear-felling and the total stand stem wood volume harvested (inclusive of volume removed at thinnings) in unthinned, lightly thinned and heavily thinned plots in the experimental plantations illustrated in Fig. 8.1 (Source—Drew and Flewelling 1979)

Plot treatment	Washington, USA		Kaingaroa, New Zealand	
	Volume per tree (m^3)	Harvested stand volume (m^3/ha)	Volume per tree (m^3)	Harvested stand volume (m^3/ha)
Unthinned	0.27	291	0.84	931
Lightly thinned	0.32	263	1.49	924
Heavily thinned	0.38	187	1.72	917

because the plantation was kept consistently within section A of the diagram. This meant there was a loss of stem wood volume harvested fromthe heavily thinned stand, which is apparent in the stand stem wood volume information in Table 8.1. On the other hand, the lighter thinning regime in Washington was too light and should have involved thinnings which positioned the stand nearer the bottom of section B of the diagram.

A density management diagram does not give the complete answer as to when, how many times and with what intensity a plantation should be thinned. However, it does serve to put some constraints on what is a feasible thinning regime for a plantation. Ultimately, the thinning regime chosen for a plantation will be that which most suits the objectives of the plantation enterprise. Usually in commercial plantation forestry, this involves finding the regime which maximises the return on the financial investment in the plantation. The factors which must be considered when settling on a thinning regime include:

- The number of thinnings which are to be done—thinning incurs expenses with logging machinery and labour. If there is a market for the logs cut from the trees removed at thinning, these expenses may be recouped through log sales and the thinning may provide a useful income part way through the plantation rotation; the thinning is then said to be a commercial thinning. Even if the wood removed at thinning cannot be sold, the acceleration of growth rate of the remaining trees that results may be sufficient to render the thinning economically worthwhile in the long term. A thinning without sale of the harvested wood is said to be non-commercial or precommercial. In that case, it may be unnecessary to fell the trees being removed; it may be sufficient simply to poison them with a herbicide and leave them to fall naturally in due course. However, whether commercial or non-commercial, the costs and/or returns involved with thinning will determine ultimately how many thinnings can be afforded during the plantation rotation.
- The timing of thinnings—the first thinning will not be carried out until the plantation has entered section B of the density management diagram. The age at which this occurs will depend on the growth rate of the plantation. If the first and/or subsequent thinnings are commercial, they may have to be delayed until the plantation has grown to an age when there is sufficient wood available for the thinning to be worthwhile commercially. The possibility of wind damage (Sect. 8.2) will become increasingly acute the longer the first thinning is delayed.
- The intensity of thinnings—if the objective of thinning is to maximise growth rates of the retained trees, the thinning will need to be sufficiently intensive to move the stand near the bottom of section B of the

density management diagram; the possibility of wind damage may limit how near the bottom of section B the thinning may approach. The more any thinning has been delayed, the further will the plantation have grown through section B of the diagram and the more intensive will the thinning need to be. Obviously, the timing, frequency and intensity of previous thinnings and the growth rate of the species concerned will all determine how far into section B of the diagram the plantation will have grown by the time of the next thinning. All these factors will interact to influence the intensity of the next thinning.

- The relative values of the different-sized log products which are to be produced ultimately by the plantation—if larger-sized logs are very much more valuable than smaller-sized logs, then it will be worthwhile thinning earlier, more frequently and more intensively to ensure that the largest-sized trees have developed at the earliest possible age to maximise the financial return from the plantation enterprise. The balance between the relative values of the various log products will be very important in determining the balance between timing, frequency and intensity of thinning.

It should be apparent from the preceding discussion that the species concerned, the productive capacity of the site on which it is growing, the objectives of the plantation enterprise and the values of the wood products which can be sold will all interact in complex ways to determine exactly what thinning regime is appropriate for any particular plantation. They will determine also what is the most appropriate age at which the plantation should be clear-felled; commercially, this will be the age at which the financial return from the plantation is maximised.

Obviously it is a complex task to consider together all these factors, and their interactions, to choose exactly what thinning regime is most appropriate for a particular plantation. In large commercial plantation enterprises, foresters have available mathematical forest growth and yield models, which are capable of predicting what wood yields of various log product sizes might be obtained under any thinning regime they consider appropriate. Forest growth and yield models were alluded to in Sect. 4.4.3. Some models include the mathematical basis of at least parts of density management diagrams (Smith 1986; Smith and Hann 1986; Seymour and Smith 1987; Tait et al. 1998; Cao et al. 2000; Mabvurira and Miina 2002; Ogawa and Hagihara 2003; Newton et al. 2004, 2005; Ogawa 2005).

In effect, growth and yield models are used to propose a number of possible thinning regimes which might be considered for a plantation. Deciding which of those regimes is most appropriate is then a complex task to which there are various approaches (Chen et al. 1980; Kao and Brodie 1980; Bullard et al. 1985; Paredes and Brodie 1987; Arthaud and Klemperer 1988; Betters et al. 1991; Pukkala and Miina 1997; Wikström and Eriksson 2000; Wikström 2001; Hyytiäinen et al. 2005). Riitters and Brodie (1984) have given an interesting example where the most appropriate thinning regimes they developed for Douglas fir forests were presented in a density management diagram, so the relationships between the various options they considered were obvious immediately.

8.5 Thinning in Practice

Thinning has a long history in forestry, extending back to the early 1800s. Zeide (2001) has given an interesting review of the developments which have led to our current view of thinning practice. There has been a trend to change from very conservative thinning practice, that is, numerous light thinnings continuing to later ages, to far more intensive practices. These changes have accelerated since the 1950s (Opie et al. 1984; Schönau and Coetzee 1989). They reflect both our increased knowledge of how trees respond to thinning and increases in the costs of thinning, which have made it economically worthwhile to undertake only a few, heavier thinnings during a rotation.

Table 8.2 is a rather arbitrary list of thinning regimes considered appropriate in various commercial plantation forestry species in various parts of the world. They are all cases where plantations were being grown for high-value, sawlog production. The list should give the reader some idea of contemporary approaches to thinning practice in large-scale, commercial plantation enterprises.

The regimes involve one to five thinnings, with around 100–300 trees per hectare left after the final thinning. Generally, clear-felling occurs at around 25–30 years of age. The trend to undertake intensive thinnings at quite young ages is apparent in the regimes for the fast-growing hardwood and softwood plantations in Australia, Costa Rica, New Zealand and South Africa (see also Kanninen et al. 2004; Ladrach 2004). The pine (*Pinus*) plantations in Spain and the USA and the hoop pine (*Auraucaria cunninghamii*) plantations in Australia are rather less productive than those other plantations and their thinning regimes tend to be less intensive.

Also apparent is that a number of the regimes involve non-commercial thinnings at younger ages. In those cases, either there is little market for small logs or the advantage gained ultimately in sawlog production at clear-felling makes the early thinning worthwhile.

Table 8.2. Thinning regimes used for sawlog production in various plantation forest species in various parts of the world. The age at which each thinning is done and the residual stocking density after thinning are shown; the stocking density at planting is shown at age 0 years. Thinnings marked with an *asterisk* are non-commercial

Species	Location	Age (years)	Residual stocking (trees/ha)	Reference
Flooded gum (*Eucalyptus grandis*)	South Africa	0	1,370	Schönau and Coetzee (1989)
		3–5	750*	
		7–9	500	
		11–13	300	
		25–30	Clear-fall	
Shining gum (*E. nitens*)	Tasmania, Australia	0	1,000	Gerrand et al. (1997)
		3–4	600*	
		10–12	250	
		30–40	Clear-fall	
Blackbutt (*E. pilularis*)	New South Wales, Australia	0	1,000	Present author, unpublished
		10	270	
		15	100	
		25–30	Clear-fall	
Teak (*Tectona grandis*)	Costa Rica	0	1,111	Pérez and Kanninen (2005)
		4	556	
		8	333	
		12	200	
		18	150	
		24	120	
		30	Clear-fall	
Honduran Caribbean pine (*Pinus caribea* var. *hondurensis*)	Queensland, Australia	0	750	Evans and Turnbull (2004)
		2–3	500*	
		22	300	
		30	Clear-fall	
Radiata pine (*P. radiata*)	New Zealand	0	1,600	Whyte (1998)
		6	600*	
		8	375	
		10	200-250	
		25–30	Clear-fall	
Radiata pine (*P. radiata*)	Spain	0	1,700	Rodríguez et al. (2002)
		12	550*	
		22	325	
		30	Clear-fall	
Loblolly pine (*P. taeda*)	Southern USA	0	1,090	Arthaud and Klemperer (1998)
		22	500	
		30	250	
		34	Clear-fall	
Hoop pine (*Auraucaria cunninghamii*)	Queensland, Australia	0	833	Hogg and Nester (1991)
		25	400	
		45–50	Clear-fall	

9 Pruning

In plantation forestry, pruning involves the removal of live or dead branches for some distance up the tree stem. Its purpose is quite different from that in horticultural crops, where plants are pruned to train them to adopt appropriate shapes and sizes, which opens their canopy allowing the sun to penetrate and stimulate the production of fruit.

The reasons for pruning in plantations include:

- Producing knot-free wood in the stem, from which high-quality timber can be sawn
- Maintaining smooth stems, where the plantation is being grown to produce posts or poles
- Harvesting the leaves for animal fodder, leaf mulch or leaf products such as essential oils
- Using branch wood as firewood
- Allowing easier access for animals or for intercropping of trees with other crops in agroforestry
- Maintaining the health of the plantation, where lack of pruning may attract damaging pests or **diseases**
- Preventing fallen leaves accumulating in the lower branches of trees, to minimise the chance of fire moving from the ground into the tree crowns.

If pruning is considered worthwhile in a forest plantation, decisions will have to be made about how it is to be done, when in the life of the plantation it should done, how far up the stem each tree should be pruned and how many of them should be pruned. This chapter will consider the issues involved in developing appropriate pruning regimes for plantations.

9.1 Natural Pruning

Section 2.2 has described how, over the first few years of the life of a plantation, individual tree crowns grow until they meet to form a continuous canopy. The time it takes for this to happen varies greatly, depending on the species being grown, how productive the site is and the stocking density at which the plantation was established. It may be very rapid (1–2 years) on highly productive sites or much longer (10 years or more) on less productive sites. If the stocking density at planting is very low (say less than about 50–100 trees per hectare), the canopies may never touch, because individual tree stems are not sufficiently strong to support the weight of the very wide spread of the branches that would be necessary to cover the gaps between the trees.

Once the canopy has closed, leaves are shed continuously from its base and are replaced progressively by new leaves at its better-lit top (Sect. 2.3.3). Once leaves have been shed from the base, the branches which held them no longer serve any purpose. They too are then shed, eventually leaving the lower part of the stem clear of branches (Fig. 9.1). Branch shedding involves the development of an abscission layer at the base of the branch, a layer of weak tissue. This eventually breaks and allows the branch to be ejected from the stem, usually leaving a small stub. The stem then produces new tissue which grows over this stub, a process known as occlusion. The final result of this self-pruning process is a knot left in the stem wood. The wood which grows beyond the knot is 'clear' (that is, knot-free), although there is usually distortion of the grain for 2–4 cm beyond the knot.

Fig. 9.1. A branch just about to be shed from the stem of a 3-year-old blackbutt (*Eucalyptus pilularis*) tree, growing in a plantation in northern New South Wales, Australia (Photo—P.W. West)

The whole process of branch death, abscission, ejection and occlusion often takes several years; however, different species vary greatly in the time it takes. Some eucalypts need only 1–2 years for the process to occur (Montagu et al. 2003). Other hardwoods take much longer; occlusion after death of branches of silver birch (*Betula pendula*) in Finland took 7–10 years (Mäkinen 2002). Softwood species often shed and occlude branches slowly, some retaining branches for decades (Petruncio et al. 1997; Mäkinen 1999); Taiwania (*Taiwania cryptomerioides*) in Taiwan took at least 9 years to occlude branches after their death (Wang et al. 2003).

9.2 Knots and Wood Quality

As discussed in Sect. 3.3.8, knots are the chief source of defect in sawn timber. Australian and South African experience with eucalypts is that knots deriving from branches which had a diameter of 2.5–3 cm or more at their junction with the tree stem are likely to cause appreciable defect in sawn timber (Jacobs 1955; Schönau and Coetzee 1989). Even knots produced from branches smaller than this can cause defects. For example, Marks et al. (1986) studied timber sawn from mountain ash (*Eucalyptus regnans*) trees and found that 85% of knots deriving from branches of 2–2.5-cm diameter were sufficiently large to produce at least some degree of defect in the timber. However, they found also that the amount of defect declined rapidly as the knot size declined; only 32% of knots deriving from branches of 0.5-cm diameter produced some defect. Very large knots, deriving from branches with a diameter in excess of 3.5 cm, may cause sufficient defect to weaken seriously the structural strength of eucalypt timber (Neilsen and Gerrand 1999).

The effects of knot size on wood quality are important in species other than eucalypts. Larger knots may be tolerable in some species, probably because the occlusion process is more efficient than it is in eucalypts. MacDonald and Hubert (2002) suggest knots deriving from branches of 5-cm diameter are tolerable in timber cut from the plantations of Sitka spruce (*Picea sitchensis*) in Great Britain. Fahey and Willits (1995) have illustrated the various grades of timber that are sold in the USA and how the presence of knots lowers the grade. Grades vary from the highest-quality 'appearance' grades, which are used for timber to be exposed in use for decoration, to 'structural' grades, which are used in building construction and must be free of defects which affect their strength, to 'factory' grades, which are to be cut up subsequently for use in mouldings or window frames and so on. The highest-quality timber in each grade may be knot-free or may have only a few, small knots. Grade, hence value, declines as both the number and the size of knots in the sawn timber increase.

Given this, if trees are being grown to produce wood of the highest quality for sawn timber production, it is obviously important that their stems should contain a minimum of knots. At least in Australian plantation forestry, this has led to the view that stem pruning should be considered if the trees in the plantation are going to develop branches larger than about 2.5–3 cm in diameter at their base. Furthermore, the earlier in the life of any tree that branches are pruned from its stem, the higher will be the proportion of completely knot-free wood it will contain when it is harvested finally. Research in Australia (see summary in Montagu et al. 2003) found that in unpruned eucalypt plantations clear-felled at 15–25 years of age, as little as 20% of the sawn timber obtained from them had a clean, virtually knot-free surface and so was of the highest-quality appearance grade. On the other hand, as much as 60% of the sawn timber was of appearance grade from plantations which had been pruned early in their life-time.

9.3 Branch Development

As discussed in Sect. 7.2.2, the stocking density of a plantation is probably the most important factor which determines branch size. The results in Fig. 7.5 show that branches with diameters at their base in excess of 2.5–3 cm will be found often on trees in plantations planted at stocking densities of around 1,000 trees per hectare, a planting density used commonly today (Sect. 7.2.4). As discussed in Sect. 9.2, this is a branch size often sufficient to lead to appreciable defects from knots in sawn timber. All the eucalypt plantations included in the experiments shown in Fig. 7.5 were growing on highly productive sites and were only 5 years of age at the time they were measured. This emphasises how early in the life of a plantation branches may reach sizes which can lead to wood defect. In fact, a large proportion of the branches measured in those eucalypt plantations were already dead by 5 years of age, so had reached sizes in excess of 2.5–3 cm diameter even earlier in the life of the plantation.

Not only will the maximum size that branches attain in a plantation increase as the stocking density of the plantation decreases, but so also will the overall **branchiness** of the trees. Branchiness has been defined as the total cross-sectional area, at their bases, of the branches on a stem, expressed as a proportion of the total surface area of the length of the stem along which they occur (Mäkelä 1997; Kellomäki et al. 1989). Thus one tree may be branchier than another because it has either more or larger branches than the other. The branchier a tree, the more work would be necessary to prune it. Both Mäkelä (1997) and Kellomäki et al. (1989) used growth models to illustrate that branchiness of Scots pine (*Pinus sylvestris*) increased as stocking density decreased. Recent developments with this type of model predict not only tree branch development, but also the properties of the wood in the stem and where knots will occur in it (Kellomäki et al. 1999). Other research has developed models to predict branch characteristics of Scots pine (Mäkinen and Colin 1998), Norway spruce (*Picea abies*) (Mäkinen et al. 2003b) and silver birch (*B. pendula*) (Mäkinen et al. 2003a) in Scandinavia and radiata pine (*Pinus radiata*) in New Zealand (Woollons et al. 2002). As research continues to develop models of these types for other species, they will be used increasingly to assist plantation growers to make decisions about if and how to prune their forests.

9.4 Effects of Pruning

Pruning trees in a plantation may involve the removal of dead limbs only. In species which maintain dead branches for many years (Sect. 9.1), their removal may limit the size of the **knotty core** of the stem, that is, the central part of the stem to which branches were attached before the tree was pruned; the smaller the knotty core of a log finally harvested from the forest, the greater is its value likely to be. Removal of dead limbs may be done also to allow easy access between the rows of the plantation, to remove limbs which catch falling needles and constitute a fire hazard or to provide firewood from the branch material.

As discussed in Sect. 9.3, live branches may reach a size sufficient to cause defects in timber at early stages of development in highly productive plantations (Sect. 9.3). If such plantations are pruned to avoid this, pruning will involve removal of live branches from the lower parts of the stem; of course this will involve the removal also of the leaves on those branches. If plantations are being grown to produce products from the leaves themselves, animal fodder or leaf oils for example, pruning of these plantations will involve removal of live branches also.

Removal of live branches and leaves by pruning has a variety of effects on subsequent tree growth and development (Briggs 1995; Maguire and Petruncio 1995; Pinkard and Beadle 2000; MacDonald and Hubert 2002). The density of stem wood may increase (Sect. 3.3.3), there may be earlier transition from juvenile to mature wood (Sect. 3.3.3), there may be more heartwood produced (Sect. 3.3.5) and stems may become less tapered (Larson 1965; Maguire and Petruncio 1995). All of these responses may be a result of the need for the stem to maintain a constant distribution of stresses along it following removal of the weight of lower branches (Sect.

8.1). They may be a result also of the need to maintain a sufficient width of sapwood along the whole length of the stem to transport water from the roots to the leaves (Sect. 2.1.1). Scientific opinion remains divided as to the exact biological mechanism which leads to these changes in stem characteristics after pruning (Mäkelä 2002); however, none of these changes seems generally to be sufficiently large to have discouraged commercial plantation growers from undertaking pruning.

Perhaps the most important effect of pruning live branches is that it may lead to a reduction in subsequent growth of the tree. The loss of leaves may reduce the total amount of photosynthesis that a tree can undertake, at least for some time following the pruning until the tree can replace the leaves which were removed. However, research results have suggested that as long as no more than about 40–50% of the leaves are removed from the lower parts of the tree crown of many hardwood species, or about 30–40% for softwood species, the growth of the tree will be unaffected (Maguire and Petruncio 1995; Pinkard and Beadle 2000; Neilsen and Pinkard 2003; Pinkard et al. 2004).

There are several reasons why trees can withstand loss of this much of their leaves without reducing their growth:

- Tree growth rate is determined by the amount of sunlight intercepted by their leaves; this is determined largely by the leaf area index of the plantation (Sect. 2.3.1). Highly productive plantations, with a leaf area index of 6–8 m^2/m^2, say, usually intercept more than 95% of the sunlight falling on their canopies. Because many leaves in the canopy normally shade other leaves, the plantation will still intercept about 80% of the sunlight even when half its leaf area is removed by pruning (Pinkard and Beadle 2000; Montagu et al. 2003). That is, a large leaf loss does not necessarily mean a correspondingly large reduction in the amount of sunlight which the canopy can intercept.
- Normally, leaves do not function at their maximum photosynthetic capacity. When some of the leaves are removed from a tree, whether by pruning or by other forms of defoliation such as insect browsing (Sects. 10.2, 10.3; Retuerto et al. 2004), the remaining leaves are capable of modifying their physiological behaviour and increasing the amount of photosynthesis they undertake to make up for the loss of other leaves. As well, the lifespan of leaves may increase to minimise further leaf loss through normal leaf shedding, until the tree is able to replace the leaves lost by pruning. So large may be these responses, that growth may actually be greater in pruned than in unpruned trees for some time following the pruning (Pinkard et al. 1999; Pinkard and Beadle 2000; Pinkard et al. 2004).

- The new leaves which develop following pruning are sometimes larger in area than new leaves which develop in an unpruned canopy (Pinkard and Beadle 1998, 2000). This allows them to intercept more of the sunlight falling on the canopy, so allowing relatively more growth to occur in pruned than in unpruned trees.
- As the trees respond to pruning by replacing the leaves which were removed, more of their growth may be assigned to leaf production rather than to stem or root production than is the case for unpruned trees (Reich et al. 1993; Pinkard and Beadle 2000). Thus, trees may respond to pruning by attempting to replace the lost leaves as rapidly as possible, so minimising any loss of growth.

Another important effect of pruning live branches may be the sprouting of epicormic shoots along the pruned stem; these develop eventually into normal leaves or leaf-bearing branches (Sect. 5.5). Re-development of branches from epicormic shoots may frustrate completely the purpose for which pruning was done. As an example, Deal et al. (2003) studied epicormic shoot development after pruning and thinning stands of Sitka spruce (*Picea sitchensis*) in Alaska. In this case the trees were growing in native stands, although Sitka spruce is used extensively in plantations in the northern hemisphere. At 6–9 years after pruning, they found profuse epicormic shoot development, with an average of 9–11 shoots developing along each metre of pruned stem. Many of the shoots were of small diameter and may not have led ultimately to large knots in the timber; however, some shoots had developed into branches which exceeded 1.5 cm in diameter at their junction with the stem and so may have been large enough to lead ultimately to some defect in sawn timber. Maguire and Petruncio (1995) have reviewed a number of experiments where epicormic shoots appeared after pruning North American conifers.

A further important effect of pruning on tree growth and development is that the wound produced by pruning may be a point of entry to the stem for wood decay fungi (Sects. 11.1, 11.2). There are many such fungi and the decay they cause may render wood quite useless for any commercial purpose. Trees may react to wounding by producing chemicals with antifungal properties and by establishing a barrier zone of tissue within the stem, beyond which decay does not spread further towards the centre of the stem (Eyles et al. 2003). If this happens, it limits the spread of the decay, but often not before considerable defect has been caused, reducing substantially the quality of the timber obtained ultimately. The season when wounding occurs may affect the extent of decay and sometimes it is more extensive on sites of high fertility (Montagu et al. 2003).

Finally, pruning may lead to a more indirect effect on tree growth and development. If the pruned branches are simply left on the ground, they can provide ideal breeding sites for insects (Sect. 10.2) which damage the live trees subsequently. Johnson et al. (1995) describe a number of examples where this has occurred in northern hemisphere conifers. Others (e.g. O'Hara et al. 1995) have pointed out that the pruned branches may constitute an increased fire hazard if they are not removed from the site after pruning.

9.5 Pruning Regime

If it has been decided to prune a plantation, the decisions necessary to establish an appropriate pruning regime include:

- When and how often during the life of the plantation pruning should be done
- How far up the stem should branches be removed at any pruning
- How many and which trees in the plantation should be pruned.

If pruning is to involve removal of dead branches only, these decisions are fairly easy to make. They are rather more difficult when live branches are to be pruned, which is the case commonly in plantation forestry today.

Pinkard and Beadle (2000) have proposed that a general approach to developing a pruning regime for any plantation involves following its leaf area index development over time. As long as it has been determined for a particular species how much leaf removal it can stand without growth loss, knowledge of leaf area index should allow determination at any time in the life of the plantation of the height to which it is reasonable to prune the trees: as mentioned in Sect. 9.4, most species can withstand removal of somewhere in the range 30–50% of their leaf area index without subsequent growth loss.

It is impossible in a book of this nature to specify the pruning regime appropriate to any particular plantation. The reasons for which pruning is being done, the biological characteristics of the species concerned and the productive capacity of the site on which the plantation is growing will all be important in determining the regime. The following discussion attempts to apply Pinkard and Beadle's principles in devising pruning regimes for plantations.

9.5.1 When to Prune

It is clear from the discussion in Sect. 9.3 that branches large enough to produce knots which cause serious defects in sawn wood can develop early in the life of a plantation. This may be within the first 5 years of growth of highly productive plantations. In less productive plantations, such as the *Pinus patula* and Scots pine plantations referred to in Fig. 7.5, it may take longer for large branches to develop.

Both Pinkard and Beadle (2000) and Montagu et al. (2003) agree that the optimum time to undertake the first pruning is when leaf area index reaches a maximum, at about the time the crowns of the trees have developed sufficiently so that the canopy is just closed (Sect. 2.3.1). Interception of sunlight by the canopy will then be at a maximum, so the amount of light reaching the ground will be minimised. Then, following the arguments of Sect. 9.4, the reduction in sunlight interception will be minimised if part of the canopy is removed at that time.

In rapidly growing eucalypt plantations, the canopy may reach its maximum leaf area index at 3–4 years of age (Beadle et al. 1995; Pinkard and Beadle 2000), or even as early as 2 years of age as in the example in Fig. 2.1. As long as pruning at these young ages does not remove more than about 40–50% of the lower part of the canopies of trees in these hardwood plantations, there should be no loss of growth and the size of the knotty core of the tree stems should be minimised. In less productive plantations, it may be some years later in their life before they reach their maximum leaf area index and, hence, the age at which pruning might first be appropriate.

If a first pruning is constrained to removing no more than a certain proportion of the leaf area index, the height to which stems are pruned may be less than the total height to which it is desired ultimately to prune (Sect. 9.5.2). This means that pruning will need to be done in several 'lifts', that is, on several different occasions to successively greater heights. Each subsequent pruning would need to be delayed until the leaf area index of the canopy has recovered to such an extent that it can stand reduction again without loss of growth; this might be when it has regained about 75–80% of its long-term, maximum value. Pinkard and Beadle (2000) refer to an example of a highly productive shining gum *(E. nitens*) plantation in Tasmania, Australia, where the recovery period was only 13 months after a first pruning lift at 3 years of age. On less productive sites, it may take longer for the canopy to recover sufficiently for subsequent prunings.

If the objectives of pruning are other than the production of high-quality sawn timber, other considerations may determine the timing of pruning. Pinkard (2000) referred to an example where leaf oil and flowers were required from a young age in plantations of Tasmanian blue gum (*E. globulus*), plantations which were intended ultimately to produce high-quality wood. She suggested that a very light pruning, before canopy closure at 2 years of age and which removed only about 20% of tree leaf area index, could be done in these plantations to provide an early yield of oil and flowers. A heavier pruning at that time would lead to a loss of growth of the plantations.

Where the objective of pruning is for other purposes again, such as to provide firewood, access for agroforestry or to prevent the accumulation of fallen leaves, the timing of pruning will often be determined by pragmatic considerations. Thus, pruning might be done when the trees are sufficiently large that animals cannot reach the remaining leaves to browse them or otherwise damage the trees. Or, it might de done when an accumulation of leaves has occurred on dead branches. However, in other cases when pruning of live branches is being done, such as to provide fodder for animals, its timing will also be limited by leaf area development to avoid excessive leaf removal and so reduce tree growth.

9.5.2 Height of Pruning

If pruning is being done to produce high-quality timber, the maximum height to which any tree will be pruned will be no more than the height up the stem from which the larger, high-quality sawlogs will be obtained when the tree is finally harvested. In many plantations today, this will be a height around 6–10 m above ground. Often, this height will be limited by the costs involved. As discussed in Sect. 9.6, the costs of pruning rise rapidly once it is done above a height which workers can reach from the ground. An economic balance, between the cost of pruning and the increased value of the wood resulting from it, will determine to what height it is reasonable to prune and, indeed, if pruning is worthwhile economically at all.

Leaves are not distributed evenly down through the canopy of trees; usually there are more leaves positioned near the middle of the canopy than near its top or bottom. Information in Pinkard and Beadle (2000) suggests that removal of 30–50% of the leaves at any pruning lift would usually involve removing branches to a height of about 20–45% of the length of the green crown (that is, from the lowest live branch on the stem to the tip of the tree). They point out also that different tree species distribute their leaves in rather different ways along the stem. A study would have to be made of the leaf distribution in the canopy of any particular species before it could be said with certainty how high up the stem it would be reasonable to prune that species at any one pruning lift.

If live branches are being pruned from a species which sheds dead branches fairly readily, it has been suggested there may be little point in wasting time and money by pruning any dead branches that are still persisting on the lower part of the stem. Research with eucalypts has found that the knots produced after natural shedding and occlusion would be little different from those produced if the dead branch had been pruned before it was shed naturally (Montagu et al. 2003). Furthermore, Pinkard (2002) reported that stubs left after pruning dead branches of eucalypts may break off and be trapped in the occluding tissue leaving kino (a gum exuded by damaged tissues) traces, which can further degrade the quality of the wood. However, in species for which the process of branch shedding and occlusion takes much longer than the few years required in many eucalypts, it may be worthwhile pruning dead branches to ensure the size of the knotty core of the stem is minimised. O'Hara et al. (1995) and Maguire and Pentruncio (1995) reviewed experimental results for some North American conifers; in those, the time to occlusion of pruned dead branches was appreciably longer than for pruned live branches and also increased appreciably as the size of the pruned branch increased.

Where the objective of pruning is for purposes other than the production of high-quality timber, pragmatic considerations will often determine the height of pruning. Thus, if animal access is required, pruning would need to be only to a little above the height of the animals. If pruning was being done to minimise the chance of fire ascending into tree crowns through dead leaves caught on dead branches, pruning would need to be to a height somewhat above the flame height of ground fires.

9.5.3
Trees to Be Pruned

Usually, it is only at the time of clear-felling of a plantation that trees will have grown to sizes where appreciable quantities of the largest, most valuable logs can be harvested from them. To achieve large tree sizes in a reasonable time, a plantation will usually have been thinned at various times during its lifetime (Chap. 8). Obviously, there is no point in spending money to prune trees which will be removed later at thinning. This means that the final-crop trees will need to be selected for pruning at the time the first pruning lift is done; as discussed in Sect. 9.5.1, this can be within the first few years of the life of highly productive plantations.

The trees selected for pruning will be those which are taller than average, hence are growing most vigorously, and those with the best form, that is, without crooked stems or excessive branchiness. These will be the trees most likely to produce the largest quantities of high-quality timber in the shortest possible time (Sect. 8.3). Neilsen and Gerrand (1999) studied the availability of suitable trees for pruning at 5 years of age in highly productive plantations of shining gum (*E. nitens*) in Tasmania. In their case, it was desired to prune 300 trees per hectare, which were to be the final-crop trees. They found that unless at least 1,000–1,100 trees per hectare had been planted, there would be insufficient trees of appropriate quality available to choose the desired 300 trees per hectare. That is to say, the availability of suitable trees for pruning may be important when deciding what the original planting density of the plantation should be (cf. Sect. 7.2).

If not all trees in the plantation are to be pruned, it is important that pruning should not reduce the subsequent growth of the pruned trees. Apart from the possibility of economic loss if tree growth is reduced, pruned trees might be overshadowed eventually by faster-growing, unpruned trees. The pruned trees would then lose their dominant position in the stand and their growth would be reduced even further (Maguire and Petruncio 1995).

When pruning for a purpose other than the production of high-quality timber, the number of trees to be pruned will often be determined by that purpose. If the purpose of pruning is to use the leaves, say for oil production or animal fodder, all the trees in the plantation might be pruned. Similarly, if pruning for access, to provide firewood from the branches or to reduce fire hazard, all trees would usually be pruned.

9.6 Pruning Method

Most commonly, pruning is carried out by hand, with shears or fine-toothed saws. Saws may be mounted on the end of a pole to extend their reach to about 4–5 m above ground. Removing branches with these tools generally leaves a neat branch stub, perhaps of 1–2-cm length. Smoothly cut, neat branch surfaces seem to minimise defect resulting from stub occlusion and the entry of fungal decay (Petruncio et al. 1997; Montagu et al. 2003).

At least amongst American foresters, there seems to be considerable controversy about whether the cut should be made vertically down through the branch, that is, parallel to the axis of the stem, or at an angle away from the vertical, perpendicular to the axis of the branch (O'Hara et al. 1995; Russell 1995). The difference is illustrated diagrammatically in Fig. 9.2. Vertical pruning allows the cut to be made as close as possible to the stem to minimise the size of the branch stub. An angled cut will minimise the size of the wound and so minimise the area available for entry of decay fungi; the level of incidence of decay fungi will probably determine which type of cut is preferred in any particular plantation. In either case, stripping of bark from the stem should be avoided as the cut is made, to minimise the area of entry for decay. Particularly with large branches, this may require that an undercut be made first, part way through the branch. The cut is then completed from the top of the branch. This avoids the branch snap-

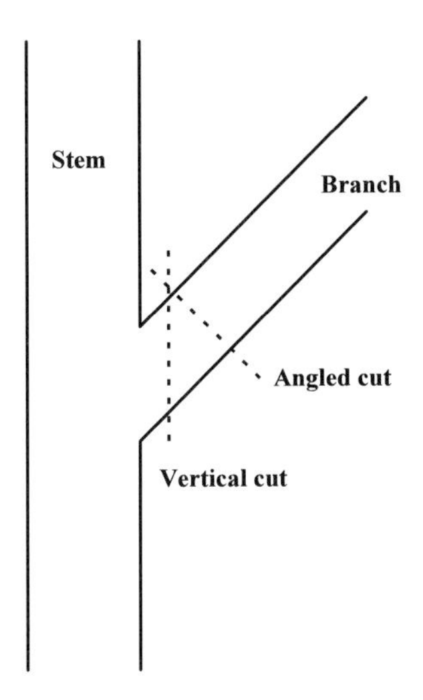

Fig. 9.2. Representation of a vertical pruning cut, made parallel to the axis of the stem, or an angled cut, made perpendicular to the axis of the branch

ping off before the cut is completed and stripping bark from the stem as it falls.

Controversy exists also (O'Hara et al. 1995; Russell 1995) about whether the cut should be made through or beyond the branch collar, which is a swollen ring of tissue where the branch emerges from the stem. In hardwoods, the collar often consists only of a swollen ridge within the crotch of the branch (Russell 1995). Cutting beyond the collar minimises the wound size, so minimising the possibility of fungal decay entry (Russell 1995); however, some studies have shown that occlusion occurs more rapidly when the cut is made through the collar (O'Hara et al. 1995), not only because the stub is smaller but also because the tree grows the occluding tissue more rapidly. Montagu et al. (2003) advocate pruning beyond the branch collar in eucalypts to minimise the possibility of decay entry. Detailed study will be necessary to determine what is appropriate for any particular plantation species.

If pruning is required to heights which cannot be reached from the ground, workers may have to climb trees, perhaps using ladders or spurs on their boots to grip the stem. This is obviously hazardous, slow and expensive. Damage to the stem from spurs may also risk the entry of decay fungi. Whether or not it is worthwhile to climb trees to prune them requires careful economic analysis of the value of the high-quality wood achieved eventually against the cost of the pruning.

Many attempts have been made to develop hand-held, powered devices for pruning. Wilkes and Bren (1986) have reviewed a number of them. They include small chainsaws, circular saws and hydraulic or pneumatic shears. They may be powered by small engines carried by the operator or by a larger, centralised power source with hoses running to several devices carried by different operators. These devices are noisy, sometimes hazardous and often unable to make cuts neatly on larger, acutely angled or closely spaced branches.

Attempts have been made also to develop machines to undertake pruning automatically to whatever height is required (Wilkes and Bren 1986; Petruncio and McFadden 1995). These are quite large, self-powered devices which are clamped around the stem base and move up the stem, cutting branches with a chainsaw or knives. As with the hand-held powered devices, they often have problems removing large, acutely angled or closely adjacent branches. Also, these machines may have difficulty with stems with lumps or bumps and can cause excessive damage to the bark as they move up the stem. These difficulties may offset the advantages of the increased rate at which these machines can prune trees when compared with hand pruning (Petruncio and McFadden 1995). The possibility of using elevating platform vehicles, to carry the operator to height with a hand-held pruning device, has been considered also. Factors such as debris on the ground, steepness of slope or wetness of the site have limited their usefulness. In general, none of the powered devices seems to have gained universal acceptance. Pruning continues to be done most commonly by hand in plantation forestry (Reutebuch and Hartsough 1995; Knowles 1995).

Where decay fungi commonly infect pruning wounds, paints or wound sealants may be applied to each pruned stub to prevent their entry (Montagu et al. 2003). Sometimes, decay entry is more common in warmer weather, so pruning in winter may be recommended (Gadgil and Bawden 1981; Petruncio et al. 1997). For some North American conifers, Russell (1995) suggests that accidental stripping of their stem bark can be minimised by pruning at the end of the dormant season (late winter) or just after new needles form (midsummer).

9.7 Examples of Pruning Regimes

Many different pruning regimes have been developed over the years for different plantations throughout the world. This chapter will conclude with several examples to give some idea of the variety of regimes that are used under different plantation circumstances.

9.7.1 Eucalypts in Australia

Detailed research conducted in Tasmania, Australia (summarised by Pinkard and Beadle 2000), has led to a regime used there for highly productive shining gum (*E. nitens*) and Tasmanian blue gum (*E. globulus*) plantations being grown to produce high-quality timber (Gerrand et al. 1997; Pinkard and Battaglia 2001; Montagu et al. 2003; Pinkard, personal communication). Plantations are established at about 1,000 trees per hectare and are thinned once or twice during their lifetime, to retain 250–300 trees per hectare until clear-felling. At the time of canopy closure, which occurs at around 3 years of age, the best 350 trees per hectare are selected for pruning; this allows that some of the pruned trees may develop poorly and be removed subsequently at thinning.

The first pruning lift removes live branches for about one third of the length of the green crown of each tree, leaving them pruned to a height of about 2.5 m; as discussed in Sect. 9.5.2, this degree of pruning should not lead to any reduction in growth rate of the pruned trees. Two further pruning lifts, each removing another third of the length of the green crown, are carried out when the trees are about 4 and 5 years of age, to heights of about 4.5 and 6.4 m, respectively. The 1-year delay between each lift has been found sufficient for the canopies of pruned trees to re-establish their leaf areas to an extent which allows subsequent pruning without loss of growth.

The entry of wood decay fungi through pruning scars has been found to be a problem in Tasmania. In shining gum trees, Barry et al. (2005) found decay columns, averaging 45 cm in length and up to 4.5 cm towards the centre of the stem, spreading in the wood above and below both branches pruned 5 years previously and branches which had been shed naturally. The decay seems to invade at any time of year, so there is no preference in Pinkard and Beadle's regime to prune in any particular season. The timing of the regime ensures that the pruned branches are generally no more than about 2 cm in diameter at their base when pruned; it has been found that the incidence of wood decay increases substantially if branches larger than 2 cm in diameter are pruned.

The aim of pruning is to leave branch stubs of no more than 1–2 cm in length and it has been found that pruning shears, rather than saws, are better able to achieve this. Also, shears have been found to cause less damage than saws to the bark surrounding the pruned branch, thus helping to minimise the possibility of decay entry. Care is taken also to avoid damage to the bark immediately above the pruned branch; in these species, this has been found to speed occlusion of the pruned branch stub.

9.7.2 Teak in Costa Rica

Pérez et al. (2003) described a pruning regime for plantations of the tropical hardwood teak (*Tectona grandis*) in Costa Rica. Teak is native to Asia, but it is used extensively in many parts of the tropics to produce high-quality timber.

In Costa Rica, teak plantations are established with about 1,100 trees per hectare and are thinned several times to retain, ultimately, about 120 trees per hectare (Pérez and Kanninen 2005; Table 8.2). Pérez et al. recommended that the trees to be retained after thinning should be pruned in three lifts, first to 2–3 m above ground when the trees are about 4–5-m tall, then to 4–5 m when they are 9–10-m tall and finally to 7 m when the trees are 12-m tall.

On more productive sites in Costa Rica, the trees would be tall enough for the first pruning lift at 2–2.5 years of age. On less productive sites the first pruning would be delayed until about 1 year later. No more than 30–40% of the leaf biomass would be removed at any of these pruning lifts, which Pérez et al. considered was appropriate to prevent any loss of growth as a result of pruning.

9.7.3 Western White Pine in Northwestern USA

Pruning to both minimise a disease problem and promote the development of clear wood is illustrated in an example for plantations of western white pine (*Pinus monticola*) in the USA (Hunt, 1998; Bishaw et al. 2003). This species is native to the western regions of North America and is of interest for plantation forestry there.

In both native forests and plantations, western white pine is affected by the disease white pine blister rust, which is caused by the fungus *Cronartium ribicola*. The disease infects needles and causes **cankers** on tree stems and branches, often leading to tree death; the worst infections occur within 2–3 m of the ground (O'Hara et al. 1995). At present, the incidence of the disease is sufficient to limit the use of western white pine for plantation forestry. However, if the lower branches of infected trees are pruned, the development of the disease can be restricted so that its effects on trees are minimal; the reasons for this are discussed in more detail in Sect. 11.3.2.

On the basis of experimental work in Washington, USA, Bishaw et al. (2003) suggested that pruning to minimise disease effects should start either at about 3–5 years of age or when the trees averaged 0.6–2-m tall. Pruning should continue in several lifts over succeeding years, perhaps to

5 m above ground to provide ultimately knot-free wood to that level. They suggested pruning should not remove more than one third of the length of the green crown.

It is clear from the information they provided that their plantations were relatively slow growing by world standards (Sect. 3.2), no doubt reflecting the rather cold climate of the mountain ranges of Washington state. In this regime, pruning at very early ages in these plantations would be principally for disease control; if production of high-quality timber was the only reason for pruning, it would probably be undertaken at a somewhat later stage of development of the plantations.

9.7.4 Spanish Red Cedar in Costa Rica

Cornelius and Watt (2003) discussed the need to prune plantations of the hardwood Spanish red cedar (*Cedrela odorata*) in Costa Rica. This species is attacked by larvae of an insect, the mahogany shoot borer (*Hypsipyla grandella*). The larvae bore into growing tips of trees, retard their growth and cause proliferation of new branches at the tips which can totally destroy the form of the tree for timber production. Members of the insect genus *Hypsipyla* are serious pests of the mahoganies (species of the genus *Swietenia*) and cedars (species of *Cedrela* and *Toona*), which are important timber species of tropical and subtropical forests in many parts of the world (Newton et al. 1993; Cunningham et al. 2005; Sects. 13.2.3, 13.2.4).

Cornelius and Watt found that the form of the trees could be restored after insect attack during the first 2 years of the growth of the plantations. This was done simply by pruning off the multiple shoots, leaving a single main shoot. So severe was the damage caused by this insect that repeated prunings had to be carried out every few months. In this example, pruning was essential to produce any trees with a form acceptable for later use as timber trees. Without it, Spanish red cedar would probably be impractical for use as a plantation species.

10 Pests

The viability of a forest plantation enterprise can be threatened by pests or diseases. A pest can be defined as a living organism (an insect for example) which damages a tree (say, by eating some of its leaves) and affects its growth or development in some way. Whereas a pest causes damage only, a disease causes some impairment to the normal functioning of the tree. Pest and disease problems overlap, for example, where a pest carries a disease-causing organism from infected to uninfected trees; the damage to the plantation may then be caused by the disease, whilst the pest causes little direct damage itself.

Many of the principles of pest management in forest plantations apply equally to diseases. In this chapter, these principles are considered in Sect. 10.1. The remainder of the chapter will consider pest problems specifically. It concentrates on both insect and mammal pests, which are the most common pest problems in plantations. Specific issues surrounding disease problems will be discussed in Chap. 11.

10.1 Principles of Pest and Disease Management

Throughout the world there is an enormous range of pests and diseases which affect agricultural crops in general and plantation forests in particular. Their study and consideration of how they may be managed are major disciplines in their own right and many books are devoted entirely to them.

This section will consider the general principles which have led to the approaches used presently to manage pest and disease problems in forest plantations.

10.1.1 Natural Occurrence of Pests and Diseases

It should be recognised at the outset that pests and diseases are a normal part of any natural **ecosystem**. For example, fungi (Sect. 11.1) and bacteria (Sect. 11.4) all have vital roles in breaking down leaf and fine-root litter in plantations, which allows recycling of nutrients (Sects. 2.3.3, 6.3). If bacteria or fungi cause disease in the trees, that disease may simply kill some of the competitively less successful trees (Sect. 2.4), trees which have been weakened physiologically by suppression making them vulnerable to the disease; this can be considered a normal way in which suppressed trees die. The larvae (Sect. 10.2) of an insect may eat the leaves of a tree and be a pest, whilst the adults of the same insect may be an important part of the natural life cycle of the tree species by spreading pollen between its flowers.

Natural forest ecosystems have evolved to maintain a balance between their various living parts. From time to time, environmental conditions, such as the weather, might vary from the norm sufficiently to favour one of the pests or diseases and allow an epidemic to occur. This can lead to substantial damage to some of the other organisms in the ecosystem and have effects which last for many years. However, in the long term usually, the natural balance between the parts of the ecosystem will be restored to normal.

In itself, the establishment of a forest plantation constitutes a disturbance to the natural environment. Establishing a large area with a single tree crop, perhaps of an **exotic** species (a species planted in a location outside its natural range of occurrence) and applying intensive silvicultural practices such as fertilisation and weed control, creates a rather unnatural ecosystem. This is typical of many agricultural enterprises and the lack of the normal natural balances in such ecosystems may encourage the outbreak of a serious pest or disease problem.

10.1.2 Control Strategies

Given that pests and diseases will occur to some extent as part of any plantation ecosystem, they need be of concern to the plantation grower only if their effects on tree growth and development are sufficiently large to frustrate the purposes for which the plantation was established. If it is decided that it is necessary to control a pest or disease, there are a wide range of strategies which can be used, including:

- Choosing to plant a species in a region where no pests or diseases to which it is susceptible occur. The success of plantations of species exotic to a particular region has often been attributed to a lack of pests and diseases of those species in their new environment (Wingfield 1999).
- Planting a mixture of species, where one of the species limits the impact of a pest or disease on another (Sect. 13.2.4).

- Undertaking a breeding programme to develop varieties of the species resistant in some way to attack by the pest or disease (Chap. 12).
- Managing other organisms in the ecosystem which control naturally the pest or disease, particularly **predators** (an animal which consumes all or part of another organism), **parasitoids** (an organism which lives at the expense of another organism and eventually kills it), **parasites** (an organism which lives at the expense of another organism, but does not kill it) or **pathogens** (disease-causing organisms).
- Employing silvicultural practices, such as thinning, pruning or soil cultivation, which limit the effects of a pest or disease (see the examples in Sects. 9.7.3, 10.3.2, 10.3.3).
- Using mechanical devices to trap a pest or deny it access to the plant, such as a metal band around a tree stem to prevent a leaf-browsing mammal from climbing the tree.
- Chemical control using any of the wide range of chemical products available for pest and disease control.

Of these various strategies, it is the last which generally causes the most concern in society. Whilst many chemicals are highly effective in controlling pests and diseases, they are often expensive, they may kill organisms additional to the pest or disease it is desired to control, they may pollute streams or soil and they may persist in the environment and have long-term adverse effects on the ecosystem generally. However, it is recognised that there are many acute pest and disease problems for which there is no alternative to chemical control. Chemical manufacturers make continuous efforts to develop new chemicals which avoid their deleterious effects.

Because of their environmental problems, it is accepted generally that there should be minimal use of chemicals in plantation forests. Instead, it is considered desirable to use, in concert, some or all of the other strategies listed above. In particular, it is considered that control measures should:

- Be based as far as possible around the natural circumstances of the pest or disease
- Aim not for a complete kill, but rather aim to maintain the pest or disease at a level where its effects are insufficient to prejudice the ultimate value of the crop
- Aim to cause as few harmful effects as possible either to other organisms, which are not pests or diseases, or to any other parts of the environment.

The combination of natural measures, with or without chemical measures, has become known as integrated pest management. Whilst this has usually been considered in the context of insect pests, its principles can be applied equally to diseases or other pests. It involves several components, which are (paraphrasing Thakur 2000):

- Understanding the ecological relationships between the pest or disease and the other plants and animals in the ecosystem.
- Identifying and using natural factors which kill the pest or disease organism. This requires careful monitoring of the populations of both the pest or disease and their enemies, to determine if the pest or disease population has reached a level where other forms of control are necessary.
- Combining and using two or more control measures to keep the pest or disease populations below some tolerable level.
- Including the pest or disease control system as an intimate part of the whole management system of the plantation.
- Undertaking specific, direct control measures only when absolutely necessary.

Because it requires a detailed understanding of the ecological relationships within a plantation ecosystem, development of an integrated pest or disease management system requires a substantial research programme. This means the system will not be available immediately, to deal with an acute pest or disease problem when it first arises. Rather, the system will evolve over some years as more and more information becomes available about the pest or disease. Some examples of integrated pest (or disease) management systems will be discussed in later sections of this and the next chapter (Sects. 10.3.1, 10.4.2, 11.3.2).

10.2 Insects

The entire animal kingdom is subdivided by taxonomists (people who classify living organisms, a discipline known as taxonomy) into about 24 major groups called phyla (singular phylum). Only two of these phyla contain animals which are serious pests of trees, namely the arthropods and the chordates. The arthropods are distinguished from other animals by having a hard, external skeleton made of a substance called chitin. They are by far the largest animal group on earth and include familiar types, such as the crabs, spiders, insects and centipedes. Of these, it is the insects which are of most concern in plantation forests. The chordates are characterised by having a stiff, supporting rod in their back. A subgroup of them is known as the vertebrates, in which the supporting rod has developed into a backbone. The vertebrates include fishes, amphibians, reptiles, birds and mammals. Of these, mammals (vertebrates which produce milk and suckle their young) are most commonly pests in plantation forests (Sect. 10.3).

Many texts discuss insect biology generally (e.g. Gullan and Cranston 1999; Thakur 2000). The insects are the most abundant animal life form on earth. About 1.2 million animal species have been described to date and over 900,000 of these are insects. It is believed that most of the insects on earth are yet to be discovered and there may be more than ten million species in total. Most are terrestrial species.

Insects (like all arthropods) have a hard, external skeleton. In their adult form, their body is divided into three parts, a head, thorax and abdomen. They have six legs, sensory organs to touch, hear, see, taste and smell and most have wings. Each species has one of a wide range of mouthparts, which can bite, chew, pierce, suck, rasp, lap or siphon, depending on what the species eats. The insect life cycle includes several stages. Females lay eggs and juveniles hatch as larvae, which have various forms as caterpillars or grubs. As the larvae feed and grow they moult several times (each stage is called an instar), until eventually they form a pupa (often in a cocoon) which develops into the adult form. Not all species go through all these stages and some have their adult form when they emerge from the egg.

The characteristics which make the insects such a successful life form on earth include (Thakur 2000):

- Their hard, external skeleton, which protects them and prevents them losing water from their bodies
- Their ability to fly, which allows them both to spread widely over the landscape and to escape their enemies
- Their small body size (the largest insects are about 15-cm long, but most are much smaller), which limits the metabolic requirements of each individual, allows them to congregate in large populations and lets them live in small spaces
- Their ability to breed rapidly and in large numbers
- Their developmental cycle through larval and adult stages, which live in different circumstances and gives them the best chance to survive and evade adverse conditions.

With such varied characteristics, insects are found occupying an enormous range of habitats. Often their life cycles involve an intimate association with plants or other animals (including other insects), which may be beneficial to both the insect and its host or harmful to its host. They have many roles as part of plant and animal ecosystems, including (Gullan and Cranston 1999):

- Breaking down leaf and wood litter, soil organic matter and animal carrion and dung, which is important as part of nutrient cycling in forests (Sects. 2.3.3, 6.3)
- Pollination of plants and dispersal of their seeds
- Maintaining the composition of plant communities by feeding on particular plants and affecting their development within the community
- Maintaining the composition of animal communities through transmission of diseases of larger animals and predation and parasitism of smaller animals
- Being food for many animal predators.

Given they are ubiquitous in natural systems, insects will inevitably be part of any plantation forest (Sect. 10.1.1). The issues for the plantation grower are whether or not the insects present damage the trees and, if so, whether or not their population size is sufficient in any year that the damage is sufficient to warrant their control. This means it is necessary to understand the ecology of insect populations to determine when and where they will be large enough to be of concern.

Study of the population ecology of insects is a scientific discipline in its own right and texts devoted to the subject are available (e.g. Schowalter 2000). At present, our understanding is limited as to exactly what factors determine the size an insect population will attain in any particular year and how that population will move around the landscape. This means it is difficult to predict if any particular plantation forest is going to suffer unacceptable damage in any particular year.

For plantation forestry, this means that populations of potential insect pests must be monitored every year; if signs are found that the population is going to develop sufficiently to cause damage, preventative measures can then be brought to bear. Inevitably, this means a substantial research programme will be necessary to understand in detail both the life cycle of the insect concerned and how it interacts with the plantation tree species.

In general, the insect pests in forest plantations can be grouped as (Gullan and Cranston 1999):

- Leaf chewers—where larvae or adults feed directly by eating leaves (Fig. 10.1).
- Sap suckers—which have long, sharp-pointed mouthparts which can be inserted into plant tissues to withdraw fluids.
- Miners and borers—which reside within the plant and tunnel their way through tissues, feeding as they go. All parts of plants may be attacked by one or other insect of this type.

Fig. 10.1. Defoliation of the tips of shining gum (*Eucalyptus nitens*) trees by the adults and larvae of the Tasmanian *Eucalyptus* leaf beetle (*Chrysophtharta bimaculata*) (Sect. 10.3.1) and the southern eucalypt leaf beetle (*C. agricola*). The plantation was growing in Tasmania, Australia (Photo courtesy of Forestry Tasmania)

- Gall formers—where damage to the plant by the insect causes the plant to react by forming a gall, a lumpy structure usually on the surface of the plant. Plant galls are extremely varied in size, shape and complexity; their study is known as cecidology.
- Seed predators—which feed on plant seeds.

To this list, might be added insects which are important in the transmission of disease to trees in plantation forests. Such insects may damage trees in some way, but are not pests themselves because the damage is only minor. Rather, it is the disease they carry which causes the problem for the plantation.

10.3 Examples of Insect Pests

So varied and numerous are insect pests of plantation forests that there are entire books devoted to them. However, for the present, a few examples will be given, to attempt to illustrate at least some problems that have been encountered in plantation forests and some approaches which have been used to deal with them.

10.3.1 Leaf Beetles in Eucalypts

This example concerns the effects of a leaf chewing beetle, the Tasmanian *Eucalyptus* leaf beetle (*Chrysophtharta bimaculata*) on eucalypt plantations in Tasmania, Australia. It provides a good example of the development of an integrated pest management system (Sect. 10.1.2). The insect is a problem in plantations of both mountain ash (*Eucalyptus regnans*) and shining gum (*E. nitens*). The economic impacts of damage by the beetle can be substantial in these plantations (Candy et al. 1992).

Leon (1989) has described the biology of the beetle. The adults are about 9 mm × 7 mm in size and hibernate over winter under the bark of trees or in cracks in dead wood. They emerge in spring and congregate on tree foliage where they mate. The females lay rafts of eggs on young foliage in two batches, the first in late spring and the second in late summer. The grublike larvae hatch in 8–11 days. They develop through four instars, each lasting 4–6 days. The larvae are very gregarious and form large groups which feed on leaves, favouring newly developed leaves (Fig. 10.2). Where feeding is heavy, the trees develop a twiggy, broom-like top (Fig. 10.1). Repeated, very severe attacks over several years can lead to tree death.

Most feeding is done by the last two instars which account for about 90% of the total consumption by the larvae. Adults continue to feed on foliage also. Late feeding in the season by adults can prevent refoliation by the trees and has particularly large impacts on tree growth (Candy et al. 1992). After they have developed for about 1 month, the larvae fall to the ground and pupate in the litter for 12–15 days. By mid-autumn, all larval activity has ceased and most adults have found hibernation sites.

There is some natural control of the beetle by other insects. Ladybirds, soldier beetles and some other insects are predators of both eggs and the early larval stages. Tachinid flies are parasitoids of the larvae and kill them when they have reached the fourth instar. In bad weather, many eggs or larvae may be shaken or washed from leaves. In many years, these natural control agents can cause as many as 95% of the eggs and larvae to be lost; most of these deaths are due to the predators, which usually account for about 80% of the losses.

Elliott et al. (1992) have described the development of an integrated pest management system which is used now to deal with the problem in some Tasmanian plantations. When the beetle population has developed to plague proportions, the only method that has been found satisfactory to

Fig. 10.2. Larvae of the Tasmanian *Eucalyptus* leaf beetle (*C. bimaculata*) feeding on eucalypt leaves. Magnification is about ×4 (Photo courtesy of Forestry Tasmania)

control them is to spray the plantation aerially with insecticide (chemicals which kill insects). However, this is expensive and there are concerns about its wider environmental effects (Loch 2005). It is done only as a last resort; the integrated pest management system is designed to avoid it if at all possible.

The key to the integrated system is careful monitoring of the development of the insect population in any year. The monitoring is carried out at about fortnightly intervals, starting in spring when the adults are mating and finishing in late summer when the larval populations have finished foliage browsing. At each inspection, the number of groups of eggs and the numbers and sizes of larvae on foliage samples are measured and the presence of any of the natural predators and parisitoids noted (J. Elek, Forestry Tasmania, personal communication). Based on results from earlier research work, the level of defoliation which would be likely to occur in the plantation can then be predicted. In turn, the long-term economic losses, which would occur as a result of that defoliation can be predicted using a model system developed specifically for the problem (Candy et al. 1992). Only after this information is available, together with the local plantation manager's assessment of 'the ability of [plantations] of differing age and productivity to recover from defoliation', is the decision made as to whether or not spraying should be done. Even then, insecticides about which there is less environmental concern can be used if the larvae are not older than the second instar (Elek, personal communication).

Although this integrated pest management system sounds quite straightforward, it was developed only after a number of years of intensive research on the problem. Detailed studies were made of the beetle life cycle and its leaf-browsing habits, to determine at exactly what stage of its life cycle it is necessary to carry out the control measures. Through study of the habits of the predators and parasitoids of the beetle, the system aimed to ensure that their natural ability to reduce the pest insect populations, at their early stage of development, is exploited to the full. Further studies were made of the effects of leaf browsing on the growth of trees and their likely long-term economic consequences. In developing the system, it was always borne in mind that the expensive and environmentally undesirable chemical control measures should be used only as the last resort.

10.3.2 Pine Weevil in Norway Spruce

The pine weevil (*Hylobius abietis*) is a bark-boring insect and is one of a large number of insects involved in breaking down stumps left after deaths of coniferous trees in European forests (Szujecki 1987). To that extent, the pine weevil is a normal part of the forest ecosystem. However, adults also browse on the bark of coniferous seedlings and may easily kill them. In large areas of native coniferous forest in northern Europe, the weevil is a serious obstacle to seedling regeneration (Örlander and Nordlander 2003).

Death of seedlings due to the pine weevil is proving to be a serious problem in plantation forests also in Europe. At present, there seems to be no way of predicting when and where they are likely to occur (Hansen et al. 2005). However, the problem is particularly acute where establishment of the plantation has occurred within 3–4 years of clear-felling of native forest, whilst the stumps of the felled trees are still being broken down; the stumps harbour high populations of the weevil (Örlander and Nordlander 2003). The weevils are attracted to seedlings by various chemicals released to the air by the trees (Szujecki 1987).

Various strategies have been used to deal with the problem. Planting is sometimes delayed for several years, until the stumps of felled trees have rotted away sufficiently that the weevil population is reduced. However, this allows very vigorous weed growth to develop, exacerbating the difficulties of plantation establishment. Spraying seedlings with insecticide to kill the weevils has also proved to be an effective control measure, but it is expensive and there are environmental concerns about it (Örlander and Nordlander 2003).

The control method that seems to be proving most effective is to simply scarify the soil around each seedling, that is, scrape away all vegetation and any plant litter on the ground surface to leave a bare soil surface. The effectiveness of this method relies on the biology of the insect. The weevil walks from stumps to seedlings, but is rather sensitive to temperature and humidity. It will move only if the temperature at the ground surface is 13–18°C and the relative humidity is 74–100% (Szujecki 1987). Temperature and humidity at the surface of bare soil are often too extreme for the insects; however, any material on the surface, weeds, mosses, leaf litter or even broken down organic matter, is sufficient to afford shade and protection so the temperature and humidity are tolerable for the insects and they can travel through the material. As well, the weevils are more likely to suffer predation on the bare surface (Örlander and Nordlander 2003).

Örlander and Nordlander (2003) carried out an experiment, in south-central Sweden, to assess the effectiveness of soil scarification in preventing weevil damage attack of newly planted Norway spruce (*Picea abies*) seedlings. The plantation site had been cleared in the previous year of an old forest, which consisted of a mixture of Norway spruce and Scots pine (*Pinus sylvestris*). A 50-cm-square area was scarified around each seedling. In some treatments in the experiment, scarified areas were kept free of any new weed growth, by spraying with a herbicide whenever weeds started to appear. Deaths of seedlings due to pine weevil attack were recorded a year after planting.

The results showed that scarification was very effective in preventing weevil damage as long as no weeds were allowed to regrow on the scarified patches. Without scarification or any weed control, 62% of planted seedlings died. With scarification, but without subsequent herbicide treatment to prevent weeds regrowing, 18–26% of seedlings died. Where scarified areas were kept free of weeds with herbicide, only 5% of seedlings died. Subsequent work (Petersson et al. 2005) has emphasised how thorough the scarification needs to be; retention of any vegetation litter on the scarified spots will increase seedling deaths.

From a practical point of view, these results show very clearly that the pine weevil problem can be controlled by careful management of the plantation. Thorough cultivation of the planting rows will be required to provide a bare soil surface at planting. Rigorous weed control will then have to follow planting, at least for several years until the stumps of the felled trees have been broken down and the pine weevil population in the plantation has been reduced.

10.3.3 Transmission of Disease of Douglas Fir by Beetles

This example concerns a case where insect pests are not the primary problem for the plantation forest, rather, it is the disease the insects carry. Thus, part of the management of the disease problem requires management of the insect pests.

The plantation tree species involved is Douglas fir (*Pseudotsuga menziesii*). The insects concerned are three bark-boring beetles (*Hylastes nigrinus*, *Pissodes fasciatus* and *Steremnius carinatus*) and the disease is black-stain root disease, which is caused by the fungus *Leptographium wageneri* (formerly called *Verticladiella wageneri*).

Black-stain root disease is distributed widely throughout western North America and is capable of killing a variety of pine (*Pinus*) species and Douglas fir through root infection (Witcosky et al. 1986). The fungus lives in the soil and normally spreads from tree to tree only slowly. It enters roots through wounds (Hessburg and Hansen 2000), although it may be transferred also from tree to tree through grafts between roots of neighbouring trees (Hessburg and Hansen 1986).

The bark beetles allow the disease to spread much further and more rapidly. Through contact with infected trees, the adult beetles pick up fungal spores on their bodies and, when they migrate to find egg-laying sites on the stems of new trees, they spread the disease. The adult beetles seem to be attracted particularly to trees where the bark has been damaged; it is believed chemicals released into the air by damaged trees attract the adult beetles to these egg-laying sites (Szujecki 1987; Schowalter 2000). Spread of the disease has been noticed particularly where there has been disturbance in the forest through road construction, logging or thinning (Witcosky et al. 1986), all disturbances which lead to tree bark damage.

Witcosky et al. (1986) carried out an experiment in 12-year-old Douglas fir plantations in Oregon, USA. They studied the effects of thinning on the occurrence of adult beetles in the forest. The thinning was quite heavy, reducing the stocking density of the plantations from 2,000–4,000 trees per hectare to 900–1,000 trees per hectare. At fortnightly intervals for a year or so following the thinning, they trapped and counted beetles moving about both the thinned and unthinned stands.

The adult beetles migrate during spring and, during that period, Witcosky et al. trapped many times more beetles in the thinned than in the unthinned stands; in fact they trapped very few beetles at all in the unthinned stands. After the migration period, the number of beetles trapped declined to relatively low levels, but was consistently slightly higher in thinned than in unthinned stands. The following autumn, they excavated some stumps of trees which had been removed at thinning. They also felled and excavated the stumps of some standing trees in both the unthinned and the thinned stands. They found large numbers of larvae and adult insects under the bark of thinned stumps and none in the still living trees.

Witcosky et al.'s results show very clearly that the stumps left after thinning in these Douglas fir plantations are very attractive to the beetles; thus, thinning plantations has great potential to aid the spread of black-stain root disease. They concluded that if thinning is to be done, it should take place only in early summer, immediately after the end of the annual migration period of the beetles. This would give a period of at least 9 months following the thinning before the next beetle migration occurred; they believed this was sufficient time to reduce greatly the attractiveness to beetles of the stumps of thinned trees.

Just as with the pine weevil example (Sect. 10.3.2), this example illustrates a case where management practices can be manipulated to minimise the detrimental effects of a plantation pest.

10.4 Mammal Pests

Insects are probably the most common pests of plantation forests. However, other types of animals are pests also, the most common group being the mammals. Generally, pest mammals feed on trees, usually on leaves or the fresh tissues immediately below the bark. Some climb larger trees to obtain their food, whilst others feed on newly planted seedlings and can easily destroy a newly established plantation. Even where trees are not destroyed, the long-term impact on growth and economic values can be substantial (Sullivan and Sullivan 1986; Sullivan et al. 1993; Wilkinson and Neilsen 1995; Montague 1996).

Table 10.1 gives information about a number of mammals which have been found causing more or less serious damage in forests in various parts of the world. The list is a rather arbitrary collection from the literature and is by no means comprehensive; however, it illustrates the range of animals which have caused problems from time to time and place to place, the types of trees they damage and the types of damage they do.

There are various features of interest about this list:

- Some of the animals have been problems more commonly in native forests (e.g. hares, elk, moose and bears), where they often prejudice regeneration by destroying young seedlings. However, they would certainly pose problems for any plantation forest established within the animals' ranges.
- Some are large, potentially dangerous animals (e.g. elk, moose or bears), whilst many are quite small (e.g. squirrels, voles or rats). Most

Table 10.1. Some mammals which are pests in forests

Animal common name	Animal scientific name	Types of trees affected	Location	Type of damage	References
Placentals					
Moose	*Alces alces*	Various conifers and hardwoods	North America, northern Europe, Russia	Foliage browsing	Schewe and Stewart (1986); Risenhoover and Maass (1987); Siipilehto and Heikkilä (2005); Koski and Rousi (2005)
Roe deer	*Capreolus capreolus*	Lodgepole pine	Sweden	Stem damage by antler cleaning	Hansson (1985)
Elk	*Cervus canadensis* (or *Cervus elaphus*)	Various conifers	Northwestern USA	Bark stripping, foliage browsing	Hanley and Taber (1980); Bishaw et al. 2003
Sika deer	*Cervus nippon*	Various conifers	Japan	Bark stripping, foliage browsing	Akashi and Terazawa (2005)
Porcupine	*Erethizon dorsatum*	Various conifers	Western North America	Bark stripping, foliage browsing	Hood and Libby (1980); Sullivan et al. (1986)
Snowshoe hare	*Lepus americanus*	Various conifers	Canada, northern USA	Bark stripping, tip browsing	Sullivan and Sullivan (1982); Bergeron and Tardif (1988); Rodgers et al. (1993); Rangen et al. (1994)
Mountain hare	*Lepus timidus*	Various conifers	Western Canada	Bark stripping, foliage browsing	Rangen et al. (1994)
Vole	*Microtus agrestis*	Lodgepole pine, Sitka spruce, Silver birch	Northern Europe	Bark stripping, seedling browsing	Hansson (1985); Nordborg and Nilsson (2003); Koski and Rousi (2005)

Table 10.1. (Continued)

Animal common name	Animal scientific name	Types of trees affected	Location	Type of damage	References
Meadow vole	*Microtus pennsylvanicus*	Various deciduous hardwoods	Alberta, Canada	Girdling of seedlings	Radvanyi (1980)
Black-tailed deer	*Odocoileus hemionus*	Various conifers	Western USA	Browsing of foliage, seedling buds and branch tips	Hanley and Taber (1980); Hood and Libby (1980); Margolis and Waring (1986)
White-tailed deer	*Odocoileus virginianus*	Various hardwoods	Central Canada	Browsing of foliage, seedling buds and branch tips	Schewe and Stewart (1986)
European rabbit	*Oryctolagus cuniculus*	Radiata pine, various eucalypts	Southern Australia	Seedling browsing and destruction	Friend (1982); McArthur and Appleton (2004)
Bush pig	*Potamochoerus porcus*	Several *Pinus* species	Zimbabwe	Uprooting trees, feeding on roots	Sniezko and Mullin (1987)
Allied rat	*Rattus assimilis*	Radiata pine	Southern Australia	Young tree browsing and destruction	McNally (1955)
Bush rat	*Rattus fuscipes*	Radiata pine	Southern Australia	Young tree browsing and destruction	Friend (1982)
Southeastern fox squirrel	*Scirius niger*	Loblolly pine, slash pine	Southeastern USA	Cone destruction in seed orchards	Asaro et al. (2003)
Red squirrel	*Tamiasciurus hudsonicus*	Various conifers	Canada and USA	Bark stripping, cone destruction	Sullivan and Sullivan (1982); Sullivan and Vyse (1987); West (1989)

Table 10.1. (Continued)

Animal common name	Animal scientific name	Types of trees affected	Location	Type of damage	References
Black bear	*Ursus americanus*	Various conifers	Western North America	Bark stripping	Sullivan (1993)
Grizzly bear	*Ursus arctos*	Various conifers	Western North America	Bark stripping	Sullivan (1993)
Marsupials					
Red-necked wallaby	*Macropus rufogriseus*	Eucalypts	Tasmania, Australia	Foliage browsing	le Mar and McArthur (2001)
Red-bellied pademelon	*Thylogale billardierri*	Eucalypts	Tasmania, Australia	Foliage browsing	le Mar and McArthur (2001)
Mountain possum	*Trichosorus caninus*	Various *Pinus* species	Victoria and New South Wales, Australia	Bark stripping near tree tip, foliage browsing	McNally (1955); Barnett et al. (1977)
Common brushtailed possum	*Trichosorus velpecula*	Various *Pinus* and eucalypt species	New Zealand and southern and subtropical Australia	Bark stripping near tree tip, foliage browsing	McNally (1955); Barnett et al. (1977); Friend (1982); Dungey and Potts (2002)
Swamp wallaby	*Wallabia bicolor*	Radiata pine, various eucalypts	Victoria, Australia	Foliage browsing	McNally (1955); Friend (1982); Montague (1996)

are placental mammals (their young grow to a relatively large size before birth, because they are nurtured by a placenta), whilst the last five are marsupial mammals (which lack a placenta and give birth to tiny, very immature young), these five all being native to Australia.

- Some are exotic species in the country where they have been a problem. The common brushtailed possum was imported to New Zealand from Australia. Plagues of the European rabbit caused enormous damage to agricultural crops and grazing lands in Australia, until they were largely controlled by the release in the 1950s of a viral disease called myxomatosis; however, many rabbits persist still in Australia.
- Some damage trees by eating foliage. Some strip bark from the stem to eat the fresh tissue below, which may girdle the tree and kill it. The southeastern fox squirrel eats the seed from cones of coniferous trees; this can be a serious problem if it occurs in a seed orchard, where the seeds are to be collected as part of a tree breeding programme (Chap. 12). Some are problems principally with seedlings (e.g. voles or European rabbits), whilst others damage older trees to some distance up the stem (e.g. snowshoe hares or mountain possums).

10.4.1 Control Measures

There are a number of approaches which can be taken to control mammal pests. Fencing to keep them out can be effective; however, fences are expensive to install and maintain. Some animals climb so well (e.g. squirrels or possums) that fences are completely ineffective; electrification can deal with this problem, but again maintenance is tedious.

Trapping and moving animals away from the plantation can be effective, just as bears in North America are trapped and moved away from places where they endanger people. For large and dangerous animals this is a specialist task. For large populations of smaller animals, trapping would simply be too expensive to countenance.

Placing protective guards around seedlings can be effective; various proprietary devices of this type are available. However it is prohibitively expensive to provide sufficient guards for large plantation programmes. For some animals they are ineffective. For example, Radvanyi (1980) considered small plantations of deciduous hardwoods being established to rehabilitate mining sites in Alberta, Canada. A guard placed around each seedling was quite ineffective in preventing damage by meadow voles, which could either dig beneath the guard or reach the top of the seedling when there was sufficient snow to bury the guard. Montague (1993) tested the effectiveness of a number of different types of guard to protect eucalypt seedlings in Australia from browsing by swamp wallabies. Only 1-m high, rigid plastic tubes effectively protected 30–40-cm high seedlings; wallabies could reach inside shorter tubes or through the holes in mesh tubes. However, Montague concluded that the cost of the effective tubes was too high for use in large plantation areas.

Poisoning programmes can be very effective. In the same mine rehabilitation sites discussed in the previous paragraph, Radvanyi found that poisoning reduced the animal pest population very adequately. He used specially designed feeding stations positioned at a number of spots around the plantation. These protected the poisoned bait from the weather; poisoned baits can often lose their effectiveness quickly if exposed to the weather. Le Mar and McArthur (2001) considered the effectiveness of a poisoning programme to control red-necked wallabies, red-bellied pademelons, common brushtailed possums and European rabbits in eucalypt plantations in Tasmania, Australia. They used a poison called 1080 (which is the chemical sodium monofluoroacetate), a poison which has been used widely in Australia for many years to protect plantations, particularly from the European rabbit. To attract the animals, it is usually put on carrots and distributed around the plantation along furrows in the ground. In their trial, le Mar and McArthur found that pademelons particularly were killed by the poison, but the effects were much less on the other species; even though they are quite small, the pademelons were particularly aggressive and tended to keep the other animals away from the baits.

Poisoning of animals is considered cruel and is becoming increasingly unacceptable to society. It may affect the **biodiversity** of the region by killing the native animals which are pests and also those which are not pests and are not the targets of the poisoning. So strong are these community pressures, many plantation growers are seeking alternatives to poisoning to deal with their mammal pest problems; in some places, governments have banned their use. As well, repeated use of poisons may lead to development of strains of the animal concerned which are resistant to the poison (Radvanyi 1980). Shooting is an alternative to poisoning, but it faces similar community pressures and is impractical for very large pest populations.

Many different compounds have been tested from time to time for their ability to repel animals, either through bad smell or bad taste. Some are sprayed onto the surface of foliage of seedlings, often together with other compounds to ensure the spray adheres to the foliage. Others may be taken up by the tree, either through its roots or leaves, and then persist within the foliage rather than being simply retained on its surface; chemicals taken up by plants in this way are known as systemic substances.

Different repellents may be more or less effective with different plant and animal species. Montague (1994) found that a systemic repellent, which had been found effective in America to deter browsing of Douglas fir (*Pseudotsuga menziesii*) seedlings by deer, was quite ineffective in deterring browsing of eucalypt seedlings by wallabies in Australia. The problem with most repellents is that their life span is limited, usually to periods of several months. Those applied to foliage usually wash off after a few months and need to be reapplied, which is likely to be prohibitively expensive. However, Knowles and Tahau (1979) made a novel suggestion, where a repellent might prove useful in agroforestry plantations in New Zealand. They tested a surface-applied repellent, which they found was effective for 3–4 months in deterring sheep from browsing radiata pine (*Pinus radiata*) seedlings in plantations in New Zealand. They suggested this would be sufficient time to ensure the pine seedlings were undamaged whilst the sheep grazed the weeds in the plantation, weeds which would otherwise have to be removed by other means.

10.4.2 Integrated Pest Management Approach

Approaches involving integrated pest management systems (Sect. 10.1.2) are now being considered for the control of mammal pests. Work done over the last few years in Tasmania, Australia, offers a good example of the type of research necessary to develop such a system.

Large areas of plantations of both Tasmanian blue gum (*E. globulus*) and shining gum (*E. nitens*) have been established in Tasmania over the last 10–20 years. The red-necked wallaby, red-bellied pademelon, common brushtailed possum and the European rabbit (Table 10.1) are all mammal pests which browse foliage in these plantations and may destroy young seedlings.

Over the first year following plantation establishment, Bulinski and McArthur (2003) studied the browsing habits of these mammals in 32 commercial shining gum plantations in Tasmania, the areas of which varied from 10 to 109 ha. They found that browsing was very variable between the different plantations, varying from less than 1% of the seedling biomass removed over the year to more than 90%. The common brushtailed possum caused most damage.

All four mammals feed in the plantations at night and shelter in adjacent forest areas during the day. Bulinski and McArthur found that this behaviour had an important influence on the extent to which the plantations were browsed. Larger plantation blocks, which have a smaller ratio of the length of their perimeter to the plantation area, tended to be browsed less because the animals had further to travel from their daytime shelter to reach the centre of those plantations; browsing tended to be heavier nearer their edges than their centres (see also Barnett et al. 1977; Bulinski and McArthur 2000). Plantations with a higher proportion of their perimeter adjacent to forested areas were browsed more heavily because of the higher populations of animals the adjoining forest was supporting. Plantations adjacent to forests with more open canopies were browsed more heavily because possums prefer to shelter in more-open forests and, hence, their numbers were higher there.

Bulinski and McArthur found also that plantations with more weed growth (most commonly grasses) tended to be more heavily browsed. The animals eat vegetation other than tree seedlings, so the presence of weeds in plantations provided an additional source of food and encouraged them to remain longer in the plantation. Thus, weed control should reduce browsing as well as encouraging tree growth (Sect. 5.4). By contrast, when one of the weeds was bracken fern (*Pteridium esculentium*), Bulinski and McArthur found that plantation browsing declined as the amount of bracken fern increased. Bracken fern is one weed species on which the animals prefer not to feed and, where it forms clumps around tree seedlings, the animals tended to avoid the seedlings. Other work in Tasmania in radiata pine (*Pinus radiata*) plantations found that tree seedlings were browsed less commonly when certain weed species were present, weeds which the animals prefer over pine seedlings (Pietrzykowski et al. 2003). These findings present somewhat of a quandary for the Tasmanian plantation grower. Controlling weeds to encourage tree growth may render some plantations subject to more browsing of the tree seedlings and others to less browsing, depending on the tree species planted, the weed species concerned and the eating preferences of the animals.

As part of an integrated pest management system, all the findings by Bulinski and McArthur will be useful in assessing the browsing risk faced by Tasmanian blue gum and shining gum plantations in Tasmania. Smaller plantations surrounded by more open-canopied forest and with a high level of weed control will be at much higher risk than larger plantation areas surrounded by open land and where weed control has not been so rigorous. Similar approaches to that of Bulinski and McArthur have been taken by others to assess the risk of mammal browsing and, hence, the likelihood that animal control measures will be necessary. For example, Sullivan et al. (1994) developed a model system to assess the risk of red squirrel damage in managed lodgepole pine (*Pinus contorta*) stands in relation to the past history and the present conditions of stands.

A second important part of the approach being taken to deal with animal browsing in Tasmanian (and other Australian) plantations is consideration of the condition of the tree seedlings which are planted. McArthur et al. (2003) found that captive red-bellied pademelons and common brushtailed possums preferred least seedlings which had been raised under partial shade and with a relatively poor supply of the nutrient element nitrogen. The leaves of those seedlings were rather more fibrous, which made them more difficult for the animals to chew. The leaves also had higher levels of certain chemical compounds (known as phenolics) which can interfere with food digestion by the animals. Many similar studies have suggested that mammals often prefer to browse seedlings more plentifully supplied with nutrients (Rodgers et al. 1993; Montague 1994). These results are perhaps unfortunate from the point of view of browsing control in Tasmanian plantations. To ensure maximum growth, seedlings richly supplied with nutrients in the nursery are preferred generally for planting (Sect. 5.2); McArthur et al.'s results suggest these seedlings will be most preferred by the browsing animals. On the other hand, McArthur and Appleton (2004) found that Tasmanian blue gum seedlings with a diameter at their base in excess of about 6 mm were too large for European rabbits to clip off at ground level with their teeth; this means there is potential to limit rabbit damage by planting seedlings larger than this.

The third approach being taken to deal with mammal browsing is to consider genetic variation in seedling palatability. Experiments with many different mammals have found that certain tree species and certain genetic strains of particular species are damaged more severely than others by the mammals (Hood and Libby 1980; Bergeron and Tardif 1988; Montague 1994, 1996; Rangen et al. 1994; Hertel and Kaetzel 1999; Lawley and Foley 1999; O'Reilly and McArthur 2000). In Tasmania, O'Reilly-Wapstra et al. (2002, 2004) found considerable variation in feeding preference by red-bellied pademelons and common brushtailed possums amongst different genetic varieties of Tasmanian blue gum, although the preferences could be influenced by how well supplied seedlings were with nutrients (O'Reilly-Wapstra et al. 2005). Their results raise the possibility of breeding Tasmanian blue gum varieties which are less susceptible to browsing.

This research on mammal browsing in eucalypt plantations in Tasmania has not yet reached the stage where a final integrated pest management strategy has been developed to deal with the problem. However, the results described here suggest that a combination of plantation circumstances, the treatments applied to seedlings in the nursery and the availability of trees bred for resistance to browsing will all be important in determining the risk mammal browsing presents in any particular plantation. Minimising that risk will minimise the need to poison animals, the control measure of last resort.

11 Diseases

At the beginning of Chap. 10, a disease was defined as something which causes impairment to the normal functioning of a plant or animal. Diseases can be either **biotic**, that is, the impairment is caused by another living organism (such as a fungus or bacterium), or **abiotic**, that is, the impairment is caused by something in the non-living part of the environment (such as an oversupply or undersupply of one or more nutrient elements).

The general principles of disease management in forest plantations were established in Sect. 10.1. This chapter will illustrate disease problems of plantations through a number of examples. It will concentrate on fungal disease problems, because these are the most common diseases which cause serious damage to plantations. At the end of the chapter, brief reference will be made to other diseases.

11.1 Fungi

Fungi make up an entire kingdom of living organisms on earth. The other kingdoms are the Monera (bacteria and blue-green algae), the Protista (other algae, protozoa and slime moulds), the Plantae (plants) and the Animalia (animals). The Monera and Protista are the simpler, mainly unicellular, forms of life on earth. The Fungi, Plantae and Animalia are the more complex life forms, being multicellular and with different types of cells performing different functions. The study of fungi is a branch of science known as mycology.

Fungi do not have the well-developed structure of plants, with stems roots and leaves. Rather, they have a branched, filamentous growth form, with a cell wall surrounding the filaments (which are called hyphae). Some fungi reproduce asexually only, whilst others reproduce sexually and produce microscopic spores, often in fruiting bodies. The common table mushroom is a typical example of a fungal fruiting body; many other fungi have only microscopic fruiting bodies. Fungi do not carry out photosynthesis. They obtain their food by breaking down organic matter and absorbing it.

Fungi are an extremely important part of natural ecosystems. Together with bacteria, their major role in forest plantations is the decomposition of leaf and fine-root litter from the trees. This allows recycling of nutrients through the soil (Sect. 6.3).

11.2 Fungal Diseases

Most fungi are free-living. However, some are parasites; that is, they live in, with or on another organism and derive benefit in some way from the other organism. Parasitic fungi often invade the tissue of plants to obtain nutrition from them; some are harmful to their hosts and these are the pathogens which may cause disease of trees in forest plantations.

There are a wide range of fungal pathogens. Different ones can cause disease or damage at any stage of tree development, from seed to fully mature, and to any part of a tree, from its fine roots to its leaves. In his excellent book on tree diseases, Manion (1991) classified tree fungal diseases as:

- Foliage diseases—which cause dead spots or patches on leaves, or even complete shrivelling and death of leaves. They reduce the photosynthetic capacity of a tree and, hence, its growth. The loss of vigour by the tree may render it more susceptible to invasion by other pests or diseases.
- Rusts—which cause diseases of leaves, branches and stems. Rust fungi are all obligate parasites, that is, they require living hosts for their normal development and have particularly complex life cycles, usually involving two quite different plant hosts.
- Cankers—which cause disease in the outer layers of the wood and the thin layers of living tissues which surround the wood (Sect. 2.1.1) of branches, stems and woody roots.
- Vascular wilts—which invade wood and interrupt the flow of water from roots to leaves, causing the leaves to wilt.
- Wood decay diseases—which digest wood and render it useless.
- Wood stain diseases—which generally discolour wood without destroying it and are generally incapable of killing trees.
- Root diseases—which may destroy root tissues generally, or just the wood of roots, or they may invade the wood of roots and interrupt the flow of water through it.

Manion's book may be consulted to learn about many of the fungal diseases which have caused problems in both native forests and plantations, particularly in North America. The comprehensive work by Keane et al. (2000) describes a wide range of fungal diseases of eucalypts.

11.3 Examples of Fungal Disease

In this section, some specific examples will be given of various fungal diseases which have been major problems in plantations and how they have been managed. The examples include a number of the different types of fungal disease recognised by Manion (1991) (Sect. 11.2).

11.3.1 Pine Needle Blight in Australasia

This example concerns a disease called pine needle blight, which has caused substantial losses in some pine plantations around the world. Pine needle blight (also called dothistroma needle blight) is a foliage disease (Sect. 11.2) which is caused by the fungus *Dothistroma septospora* (there is confusion presently about its name and it will be seen referred to also as *D. pini*). It is believed to be Central American and Himalayan in origin (Bradshaw 2004). The disease occurs extensively in North America, where it infects more than 30 pine (*Pinus*) species as well as Douglas fir (*Pseudotsuga menziesii*) and European larch (*Larix decidua*) (Manion 1991; Bradshaw 2004). It has been found also affecting Sitka spruce (*Picea sitchensis*) in exotic plantations (Neumann and Marks 1990).

One of the pine species affected in North America by pine needle blight is radiata pine (*Pinus radiata*). Radiata pine has a very limited natural distribution over a total of only 5,000–8,000 ha on the Monterey Peninsula in California, USA and some small islands to the south, lying just off the Mexican coast (Ohmart 1982; Pederick and Eldridge 1983; Piirto and Valkonen 2005). Despite this small native occurrence, there are now over 3 million ha of radiata pine in plantations across the world, particularly in the southern hemisphere; it is probably the pine species grown most extensively in exotic plantations around the world.

Wingfield (1991, 2003) suggests that one reason for species growing well when planted as exotics, outside their range of natural occurrence, is that they are free of the pests and diseases which occur in their native habitat. However, for all exotics there is a risk that those native pests or diseases will eventually be introduced into their exotic habitat; Waring and O'Hara (2005) have reviewed the problems that have been encountered around the world when this has occurred. It certainly happened with pine needle blight and radiata pine. In the 1950s, the disease was found to be causing serious damage of radiata pine plantations in Tanzania, Africa. By the early 1960s it was found in New Zealand and then spread to both Australia and South America. This example will concentrate on how the introduction of the disease has been handled in Australasia (Australia and New Zealand).

Pine needle blight is spread from tree to tree by airborne spores in water droplets (Neumann et al. 1993). Infection causes yellow and tan spots or bands on needles, which may gradually turn brown, then red. The needles eventually die and are shed (Power and Dodd 1984). Repeated infections over several years may be sufficient to kill trees (Neumann et al. 1993). It has even been suggested that the presence of the disease is a danger for forest workers, because *D. septospora* releases chemicals into the atmosphere which may be hazardous to human health (Elliott et al. 1989; Bradshaw et al. 2000; Bradshaw 2004). The loss of needles by infected trees reduces the amount of photosynthesis they can carry out and, hence, their growth. Studies in New Zealand found that growth losses were sufficiently large to be of great concern to radiata pine plantation growers there (van der Pas 1981; Woollons and Hayward 1984).

Until they are about 20 years of age, radiata pine trees are much more susceptible to infection than when they are older (Power and Dodd 1984; Neumann and Marks 1990). The reason is not understood, although older radiata pine trees show a build up of wax within their stomata, which may prevent the fungus entering the needle (Power and Dodd 1984). On sites which are particularly favourable for development of the disease, much older trees have been severely affected (Simpson and Ades 1990).

Studies of the disease in Australia and New Zealand have shown that rainfall, humidity and temperature are the most important factors determining the likelihood of it occurring at any particular site (Marks et al. 1989). Spores of *D. septospora* need leaves to be wet for at least 10 h for infection to occur; this is likely to happen on wetter and more humid sites. These studies led to development of a hazard index for occurrence of the disease (Marks et al. 1989). The index is determined as the number of months in the year with both a rainfall in excess of 100 mm and an average temperature in excess of 10°C. On sites where the index is less than 3 months per year, there is little risk of the disease occurring. As the index rises above this, the risk increases progressively. However, the index is useful as a guide only. Even within a region which the index indicates is generally at low risk of the disease, there may be local spots at high risk, where the weather is affected by the local topography (Marks et al. 1989).

In areas which are at relatively high risk of the disease, it may be possible to minimise its effects by pruning and thinning (van der Pas et al. 1984; Marks and Smith 1987). Pruning may remove heavily infected lower branches and so reduce the amount of infective material from which the disease may spread. Thinning opens the stand and may reduce humidity, so reducing the likelihood of infection. Other work has shown that the risk of infection is higher on soils where the availability of the nutrient element sulphur is low (Eldridge et al. 1981; Lambert 1986).

Where the disease has become established, it can be controlled by aerial spraying of the plantation with copper oxychloride, a chemical compound which is quite benign in its environmental effects. Copper is very toxic to *D. septospora*, preventing the production and germination of its spores and inhibiting its growth (Franich 1988). Spraying in New Zealand was usually considered appropriate when the level of infection had caused more than 25% loss of foliage from a plantation, although Woollons and Hayward (1984) showed that even at this level of infection there was appreciable loss of wood volume growth of radiata pine which could lead to economic losses for the industry (New 1989). In one severely affected region of New Zealand, it was found necessary to spray an average of five times during the first 16 years of the life of plantations, that is, over the ages when radiata pine is most susceptible to the disease (Dick 1989). Such repeated aerial spraying of plantations is very expensive (van der Pas et al. 1984; Dick 1989; New 1989).

It has been found that radiata pine can be bred for resistance to *D. septospora*. Resistant trees are now planted in New Zealand (Carson 1989; Simpson and Ades 1990) and the breeding programme continues there (Jayawickrama and Carson 2000; Ganley and Bradshaw 2001; Bradshaw et al. 2002). Methods are being developed to locate areas affected by the disease from light sensors mounted in aircraft (Coops et al. 2003; Stone et al. 2003).

The history of pine needle blight of radiata pine is a good example of the risk that a fungal disease may pose to a forest plantation enterprise and the effort necessary to deal with it. In the first appearance of the blight in exotic radiata pine plantations, in Africa in the 1950s, so severe were its effects that planting there of radiata pine ceased (Dick 1989). New Zealand has a climate suited generally to occurrence of the disease and it caused extensive damage after it appeared there in the 1960s. Over the next 20 years or so, it required a major research effort both to understand the disease and to develop methods to deal with it. At first, expensive spraying programmes were used to control it, until breeding programmes could develop trees resistant to the disease. Australia was a little more fortunate when the disease arrived there in the 1970s. Most of its radiata pine plantations are established in regions with climates less suited to the development of the disease than New Zealand (Marks et al. 1989). In Australia now, areas susceptible to the disease tend to be avoided when establishing plantations of radiata pine.

11.3.2
White Pine Blister Rust in North America

There are many rust fungi which are serious pathogens of trees and of agricultural crops. They usually infect foliage initially and may eventually invade the cambium and phloem of branches and stems. Most have a complex life cycle which involves two quite different plants as hosts.

Manion (1991) has summarised the experience in North America with the disease white pine blister rust, which is caused by the fungus *Cronartium ribicola*. This fungus is a serious pathogen of various native pine species in America, including eastern white pine (*Pinus strobus*), western white pine (*Pinus monticola*) and sugar pine (*Pinus lambertiana*). Mention has been made already of this disease in Sect. 9.7.3, as an example where pruning in plantations may aid in its control. White pine blister rust has been principally a problem of native forests in America, but its presence has restricted opportunities to establish plantations of western white pine there (and in other parts of the world; Kinloch 2003). Manion's discussion of the disease will be summarised here, as a second example of a major fungal disease problem to which considerable effort has been dedicated in attempts to control it (for a more recent review, see Kinloch 2003).

White pine blister rust is an introduced disease in American forests, deriving originally from its native occurrence in Asia. In America, its life cycle involves both the pine trees and understorey bushes of the genus *Ribes*. Spores from the fungus on the *Ribes* infect needles of the pine trees. The fungus then develops in the needles and grows down into the twigs to which the needles are attached. It then develops in the cambium of the twig, eventually killing it. It may even grow down into the cambium of the stem. Cankers may form on the twigs or stems (Manion distinguishes between rust canker diseases and canker diseases, because rusts need two quite different hosts to complete their life cycle). If the infection is sufficiently severe, the fungus may eventually kill the tree. Spores released from the infected pine tree infect *Ribes* and the fungal life cycle recommences. The fungus develops differently on the trees and on *Ribes*; both hosts must be present or the fungus is unable to complete its life cycle.

Towards the middle of the twentieth century, white pine blister rust was established widely across the USA, over the entire natural range of eastern and western white pines and sugar pine. So damaging were its effects, that it became the subject of various programmes aimed at eradicating it, or at least containing it within tolerable limits.

The first control programme started in the 1930s and did not cease completely until the 1970s. It aimed to control the disease by removing *Ribes* from the forests; it was argued quite correctly, that if one of the hosts necessary for the life cycle of the fungus was absent the disease would disappear. However, experience proved that *Ribes* was very difficult to eradicate. It would resprout readily, if digging and pulling the bush did not remove all the roots. Its seeds can lie dormant in the soil and sites thought to be free of *Ribes* may be recolonised from dormant seeds after many years. Removal of *Ribes* was found to be an effective control strategy only for relatively small areas which were thoroughly and repeatedly treated. Ultimately, the programme proved too expensive and too difficult to provide complete control of the disease.

During the 1950s and 1960s, broad-scale aerial spraying with fungicides was introduced; however, the fungicides proved to be only partially effective at killing the disease and it was difficult to get adequate coverage of trees with aerial spraying. Also, there were environmental concerns about broad-scale spraying of chemicals over large areas.

From the 1960s a more integrated management approach (Sect. 10.1.1) was adopted to control the disease. This was based on research which considered the environmental factors which encourage its development (for some recent work, see Smith and Hoffman 2001). It was found that infective spores could develop on *Ribes* only if there were 2 weeks of relatively cool, late summer weather. Further, there had to be 2 days with temperatures below about 20°C and with 100% relative humidity, or the spores released from the *Ribes* were unable to infect pine needles. Identifying regions where these environmental circumstances often apply has allowed maps to be drawn indicating where disease is most likely to occur. As well, it has been found that very high humidity is likely on days with heavy dew. Avoiding planting trees in small openings, pruning lower branches or planting trees below a 'nurse' crop (Sect. 13.1) with a thin crown have all been found to be effective in minimising the occurrence of dew and so minimising the effects of the disease. The pruning programme discussed in Sect. 9.7.3 arose from this understanding of the biology of the disease. As well, tree breeding programmes have been undertaken to develop disease-resistant strains of both eastern and western white pines (Hamelin et al. 2000; Kim et al. 2003; Kinloch 2003; Pike at al. 2003; Lu et al. 2005), which has improved the prospects for growing these species in plantations (Bishaw et al. 2003). Research continues to find other ways of protecting trees from the disease (e.g. Hunt 2002).

As with pine needle blight in Australasia (Sect. 11.3.1), the story of white pine blister rust in America is a good example of the complexity of dealing with a serious pest or disease problem. It has required a major research effort over many years, both to understand the biology of the rust and to develop control mechanisms for it. It illustrates also that attempts to eradicate a pest or disease can often be extremely expensive and ultimately ineffective. Finally, it shows how there was acceptance in America that it would not be possible to eliminate the disease; eventually, control measures were adopted which at least will keep its impacts within tolerable limits.

11.3.3 Chryphonectria Canker of Eucalypts in South Africa

Canker diseases cause a spreading wound of dead tissue and can affect living tissues in the outer part of stems, branches and woody roots of trees. They usually infect the bark, phloem, cambium and the outermost layers of the xylem (Manion 1991; Old and Davison 2000). They may cause only small irregularities on the surface of the affected part of the tree, or the tree may respond by producing a very large woody mass around the affected area. Most commonly, canker diseases enter the tree through wounds, although some may enter through leaves (Manion 1991). Some cankers may persist in trees for many years, whilst others may kill the infected tree. Manion (1991) describes a number of canker diseases in North America, whilst Old and Davison (2000) discuss various canker diseases of eucalypts.

This example concerns the disease Chryphonectria canker, which is caused by the fungus *Cryphonectria cubensis*. The disease is a serious problem in many eucalypt plantations around the world (Wingfield 2003). It reduces growth rates and wood quality (Old and Davison 2000). An interesting approach taken to deal with the disease is described by van Staden et al. (2004) for eucalypt plantations in South Africa. Commercial forestry in South Africa relies largely on plantations of exotic species, principally eucalypts, pines and wattles (*Acacia*); Chryphonectria canker has been causing serious losses in eucalypt plantations there (Wingfield 2003).

The disease was first found in South African eucalypt plantations in 1988. It forms cankers in stems and around branch stubs. It causes death of young trees and may weaken the stems of older trees, making them prone to breakage by the wind. Its development is favoured by high rainfall and humidity and temperatures above 23ºC. Attempts in South Africa to control the disease with fungicides have proved expensive and unreliable. Breeding programmes have been undertaken with some success and are continuing to develop tree varieties resistant to the disease (van Zyl and Wingfield 1998, 1999; van Heerden and Wingfield 2001, 2002).

The work on the disease by van Staden et al. aimed to determine which parts of South Africa have a climate suited to its development. For all the locations in South Africa where the disease has occurred, they determined the altitude, and from historical weather records, the average rainfalls of the wettest and driest months and the average temperatures of the hottest and coldest months. Using a complex modelling system, they then identified which other parts of South Africa have a similar climate and, hence, where it is likely the disease would be a problem in eucalypt plantations.

Their results are shown as the regions shaded in black in Fig. 11.1. Eucalypt plantations have been established over some of that region, although there are also large plantation areas to the south of it which, the results suggest, should be relatively safe from the disease.

Also shown in Fig. 11.1 are the regions of South Africa which Booth et al. (1989) identified as having a climate similar to those parts of Australia where flooded gum (*Eucalyptus grandis*) occurs naturally (Sect. 4.1). Flooded gum is one of the eucalypt species grown in plantations in South Africa. To obtain their results, Booth et al. used a model system similar to that used by van Staden et al. To date, eucalypt plantations have been established in South Africa only in more easterly parts of the country. The

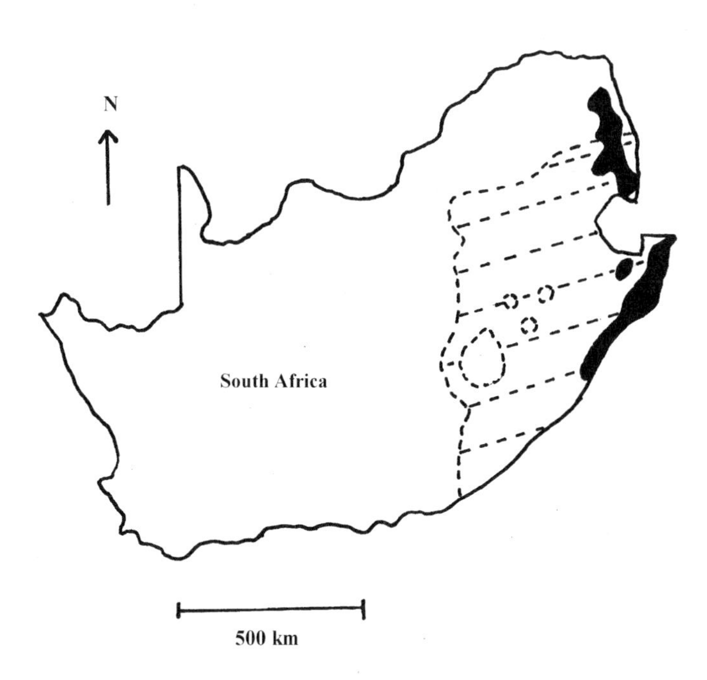

Fig. 11.1. Map of South Africa, showing regions (shaded in *black*) which were predicted by van Staden et al. (2004) to have at least a 25% chance that the canker fungus *Cryphonectria cubensis* would be a damaging disease in eucalypt plantation forests. The regions delimited with *dashed lines* were predicted by Booth et al. (1989) as having a climate equivalent to those parts of Australia where flooded gum (*Eucalyptus grandis*) occurs naturally (adapted from van Staden et al. 2004; Booth et al. 1989)

areas identified by Booth et al. as having a climate suited to flooded gum extend much further to the west; those more westerly regions are much drier than the more easterly regions and, whilst Booth et al.'s results suggest flooded gum might grow in the more easterly parts, its growth rate is likely to be too low for commercial plantation forestry.

These results illustrate the approach being taken in South Africa to contend with the problem of Chryphonectria canker. Eucalypt trees being bred for resistance to the disease will have to be planted in the easterly regions of the country, where the disease is likely to occur. However, it appears that much less concern will have to be paid to the disease in southeastern parts of the country, where the climate is still suitable for eucalypt plantations but much less suited to the disease.

11.3.4 Damping-Off in Nurseries

There are many fungal root diseases which are major problems in both agricultural crops and forests. Individual diseases can be found which attack trees at any developmental stage, from seedlings right through to fully mature trees. Some affect live roots and some cause decay of woody roots. Some are spread by spores blown by the wind and some live entirely below ground. Manion (1991) has summarised the issues surrounding fungal root diseases generally in forests, whilst Keane et al. (2000) devoted several chapters of their book to root diseases of eucalypts.

One of the most common disease problems in forest nurseries is damping-off. It affects many different tree species and occurs in nurseries all around the world. It is caused by root disease fungi, which invade and destroy tissues at the base of the stem, weakening the stem so that the seedling collapses. A number of different fungi cause the disease, including species of the genera *Fusarium, Phytophthora*, *Pythium*, *Rhizoctonia* and *Sclerotium*.

The circumstances of forest nurseries make them highly vulnerable to a disease like damping off, and indeed to other fungal diseases. The tree seedlings are small and immature, making them susceptible to infection. They are planted in close proximity and in very large numbers, allowing disease to spread readily amongst them. When they are grown open-rooted, the nursery beds are reused many times for successive seedling crops; this may allow disease populations to build up in the soil. If seedlings are grown in greenhouses, they often experience warm temperatures, high humidity and dampness, all factors which generally favour the development of fungal disease.

The key to dealing with damping-off and other fungal diseases in nurseries is hygiene. This must include all parts of the nursery system, from the soil in which the seedlings are grown and the water with which they are irrigated, to the greenhouses in which they are raised and even to the hands and clothing of the people who tend them. Brown (2000) has summarised methods which have been used in nurseries around the world to minimise damping-off and other disease problems, including:

- Maintaining free of fungal pathogens the medium in which seedlings are being germinated or raised. This is done by sterilising the soil or potting mix (chemically or by heat treatment), by using composts which are well rotted so that fungi have been killed, by avoiding reuse of the medium or by using sterile media such as vermiculite.
- Growing seedlings in containers which are supported above ground, to avoid contamination of the containers by water-splashed soil which may contain fungal pathogens (Fig. 5.2).
- Using soil or container mixes which are lightly textured and slightly acidic, as this tends to inhibit fungal development.
- Avoiding container mixtures high in nitrogen, such as certain manures.
- Avoiding sowing tree seeds when rainfall has been high or humidity is excessive.
- Avoiding excessive shading or watering of seedlings.
- Adding other microorganisms antagonistic to fungi, such as the fungus *Trichoderma viride* which has been found to inhibit growth of damping-off pathogens of eucalypts.
- Avoiding excessive density of planting of seedlings.

If measures such as these fail, damping-off is usually controlled readily with fungicides, applied either by coating seeds with fungicide powder before sowing or by drenching seedlings with fungicide solution after germination.

11.4 Other Diseases

Manion (1991) defines three classes of disease of plants, biotic, abiotic and declines. Biotic diseases are caused by living organisms which interact with the plant concerned. Fungal diseases are biotic, but there are other organisms which cause biotic diseases, listed by Manion as:

- Bacteria—these are simple, microscopic, single-celled organisms which first developed during the early evolution of life on earth. Together with the blue-green algae, they make up the kingdom Monera of living organisms (Sect. 11.1). Manion mentions three different bacterial groups in particular in his list of biotic pathogens, spiroplasmas and mycoplasmas (which are known jointly as phytoplasmas) and rickettsia.

- **Viruses**—these are very small (you need an electron microscope to see them), biologically very simple, infectious agents which can reproduce only within living cells of animals, plants, or bacteria. They are almost considered not to be living organisms, because they are unable to carry on metabolic processes and reproduce without a host cell; if they are considered living, they make up a kingdom of their own. The role of viruses in tree disease has not yet been studied well.
- Insects—these are animals and have been described in Sect. 10.2. They are often forest pests, but may cause disease also.
- Mites—these are tiny animals (0.1–6-mm long) which are a subgroup of the arthropods (Sect. 10.2). Mites may be pests of plants, causing damage by feeding on leaves, or they may cause disease, especially by transmitting viruses.
- Nematodes—also called roundworms, which are microscopic worms. They live in a wide range of terrestrial and aquatic environments and are amongst the most abundant animals on earth.
- Higher plants—a number of higher plants, such as mistletoes, are parasites of other plants.

None of these biotic agents causes disease in forest plantations as commonly as fungi. Wardlaw et al. (2000) have summarised the few known cases where viruses, phytoplasma, other bacteria or nematodes have caused disease in eucalypts. Only one of the diseases they discuss has been a serious problem for plantation forests. This was a bacterial wilt, caused by the bacterium *Ralstonia solanacearum*, which killed eucalypt seedlings in plantations in several parts of world. Manion (1991) mentions a number of bacterial diseases of trees, but none has been a serious problem in forest plantations. He refers also to a viral disease of poplar (*Populus*), which has caused problems in plantation forests in Europe. He discusses also a disease caused by the nematode *Bursaphelenchus xylophilius*, which leads to wilting of pines in native forests in Asia and North America and may become a serious problem in plantations.Yeates (1990) examined nematode occurrence in forest nurseries in New Zealand; whilst some potentially damaging nematodes were present, the management practices used in the nurseries seemed to be sufficient to prevent unacceptable damage occurring to seedlings.

As abiotic agents of plant disease, Manion (1991) lists air or soil pollution, high or freezing temperatures, pesticides, drought, salt, poor soil aeration, nutrient deficiency and mechanical damage. Some of these are discussed in more or less detail in other parts of this book. Some are induced by man. A controversial example is acid rain. Cars and some large industrial plants emit sulphur dioxide and nitrous oxide into the atmosphere; these combine with water vapour in the air to form sulphuric and nitric acids, which then fall to ground in rain. The acid causes disease of tree leaves and there has been serious damage of forests near heavily industrialised areas in parts of Europe, Asia and North America. These tend to be localised effects; large areas of plantation forests are not generally located near industrial regions. Manion is of the opinion that the effects of acid rain on forests have been overstated in many cases; he suggests that diseases attributed to acid rain often may have been due to other factors. As an interesting contrast to the acid rain issue, increased growth over recent times in some European forests has been attributed to increased availability of nitrogen, resulting from nitrogen deposition in rain from air pollution (Spiecker et al. 1996; Solberg et al. 2004).

The last class of plant disease Manion recognises is termed declines. These are diseases with complex causes, which include both biotic and abiotic agents. Manion suggests that such diseases often arise from biotic and/or abiotic factors which put a forest under stress, factors such as old age, poor soil fertility or drought. These predispose the trees to attack by one or more biotic factors, which lead eventually to the death of individual trees and even of large forest areas. Acid rain may well be one such predisposing factor. Most of the examples Manion gives of declines have occurred in native forests, but one, spruce decline, has occurred in plantation forests in Europe.

12 Tree Breeding

One of the most powerful tools available to growers of forest plantations is the ability to modify the characteristics of the trees through breeding. In agriculture, breeding has a long and very successful history in producing plants or animals with desirable characteristics (or **traits** as they are called in the terminology of breeding). Many different traits can be modified through breeding, such as more vigorous growth, resistance to damage by pests or diseases or the quality of the product (perhaps meat quality in animals or wood quality in trees).

In the past, plant breeding programmes have involved mating parents with desirable traits to produce more desirable offspring. This process can continue through many generations. These are known as conventional breeding programmes. In recent times, advances in techniques known as genetic engineering are offering new possibilities to make substantial advances in breeding, at least for some traits.

Most major forest plantation programmes around the world have a breeding programme associated with them. As will become evident in this chapter, such programmes are large and expensive, continue over many decades and involve highly skilled and specialised staff.

This chapter introduces first the principles of genetics. This is the study of why individuals of any species vary each from the other and how traits are passed from parents to offspring; all breeding programmes are based on these principles. It will then discuss how conventional breeding programmes are conducted with forest plantation trees. Finally, it will discuss briefly some of the opportunities and challenges being presented to forestry by genetic engineering.

12.1 Genetics

Two things determine the traits of any individual living organism, its genetic makeup and the environmental circumstances in which it lives. Its genetic makeup, known as its **genotype**, determines, firstly, what type of organism it is (a fungus, a tree, a cow, a person, etc.) and, secondly, the individual traits of that organism, which distinguish it from any other individual of the same species. However, at least to some extent, those individual traits can be modified by the environment in which the organism lives. The combination of genotype and environmental effects produces the final living organism which we see; this combination is known as its **phenotype**.

This section will describe exactly what is meant by the genotype and phenotype of an individual and how they are important to a breeding programme.

12.1.1 Genotype and Phenotype

By genotype (or genetic makeup) is meant the set of **genes** which an individual has. All basic textbooks on biology include considerable discussion of what genes are and how they are transmitted between parents and offspring; only a brief account will be given here.

All the living cells which make up any individual plant or animal contain an identical set of **chromosomes**. These are located in a central part of each cell, known as its nucleus. Chromosomes are very long (by the standards of chemistry) strands of a chemical substance called **deoxyribonucleic acid**, or **DNA** for short. Each chromosome is divided into many sections along its length, each of which is an individual gene. DNA has a particular chemical structure, which allows that any one gene of an individual organism may differ in chemical detail from the corresponding gene of any other individual.

Living organisms have millions of genes arranged along their chromosomes. The number of chromosomes varies amongst species. Humans have 46. Tree species vary, commonly in the range 20–40; the eucalypts have 22, the pines have 24. Chromosomes actually come in pairs, each pair known as a **chromatid**; organisms have half as many chromatids as they have chromosomes. In reproduction, an individual inherits chromosomes (hence genes) from its parents; one member of each chromatid is inherited from the father (through his sperm or pollen) and the other from the mother (through her eggs). Each chromosome of a chromatid contains a set of genes which have similar functions. Thus, genes come also in pairs, each pair of genes being known as an **allele**. However, each member of an allele may differ slightly in its DNA structure, so each may affect somewhat differently the particular trait of the organism which that allele controls.

It is determined by chance which chromosome of the parents' chromatid pairs any offspring inherits; hence, the set of genes which any individual inherits differs from the set which any other individual inherits, even when they have the same parents. The differences between the gene sets of any two individuals become less the more closely they are related. However, it is only in the case of identical twins that two individuals have identical sets of genes; in that case, both individuals develop from the same fertilised egg.

The function of a gene is to provide the cell in which it resides with a code to make a chemical substance known as a **protein**. Proteins determine what chemical reactions occur in a cell and so, ultimately, how an organism functions (the proteins which do this in living organisms are known as **enzymes**). Slight variations in the chemical structure of genes lead to slight variations in the structure of the proteins produced from them. In turn, this leads to differences between the functioning of individuals and, hence, the traits which each displays. However the environment in which an individual lives can influence also how its traits are displayed and, hence, its phenotype.

This interaction between genotype and environment is very important for breeding programmes. It means that a programme will need to consider the environment in which it is intended to grow the individuals which have been bred. For example, if trees are to be grown in plantations on relatively dry or infertile soils, the breeding programme will need to produce individuals which have genotypes which are well adapted to those types of sites. These matters are discussed in Sect. 12.2.6.

12.1.2 Qualitative and Quantitative Traits

Whilst the genotype and the environment together determine the traits of an organism that can be seen and measured (its phenotype), the number of gene pairs (alleles) involved in determining any one trait differs from trait to trait.

Some traits are determined by a single allele only. These are known as **qualitative** traits. The two genes of any allele may be identical, or they may differ slightly in the chemical structure of their DNA, so that the proteins produced from them will differ slightly also. When the two genes differ from each other, one of them is called dominant and the other recessive. When both these types make up an allele, the dominant type will prevail and determine how the trait concerned is displayed by the individual. Only if both genes of an allele are of the recessive type will the trait be displayed as determined by the recessive type. This means that any qualitative trait can be displayed in only two ways, depending on which of the dominant or recessive gene types the individual contains. Examples of qualitative traits include offspring which are either normal size or dwarf or with flowers of either one colour or another. In such cases; none of the offspring would have sizes intermediate between normal or dwarf or flowers with shades which are a mixture of the two colours. The principles of genetics were discovered in 1865 by the Austrian monk Gregor Mendel, who studied the inheritance of qualitative traits of peas.

Because only one allele is involved in determining a qualitative trait, it is usually quite straightforward to breed individual offspring which have either one or other of the two possible types of that trait. Because they are passed from generation to generation simply and unequivocally, qualitative traits are said to be highly heritable.

There are many traits of plants and animals which are determined by many alleles. These may occur on the same or several different chromatids. These are known as **quantitative** traits. Because many alleles are involved, individuals show more or less continuous variation in the trait which those many alleles control. A trait such as the height of adult humans is a quantitative trait; people have heights which vary across a wide range from rather short to rather tall. In trees, quantitative traits include tree growth rate, density of the stem wood or resistance to attack by some pests and diseases.

Because of the many alleles involved, quantitative traits are far less heritable than qualitative traits. For a breeding programme which aims to influence a quantitative trait, this means that the best that can be hoped for, is that the population of offspring will show at least some shift in the desired direction of their average of that trait. Most of the traits which are of importance in tree breeding programmes are quantitative traits. The discussion of breeding programmes, which follows in Sect. 12.2, will concentrate on breeding to influence quantitative traits.

12.2 Breeding Programme Strategy

At its outset, a breeding programme will need to set its breeding objective, that is, 'what the breeder seeks to maximise...' (Greaves et al. 1997). The objective will be determined by the circumstances of the plantation programme and the requirements of the markets it aims to serve. Inevitably, it will change from time to time, reflecting changes in things such as:

- Market requirements for the products
- Increased technological capabilities of the wood products industry to produce high-quality wood products from smaller or lower-quality trees
- The pest or disease problems which are affecting the programme
- The range of sites on which plantations are being grown

- Broad changes in environmental circumstances, such as those resulting from climate change.

To allow for all these changes, it will be important that the breeding programme never becomes focused too narrowly on specific traits. Rather, it will need to ensure there is always available to it a wide variety of tree genotypes to provide plenty of opportunity to breed for new traits as circumstances demand (Burley 2001).

Once a breeding objective has been chosen, the programme will then need a strategy to achieve it. This will include making decisions about all of the following:

- For which and how many traits the programme will aim to breed.
- How many trees will be selected for inclusion in the programme.
- How will those trees be selected.
- How they will be mated with each other.
- How will the offspring be assessed for their desirability.

There are many examples of the specific breeding objectives and strategies used in particular tree breeding programmes (e.g. Eldridge et al. 1994; Jayawickrama and Carson 2000; Gapare et al. 2003; Burley and Kanowski 2005; Hubert and Lee 2005; Koski and Rousi 2005; Savill et al. 2005; Arnold et al. 2005).

This section will discuss the strategies which are used commonly to achieve the objective of a tree breeding programme. As discussed in Sect. 12.1.2, it will be concerned principally with programmes for quantitative traits. In this book, there is sufficient space to describe these strategies only generally. Far more comprehensive texts are available which discuss them in detail, for plant breeding in general and tree breeding in particular (e.g. Allard 1960; Wright 1976; Zobel and Talbert 1984; Mayo 1987; Cotterill and Dean 1990; Eldridge et al. 1994; Falconer and Mackay 1996; Williams et al. 2002).

12.2.1 Principles

In essence, the strategy of any breeding programme involves three steps. These are repeated again and again, over many generations of the programme. Because of the time taken for trees to grow and produce offspring, it can take as long as 10–20 years to complete each generation. The steps are:

- From the population of trees available to the programme, individual trees are selected with favourable genotypes for the traits being considered in the programme.
- The selected individuals are mated to produce a new generation of offspring which, on average, will display traits closer to those desired by the programme.
- The offspring are planted out, either as part of the general plantation programme or in experimental plantings called progeny tests. The offspring then form a new population with which the steps of the programme are repeated.

As successive generations of a breeding programme are completed a hierarchy will be developed of different populations of individual trees of the species concerned. These populations are (Shelbourne et al. 1989; Eldridge et al. 1994):

- The base population—(also called the original variability, gene resource, or external population). This contains the entire resource of genotypes available to the programme. It consists generally of all the native forests and existing plantation areas of the species concerned and to which the breeding programme has access. A base population will consist of many thousands or millions of individuals.
- The breeding population—(also called the selection or selected population). This consists of the individuals with 'superior' genotypes, which have been selected from time to time in successive generations of the breeding programme. They will be used for further selections and matings. This population may consist of perhaps 500 families (a family consists of the offspring from one individual tree) with perhaps as many as 100–200 individuals within each family. It is the population with which tree breeders work most intensively in their breeding programme. As 'superior' trees are identified in each generation of the breeding programme, or even in the base population, they will be included in the breeding population. As 'inferior' trees are identified, they will be removed from the population.
- The propagation population—(also called the packaging, seed production or production population). This consists of the individual trees (perhaps fewer than 100) which have been found from time to time to produce offspring best suited for the purposes of the general plantation programme. Large numbers of offspring are propagated in various ways (Sect. 12.3) from the members of this population for planting in the general plantations. As the breeding programme develops, the members of this population change also from time to time.

At the very start of a breeding programme, the only resource of trees available is the base population. From this, the breeder needs to select trees to constitute an initial breeding population. From time to time in later generations of the breeding programme, especially as the breeding objective changes, it may be desired to supplement the breeding population with new individuals from the base population. Selection of trees from the base population is known as provenance selection and is discussed in Sect. 12.2.2.

Once a breeding population has been established, there are a variety of ways of selecting the most desirable individuals from it to be mated in the next step of any generation of the breeding programme; selection methods are discussed in Sect. 12.2.3. There are then many possible choices as to which of the selected trees should be mated with which others; Sect. 12.2.4 discusses how those matings may be done. The methods of selection and mating will vary as the breeding programme develops.

The **gain** (or genetic gain) achieved in each generation of the programme depends on how selections are made and matings done. Gain is defined as the difference between the average of a desired trait in the offspring of each generation of the programme and its average in the individuals in the breeding population from which the offspring originated. It is the role of tree breeders to adopt a strategy for the breeding programme which achieves the largest gains possible, over as few generations of the breeding programme as possible.

One example which illustrates the magnitude of the gains which have been achieved in plantation breeding programmes comes from Li et al. (1999). They discussed a programme for loblolly pine (*Pinus taeda*), which is a major plantation species grown in southern USA. The programme has now completed two generations and gains of up to 35% in stand stem wood volume have been achieved in plantations established using seed produced from the programme. Also, the plantations show improved stem form, wood quality and increased resistance to a fungal disease which is a problem in the region. Vergara et al. (2004) have reported that one generation of a breeding programme for slash pine (*Pinus elliottii*), another important plantation species in southern USA, achieved gains of 10% in stand stem wood volume production as well as increased resistance to fungal disease.

As a breeding programme progresses through more and more generations, tree breeders will accumulate more and more information about the trees in their breeding population. This will allow them to make increasingly sophisticated selections and to employ increasingly sophisticated mating strategies to achieve increasing gains with each generation of the programme. As a programme develops, the distinction between separate generations may become blurred as various mating strategies are attempted within and between the generations (Potts 2004). Jayawickrama and Carson (2000) have provided an interesting example of how the strategies of a breeding programme for radiata pine (*Pinus radiata*) in New Zealand have developed over the 50 years since the programme started.

12.2.2 Provenance Selection and Testing

The place where any tree in the base population grows is known as its **provenance**. When establishing an initial breeding population (or supplementing an existing breeding population), seeds are collected from individual trees of different provenance. Seedlings are raised from the seed and are planted out in progeny tests, called provenance tests, in the region where the plantation programme is being established. These will give some idea as to which provenances are most likely to provide trees best suited to the purposes of the plantation programme. The better performing trees will become the initial breeding population or may be added to the existing breeding population.

A good example of the scope and scale of provenance testing is given by Sierra-Lucero et al. (2002) for loblolly pine (*Pinus taeda*) in the southern states of the USA. No fewer than seven separate provenance tests were established across the region where loblolly pine is planted. The results suggested that loblolly pine with its provenance in Florida was likely to be most appropriate for plantation programmes along the coastal plain of southeastern USA. These provenance tests were aiming to find suitable trees to supplement an existing breeding population for loblolly pine, which was in its third generation of a breeding programme (Xiang et al. 2003a); earlier provenance testing had not been as thorough as it might have been and Florida provenances of loblolly pine had been overlooked in earlier provenance tests.

12.2.3 Selection

Once a breeding population has been established, the first step in each generation of the breeding programme is to select individuals from it to be mated. This is done by selecting those individuals with the highest **breeding value** (also called additive genotype) for the trait it is desired to influence in the breeding programme.

The breeding value of an individual is a measure of the extent to which its offspring show a gain in a trait. In effect, it is a measure of the extent to which that individual has a 'superior' genotype favouring that trait. Breeding value can be determined by mating the individual with many other individuals from the breeding population, growing the offspring from those matings in progeny tests and then measuring them. Those measurements will show how much larger is the average of the desired trait in the offspring of the individual than was the average of the trait over the individuals of the breeding population; that is then the breeding value of the individual.

In the first generation of a breeding programme, there is little information available about the individuals in the breeding population from which their breeding values can be assessed properly. Certainly no progeny tests will have been carried out to allow their breeding values to have been determined. At that stage of the programme, it can only be assumed that the phenotype of an individual is a reasonable measure of its breeding value. The trees selected from the breeding population will then simply be those with the highest phenotypic values of the trait concerned.

As each generation of the breeding programme is completed, progeny tests from previous generations will provide information to better assess breeding values of individuals in the present breeding population. Specialist quantitative geneticists employ complex mathematical techniques to use this information to assess breeding values (White and Hodge 1989); quantitative geneticists are important members of the team involved in a breeding programme. Some little discussion of their work will be given in Sect. 12.2.5; the advanced texts, referred to in the paragraph preceding the start of Sect. 12.2.1, will need to be consulted to learn about their work in more detail.

A further problem with selection arises because the objective of most breeding programmes is to achieve gains in each generation in more than one trait. For example, it might be desired to select individual trees with high breeding values for all of high stem wood density, fast growth rate, small branches and a straight stem. Usually it is practical to consider no more than about three or four traits at any one time in a breeding programme, or else the number of genes involved in controlling all the traits is so large that the selection process ends up achieving little gain in any of the traits.

Cotterill and Dean (1990) and Falconer and Mackay (1996) have discussed various ways in which selection to favour multiple traits can be achieved. The most sophisticated method is known as index selection. This considers all the desired traits together and, in effect, finds a compromise between them to determine which individuals should be selected; the details of index selection will not be discussed further here.

Whilst selections are generally based on the phenotypes of individuals, it would obviously be desirable if the genes of an individual could be 'seen' directly. That would determine unequivocally if the individual contained the particular genes which favoured development of a desired trait and so had a high breeding value for it. In recent times, research has been developing methods to identify and isolate individual genes; however, quantitative traits are controlled by many genes and, as yet, it has proved impossible either to identify all the genes involved in determining a particular trait or to evaluate the relative influences of each of those genes on the trait.

Instead, methods are being developed to identify what are called quantitative trait loci (often abbreviated as QTL). These are segments of chromosomes which contain genes important in determining the quantitative trait concerned. In turn, quantitative trait loci are identified using what are called **molecular markers**, which are segments of DNA. There are a number of different types of molecular marker, each extracted from tree tissue using different and complex chemical methods; they have names such as amplified fragment length polymorphism (abbreviated as AFLP), restricted fragment length polymorphism (RFLP), random amplified polymorphic DNA (RAPD) or simple sequence repeat (SSR). Research is continuing on the use of molecular markers to aid in selection in breeding programmes (e.g. Falconer and Mackay 1996; Boyle et al. 1997; Walter et al. 1998; Burley 2001; Kumar and Garrick 2001; van Buijtenen 2001; Potts 2004; Tani et al. 2004; Lörz and Wenzel 2005; Srivastava and Narula 2005).

Whatever method is used to select individuals from the breeding population, there are several issues which need to be borne in mind:

- A decision needs to be made as to how many individuals should be selected. The fewer selected, that is, the more intensive the selection, the higher will be the average breeding value of the selected individuals; thus, the bigger will be the gains in the desired traits in the offspring of their mating. However, the fewer selected, the more restricted will be the range of genotypes represented amongst the offspring. This will limit opportunities for further selection, particularly as the breeding objective of the programme changes over time; by chance, genes desirable for some traits may have been lost from the breeding population if the breeding programme has concentrated too intensively on selection of genes favouring one particular trait.
- Each generation of the breeding programme can take many years. This is more a problem for tree breeding programmes than for many other agricultural crops, where only 1 year is often required for each generation. It means that tree breeding programmes over many generations are very long-term projects.

  To minimise the time involved in each generation, it is necessary to evaluate the traits of offspring as early in their life as possible, so that the next generation of selection and mating can be started. However, when it is young, the phenotypic traits displayed by an individual tree may be rather different from those it will have when it has grown to the age at which it would normally be harvested in a plantation.

To deal with this problem, tree breeders have devoted considerable effort to finding the earliest age at which a trait can be measured on an individual and still be a reliable indicator of that trait at a much later age. For example, Raymond (2002) has summarised research work with many eucalypt species and suggested that offspring must be at least 3–8 years old before a measurement of their wood traits will indicate reliably the values of those traits in mature trees. Shelbourne et al. (1989) suggest that offspring of radiata pine (*Pinus radiata*) in New Zealand need to be at least 4–5 years old before any reliable selections can be made from them and up to 15 years old before the most reliable selections are possible. McKeand (1988) found that offspring of loblolly pine (*Pinus taeda*) in the southern USA can be selected on the basis of their height at 6 years of age and this will predict reliably their stem wood volumes at maturity; more recent work has suggested this age can be reduced to 3–5 years of age (Xiang et al. 2003b). Stener and Jansson (2005) found that selection for height growth of silver birch (*Betula pendula*) in Sweden could be done reliably at 4–6 years of age.

The identification of quantitative trait loci through molecular markers offers particular advantages in this respect. DNA can be extracted from the tissue of very young seedlings and superior offspring might be identified at very young ages.

- Difficulties may arise when a breeding programme is considering more than one trait. If some or all of the genes which determine one trait are carried on the same chromosome as those which determine another trait, both traits will be passed together from one generation to the next. In such cases breeding to favour one trait may disadvantage the other; this can frustrate the aims of the breeding programme.

The solutions to the various problems raised by these issues are complex and beyond the scope of the present discussion.

12.2.4 Mating

Once desirable individuals have been selected from the breeding population, the next step is to mate those individuals with other selected individuals. This may provide the seed to establish the next generation of the general plantation programme. Instead, or as well, the offspring may be planted in progeny tests; depending on their subsequent performance, some of the offspring may eventually become new members of the breeding population. The offspring may also provide information about the performance of their parents, which will assist in making decisions about which of those parents should be removed from the breeding population.

There are a number of ways in which mating of the trees selected in any generation of the programme may be carried out. As the generations of the programme advance, and more detailed information is available about the breeding population, the matings are undertaken in ever more complex ways (or in ever more complex mating designs as they are called). The more complex mating designs aim to produce larger gains in the desired traits in the offspring than would occur with less complex designs.

There are two principal mating designs used in tree breeding programmes, mass selection and recurrent selection. Shelbourne (1969) has provided a relatively straightforward description of their principles and his discussion will be followed here. The steps involved in these designs are shown diagrammatically in Fig. 12.1.

The principal difference between mass selection and recurrent selection is the way in which offspring are obtained from the trees selected from the breeding population. In mass selection, seed is collected from the selected trees and seedlings raised from them. Each selected tree will be the mother of the seed collected from it; hence, that seed will have inherited half of its genes from a selected tree. However, the selected trees may be scattered quite widely around the plantation estate; some may be in previously established progeny tests, some may be in the general plantation areas and some may even be in the native populations of the species concerned. Hence, the other half of the genes in the seed from the selected trees will have been inherited from the pollen of fathers which are not one of the other selected trees. It is then said that there was open-pollination of the selected trees.

In recurrent selection, seedlings are raised from the selected trees as **clones** of them. A clone is an offspring of an individual organism which has been reproduced in such a way that it is genetically identical to its parent. As discussed in Sect. 12.3, there are various ways of raising cloned offspring from trees. Because genes from unselected trees will not be included amongst them, the range of genotypes in seedlings raised in recurrent selection should, on average, better suit the requirements of the breeding programme than those raised in mass selection.

The subsequent steps in both mating designs can then be followed in Fig. 12.1. Shown there are two variations on mass selection, termed with or without progeny testing. Mass selection *without* progeny testing is the simplest and quickest way to achieve at least some gain in the next generation of the plantation programme, because there is no delay involved by conducting progeny tests. It is used commonly at the early stages of a

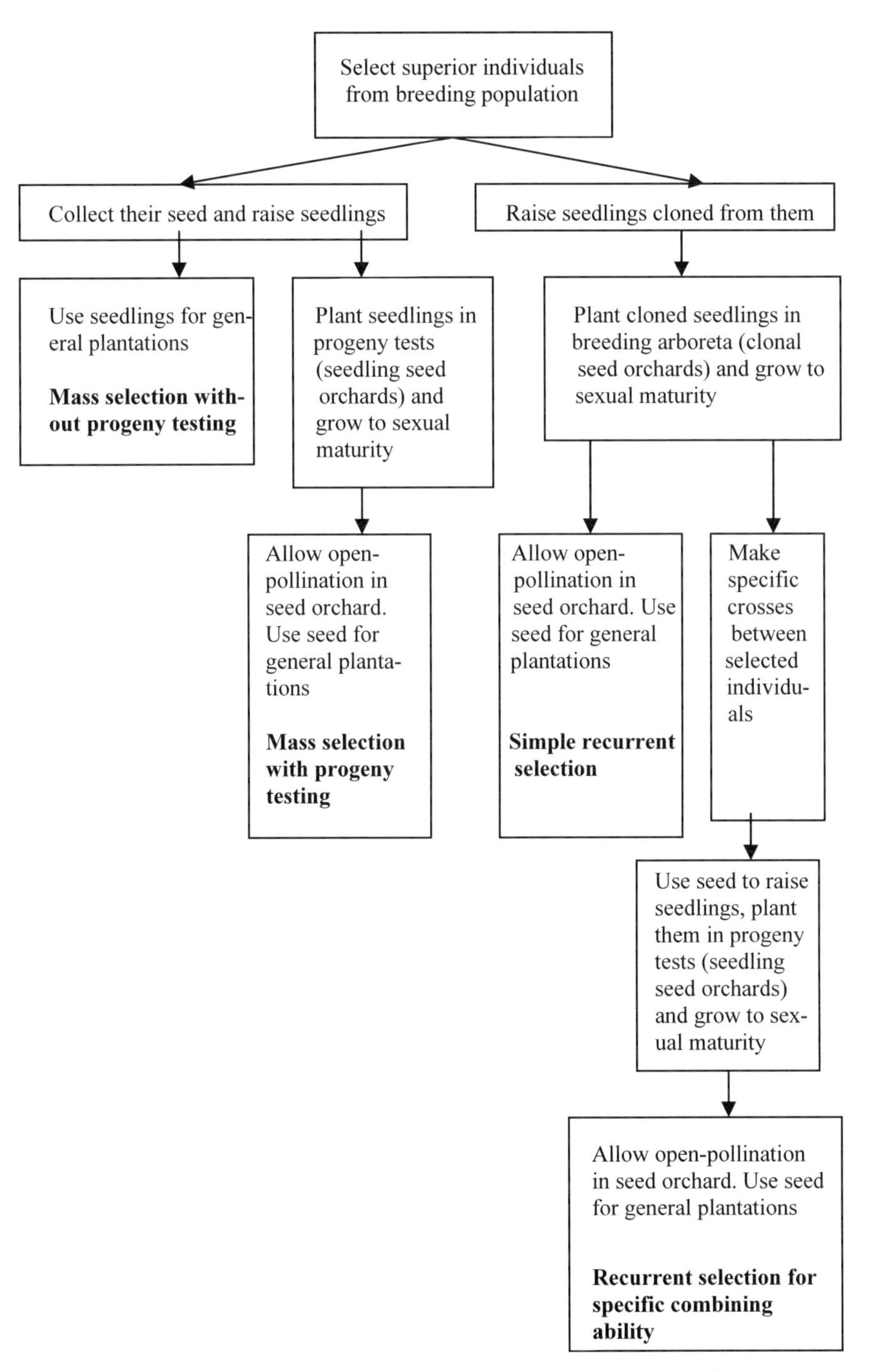

Fig. 12.1. Principals of four mating designs used in tree breeding programmes (adapted from the discussion in Shelbourne 1969)

breeding programme to obtain a supply of seed quickly for the general plantations.

The supply of seed is delayed using mass selection *with* progeny testing, because progeny tests have to be grown to sexual maturity. These progeny tests are called seedling seed orchards and are grown in a location isolated from any other trees, so that only trees in the seed orchard can pollinate others there. This will mean that the seed produced *with* progeny testing will, on average, contain a higher proportion of genes derived originally from the selected mother trees than seed produced *without* progeny testing. This means that the gains from mass selection *with* progeny testing will be higher than those from mass selection *without* progeny testing.

Gains from recurrent selection can be expected to be higher still than those from mass selection, because only genes from the selected trees will occur in the seed used eventually in the general plantation programme. As shown in Fig. 12.1, there are two forms of recurrent selection, simple recurrent selection and recurrent selection for specific combining ability.

The latter involves making specific matings (or crosses as they are called) between particular individuals in the breeding arboretum (also referred to commonly as a clonal seed orchard). That is, pollen from particularly chosen individuals is used to fertilise flowers of other particularly chosen individuals. This is the mating design used most commonly today in advanced tree breeding programmes around the world. Quantitative geneticists decide what specific crosses should be carried out; they choose crosses which should lead to the largest possible gains.

Making specific crosses is rather labour intensive, hence expensive. Pollen is collected from individual flowers of what are to be male parent trees. To avoid any possibility of self-pollination, stamens are removed from the flowers of trees which are to be female parents. Pollen from the desired male parent is then applied to the flower of the desired female parent. The flower is then enclosed in a bag, to prevent pollen from any other nearby tree fertilising it. The techniques involved in making specific crosses in forest trees have been discussed by various authors (e.g. Wright 1976; Eldridge et al. 1994; Potts 2004; Assis et al. 2005). The seed produced from the specific crosses is used to establish progeny tests and/or seedling seed orchards (Fig. 12.1). To allow quantitative geneticists to make best use of the information from these progeny tests, it is important that they be very carefully laid out; books such as Williams et al. (2002) are devoted entirely to discussion of their design and the analysis of the data obtained from them.

The time, expense and expertise necessary for these various mating designs increase progressively from mass selection without progeny testing through to recurrent selection for specific combining ability. The gains should increase in the same order. For any breeding programme, it requires careful analysis to find the most cost effective balance between the gains achieved and the expense involved when choosing the mating design.

12.2.5 Gains

Quantitative geneticists have available mathematical tools which allow them to predict what gains (defined in Sect. 12.2.1) are likely to be achieved from each generation of a breeding programme. An example will be used to illustrate how this is done. It comes from Tibbits and Hodge (1998), who were involved with a breeding programme for plantations of shining gum (*Eucalyptus nitens*) in northern Tasmania, Australia. The plantations were being grown to produce pulpwood for paper-making.

Tibbits and Hodge's breeding programme had been through an early generation and they had available results from progeny tests from that generation. Their breeding population contained about 13,000 individual trees from about 300 shining gum families. The trees in this population were growing in eight different experimental plantations, established across a range of sites in northern Tasmania.

The breeding programme was concerned with three traits, all important for pulpwood production, namely tree growth rate, wood basic density and wood pulp yield. The importance of wood basic density for pulp production was mentioned in Sect. 3.3.3. Pulp yield is the weight of oven-dry paper pulp produced per unit weight of oven-dry wood. Measurements representing the three traits had been made at 6–9 years of age on the individuals in the breeding population; that is, the phenotypes for each of these traits were measured. Tree growth rate was represented by stem **basal area** (stem cross-sectional area at breast height, cf. stand basal area). It was assumed that trees with larger basal areas were faster-growing trees. Rather more complex methods were used to measure the average basic density of the wood of the entire stem of each individual tree and its pulp yield.

Tibbits and Hodge's next step was to determine values for their breeding population of what are known as genetic parameters. These are computed routinely for a breeding population by quantitative geneticists, using information gathered from progeny tests. Their calculation is mathematically complex and will not be discussed further here. In effect, they allow estimation of the breeding values of the individual trees in the breeding population. The genetic parameter values Tibbits and Hodge determined for their shining gum breeding population are shown in Table 12.1.

The first parameter listed in the table is simply the average over all the individual trees of the breeding population of the measured (phenotypic) values of basal area, basic density and pulp yield. The phenotypic standard deviation is a variable which measures the amount of variation, over the

Table 12.1. Genetic parameters for three traits, of the trees in a breeding population of shining gum (*Eucalyptus nitens*) in northern Tasmania, Australia (derived from information in Tibbits and Hodge 1998)

Trait	Average	Phenotypic standard deviation	Narrow sense heritability	Genotypic (bold face) and phenotypic (normal face) correlations		
				Basal area	Basic density	Pulp yield
Basal area (cm^2)	250	23	0.15	–	**−0.24**	**0.24**
Basic density (kg/m^3)	430	13	0.34	−0.09	–	**0.33**
Pulp yield (%)	50	1	0.27	0.27	0.23	–

whole population, of each of those traits; influences of both genes and the environment will determine that variation.

Narrow sense heritability 'expresses the extent to which phenotypes [of offspring] are determined by the genes transmitted from the parents' (Falconer and Mackay 1996). It is commonly known simply as **heritability**. The higher the heritability, the fewer generations of a breeding programme will be necessary to make substantial gains in the desired trait. In general, quantitative traits with a heritability less than 0.1 can be expected to produce low gains in that trait from generation to generation. Traits with heritabilities in the range 0.1–0.3 can be expected to produce moderate gains, whilst traits with heritabilities above 0.3 will produce appreciable gains (Cotterill and Dean 1990). You will find reference in texts on quantitative genetics to another heritability measure, known as **broad sense heritability** (also known as degree of genetic determination). This expresses the extent to which the phenotype of an individual tree is determined by its genotype (Falconer and Mackay 1996). Broad sense heritability will play no further part in the present discussion.

The last columns of Table 12.1 show correlations between the three traits in the breeding population. Correlations take values in the range −1 to +1. The genotypic (in **bold face)** correlations show, in effect, the extent to which the genes determining one of the traits are carried on the same chromosomes as those which determine another of the traits. If the correlation is positive, it shows that breeding to favour one of the traits will also favour the other, at least to some extent. If the correlation is negative, breeding to favour one trait will tend to lead to a decline in the value of the other. If the correlation was zero, it would indicate that all the genes controlling each of the two traits were carried on different chromosomes. Thus, the genotypic correlations in Table 12.1 suggest that breeding to favour faster growth (basal area) could, at the same time, lead to lower basic density and higher pulp yield.

The phenotypic correlations in Table 12.1 show simply the correlations between the phenotypic measurements of the individuals. They indicate how phenotypes of one trait will tend to change as phenotypes of another trait change during breeding.

With this information, Tibbits and Hodge could now estimate the gains they would achieve in offspring from their breeding population, for any particular selection method (Sect. 12.2.3) and mating design (Sect. 12.2.4). Suppose they decided to use an index selection method to select individuals from their breeding population and chose to use the Smith–Hazel selection index; this is a quite sophisticated index and details of it can be found in Cotterill and Dean (1990). Suppose also they chose to use simple recurrent selection as their mating design. As well, suppose they decided to select and mate 130 trees from the 13,000 in their breeding population, that is, they used use a selection intensity of 1% (Sect. 12.2.3). Using quite complex mathematical techniques (Cotterill and Dean 1990), Tibbits and Hodge could then have predicted that the offspring from the mating would have an average stem basal area of 254 cm^2, wood basic density of 438 kg/m^3 and pulp yield of 50.4%, representing gains over the corresponding average values (250 cm^2, 430 kg/m^3 and 50%, respectively; Table 12.1) for the original 13,000 trees in the breeding population.

Tibbits and Hodge would then need to consider if the gains they were predicting were in fact large enough to be worthwhile commercially for the breeding programme to continue and actually carry out the recurrent selection mating. If it was not, they might have to consider the gains which could be achieved using different selection intensities or with other mating designs, or even reconsider completely the breeding objective of their programme.

Quantitative geneticists have mathematical tools available to them to predict the gains which might be achieved with any selection method and mating design. In any major tree breeding programme, these tools are used to help make decisions about what traits should be considered, and what selection methods and mating designs should be used in each successive generation of a breeding programme.

12.2.6 Genotype × Environment Interactions

In breeding programmes it must always be borne in mind that environmental effects will modify the extent to which any trait is displayed by an individual tree. That is to say, breeding for a particular trait will only be effective if that trait can be expressed satisfactorily in the environment in which the offspring of the tree is to be planted. In breeding programmes, the breeding population may be subdivided into different breeding lines which aim to produce trees suited to particular environmental circumstances (Sect. 12.2.1).

There are many reports of trees of a particular genotype performing relatively well on some sites and relatively poorly on others, when compared with trees of another genotype. It is then said to be displaying a genotype × environment interaction. Examples of this have been seen in radiata pine (*Pinus radiata*) in Australia and New Zealand in tree growth rate, stem form, branch form (Matheson and Raymond 1984), rooting ability (Theodorou et al 1991), distance between branch whorls (known as internode length) (Carson and Inglis 1988) and susceptibility to infection by pine needle blight (Carson 1989). Examples for other species include the level of an anticancer compound in Pacific yew (*Taxus brevifolia*) in western North America (Wheeler et al. 1995), growth of mountain ash (*E. regnans*) in southeastern Australia (Raymond et al. 1997), height growth of interior spruce (*Picea glauca* and *Picea engelmannii*) in western Canada (Xie 2003) and black spruce (*Picea mariana*) in eastern Canada (Johnsen and Major 1995) and survival, growth and rooting ability of poplar in northern USA (Hansen et al. 1992; Zalesny et al. 2005).

12.2.7 Interspecific Hybrids

Discussion to this point has concerned breeding programmes for trees of one particular species. However, it is worthwhile mentioning that considerable advantages have been gained in some plantation programmes by planting interspecific hybrids, which are crosses between two different species. Biology defines a species as consisting of related organisms, capable of interbreeding in natural circumstances; however, it is not uncommon to find that individuals of closely related species (of plants or animals) can interbreed when they are living in close proximity to each other, sometimes under natural conditions and sometimes when they are brought together under controlled conditions.

Various interspecific hybrids of both softwood and hardwood species have been used for plantation forestry (Wright 1976; Potts and Dungey 2004; Potts 2004; Arnold et al. 2005). The hybrid between two species of poplar, *Populus trichocarpa* and *Populus deltoides*, shows particularly high growth rates and is used extensively in North America in plantations grown to produce wood for paper production (Heilman et al. 1994). Hybrids of various eucalypt species also show particularly high growth rates and extensive plantations of them have been established in Africa, China, Brazil and Indonesia (Potts and Dungey 2004). The hybrid of a canker disease resistant eucalypt species, *E. urophylla*, with the fast-growing but canker-susceptible eucalypt, *E. grandis*, has both fast growth and canker resistance in plantations in the Congo (Potts and Dungey 2004). Hybrids of a species of willow (*Salix viminalis*) with other willows (*S. shwerinii* or *S. burjatica*) are showing potential to produce high wood yields in bioenergy plantations in Europe (Robinson et al. 2004). The hybrid of slash pine (*Pinus elliottii*) with Honduran Caribbean pine (*Pinus caribea* var. *hondurensis*) shows high growth rates and improved stem form; it is being grown in plantations in Australia and is being considered for use in North and South America, China and Africa (Kerr et al. 2004). The hybrid of two species of larch, *Larix occidentalis* and *L. lyallii*, has traits which might make it suitable for afforestation of sites with a climate intermediate between that of sites where the two species occur naturally (Carlson 1994).

As discussed by Volker (2002) most of the interspecific hybrids used to date in plantation forests have derived from individual hybrid trees which were found occurring naturally. There certainly appears to be considerable potential for greater use of hybrids and to undertake breeding programmes to develop them further. However, there are particular problems with their breeding, many of which arise because individuals which would not normally be expected to interbreed are being forced to do so. Potts and Dungey (2004) have reviewed the potential for and problems associated with interspecific hybrids in plantation forestry.

12.3 Propagation

Once a breeding programme has identified individuals with desirable traits, those individuals become part of the propagation population. Large numbers of offspring are produced from this population to be planted out in the next generation of the general plantation programme.

12.3.1 Seed Orchards

One of the most common ways of raising individuals from the propagation population is by growing seed orchards. That is, members of the propagation population are grown together and allowed to open-pollinate amongst themselves to produce seed for the general plantation programme.

As discussed in Sect. 12.2.4 and shown in Fig. 12.1, progeny tests conducted as part of the breeding programme may double as seed orchards. Often, their life as a seed orchard will continue well past the time they have provided the information required from the progeny test.

Sometimes, seed orchards may be established specifically for seed production. In that case, they are established using individuals from the propagation population. In deciding which individuals should make up this population, it must be remembered that some individual trees are not as profligate seed producers as others; if seed production is the principle purpose for establishing a seed orchard, a balance must be drawn between including the most desirable individuals from the breeding population and those which produce the most seed (Lindgren et al. 2004).

For major plantation programmes, seed orchards may be quite large, occupying perhaps tens of hectares. Their management is a specialised task, to ensure they provide both the quality and quantity of seed required for the general plantation programme. They need to be grown in a location which is isolated from other plantations, to avoid the possibility of pollination by trees outside the orchard. They are usually established at wide spacings, perhaps with 9–10 m between the trees. This ensures that low branches will remain alive, from which it is easy to collect seed, and so that machinery can move freely about the orchard. Seed orchards which were established originally as progeny tests will usually need thinning to these wider spacings and only the most superior individuals will be retained at the thinning.

It takes some years from the time a seed orchard is established until the trees become sexually mature, flower and produce seed. In some tree species, it can be as long as 20 years before appreciable quantities of seed become available, although many species will produce appreciable amounts after 5–6 years. From the point of view of the general plantation programme, this time should be kept to a minimum so that improved seed from the breeding programme is available as soon as possible.

An interesting possibility to speed production in seed orchards is treatment of the trees with chemical substances which promote flowering at young ages. One such substance, paclobutrazol, has been used to increase flowering (hence fruit production) in horticultural crops. It has been used successfully in seed orchards of shining gum (*E. nitens*) in Tasmania, Australia (Moncur et al. 1994) and Tasmanian blue gum (*E. globulus*) in Portugal (Araújo et al. 1995) to obtain seeds from trees as young as 1–2 years old. Other substances and methods, such as raising seedlings in a heated greenhouse, have been found to promote early flowering in some softwood species (Philipson 1995). The use of such substances and techniques remains an interesting area of research for tree breeding programmes (Reid et al. 1995; Potts 2004).

If the trees in the seed orchard are allowed to grow too tall, it can be difficult to collect the seed from them. Elevating platform vehicles are often used to reach into the canopies of tall trees. Often trees are lopped and/or their branches trained near the ground to facilitate seed collection. Some seed orchards are used for only a few years and are then abandoned, so the trees never reach heights above which it is difficult to collect seed. Some early-flowering eucalypt species have been grown in seed orchards for only 4–8 years and the trees have then been felled and allowed to regrow by coppice (Eldridge et al. 1994).

Further reading on the different types of seed orchard that are in common use for plantation forestry and on many of the issues involved in their management can be found in Shelbourne et al. (1989), Wright (1976) and Eldridge et al. (1994).

12.3.2
Clonal Propagation

The reasons why it may be desired to produce offspring which are genotypically identical to their parents, that is, clones of their parents, are:

- If recurrent selection mating designs (Sect. 12.2.4) are being used, they require that clonal offspring be raised from individuals selected from the breeding population.
- It may be desired to raise large numbers of clonal offspring, from a few highly desirable individuals, and plant those offspring out as part of the general plantation programme.

The use of clonal offspring in the general plantation programme seems to be a very attractive idea for plantation forestry. It should lead to the largest possible gains over any generation of the breeding programme, because only the most superior individuals will be used. As well, the methods of clonal propagation produce seedlings much more rapidly than seed orchards, which must wait until the trees have grown to sexual maturity. Thus, the results of a breeding programme may be exploited more rapidly when clonal propagation methods are used.

However, the establishment of large plantation areas with clonal offspring from only a few, highly superior individuals is a somewhat controversial issue in plantation forestry. It is argued that if some unanticipated, damaging environmental disturbance occurs, then one or other of the few genotypes involved might be highly susceptible to the damage and the whole plantation might be lost. For example, an insect pest that has not previously been a serious problem for the plantation programme may find the trees of one or other clones so attractive that its population rises rapidly to very damaging levels for the whole plantation. On the other hand, if seed produced in a seed orchard is used to establish the plantation, gains may not be as large, but the much wider range of genotypes present in the plantation may prevent such massive damage occurring. To date, there do not seem to have been any cases of large-scale disasters of this nature resulting from clonal plantings. However, current wisdom suggests that no fewer than about 20 clones should be mixed in clonal plantings to minimise the risk. Robison (2002) and Coyle et al. (2002) have reviewed the issues surrounding this controversy (see also Zwolinski and Bayley 2001).

Whatever the reason for producing clonal offspring, there are five principal methods of clonal propagation used with forest trees, namely grafting, cuttings, air-layering, tissue culture and somatic embryogenesis. These techniques, and many variations on them, are used extensively in agriculture and horticulture. Substantial texts are available which discuss them in detail (e.g. Hartmann et al. 2001; Gupta and Ibaraki 2005).

Grafting involves joining two pieces of living plant tissue from different plants so that they grow into one complete plant. The two pieces may be of the same or of different species. Usually a stem or branch section, or even a single bud, (called the scion) of one plant is joined to the stem, with root system attached, of another (called the rootstock). The join is usually made through a cut and the various tissues of the two plants must be aligned carefully so they will grow together to form a complete plant. The join between the scion and the rootstock is usually bound with tape, to hold the two sections firmly, whilst the plants grow together to become one; the resulting plant is called a ramet.

Various substances, often closely guarded proprietary secrets, may be applied to encourage the two grafted sections to fuse. Grafted plants are usually raised in pots in a greenhouse, under environmental conditions which have been found most suited to the requirements of the species being grafted. When the grafted plant is fully grown, the rootstock and the scion retain their respective genotypes. Most of the tree species used in plantation forestry are quite amenable to grafting (Wright 1976; Eldridge et al. 1994; Potts 2004). It is one of the most common methods used to raise clonal seedlings for the establishment of breeding arboreta (Fig. 12.1). It is a rather labour intensive process, so is not suitable for raising very large numbers of clonal individuals from a single parent.

Mass production of large numbers of clonal seedlings can be done with cuttings, air-layering or tissue culture. These methods require that an entire, fully functioning individual develops from a small piece of the parent.

Cuttings are raised by taking a small piece of stem, branch, leaf or bud from the parent and planting it in soil, or some form of potting mix or other culture medium. Roots and new shoots develop on the cutting and eventually a whole new plant forms (Fig. 12.2). In some species, cuttings are raised quite easily out-of-doors in a nursery bed. Other species require cuttings to be raised in pots in well-controlled environmental conditions in a greenhouse. Various substances, such as plant **hormones** (chemical substances produced naturally by plants and which control certain of their growth processes), can be applied to cuttings to encourage rooting.

Air-layering involves sharply bending, or making a small incision in, the stem of the parent, which causes roots to form on the stem. The rooted

a

b

Fig. 12.2. (Continued over)

c

Fig. 12.2. (This page and previous). An example of the stages involved in raising clonal seedlings from cuttings. In this case, the seedlings were to be planted in a clonal plantation in subtropical eastern Australia. They are clones of a fast-growing, superior tree, which was a hybrid of two eucalypt species, flooded gum and river red gum (*Eucalyptus grandis* × *E. camaldulensis*). **a** A 2-year-old clonal seedling which had been raised from a cutting taken from the superior tree. In the background are a large number of similarly cloned seedlings of the same tree. These are kept permanently in a greenhouse at the nursery and large numbers of new cuttings are taken from them each year. Small branch segments, with one or two leaves attached, are snipped from them to make new cuttings; you can see where pieces have been snipped from this seedling. **b** A tray of these new cuttings, 3 weeks after they were taken. They have been planted in small containers containing potting mix. To encourage rooting, they were treated with plant hormones and were grown for 4–5 weeks in the greenhouse, with very high humidity and frequent watering. In **c**, they have been transferred to root trainer containers (Fig. 5.3) and moved outside to be raised in the open air (Photos—P.W. West)

stem section is later detached from the parent and raised in a culture medium to form a new plant.

Tissue culture (often also called micropropagation) involves excising a small piece of tissue from the parent and raising it in vitro (a term from modern Latin, meaning literally 'in glass', and used to refer to an activity which takes place in an artificially created environment outside a living organism). In tissue culture, the excised tissue is raised in a sterile culture medium contained in a small, closed glass or plastic container.

Cuttings, air-layering and tissue culture all require that one type of plant tissue, say a stem section in a cutting or a small piece of a shoot tip in tissue culture, is capable of developing into an entire plant. That this is possible relies on the fact that every cell in a plant (or animal for that matter) contains all the genes necessary to control all the functions of the entire organism. Normally, the only genes which are active within any one cell of a plant (or animal) are those essential to the function of the type of tissue of which that cell is a part (a leaf cell, a root cell, a cambial cell, etc.). Genes unnecessary for the functioning of that type of tissue are 'switched off' (that is, they are present, but remain inactive). Thus, a fine-root cell cannot carry out photosynthesis, because the genes which control the chemical reactions of photosynthesis are switched off in that type of cell.

Just how and when genes in a cell are switched on or off is an area of intense research interest at present. Some chemical substances, such as certain nutrient elements (Sect. 2.1.4) or plant hormones, and certain environmental conditions (temperature, humidity, etc.) have been found to affect the switching on and off of genes; exactly how they do so has yet to be determined fully.

Nevertheless, for most plant species, at least one of cuttings, air-layering or tissue culture can be used to induce the growth and development into complete plants of many clonal individuals from a single parent. However, because our understanding of how genes in tissues are switched on or off is inadequate, it is very difficult to predict which of these practices will work successfully with any particular plant species and what environmental conditions are necessary to raise entire plants of that species. At present, research on these issues is often a rather hit-or-miss procedure; various practices and a wide range of environmental conditions are tested until some combination is found which works successfully for the species concerned. Often, growers will closely guard, as proprietary secrets, the successful results they have obtained with a particular species.

The final method of clonal propagation, somatic embryogenesis, is a form of micropropagation. It involves excising the embryo from a seed, placing it in a suitable culture medium where it reproduces many embryos, which can then be split and raised individually. Seedlings produced ultimately from somatic embryogenesis are known as emblings and can, when eventually planted out, show slightly different growth behaviour from seedlings raised from seed (Grossnickle et al. 1994; Grossnickle and Major 1994a, b).

Many of the tree species important for plantation forestry can be successfully raised by one form or other of clonal propagation. Research continues with those species for which successful techniques have yet to be developed.

12.4 Genetic Engineering

Genetic engineering involves artificial manipulation of the genes that an organism contains. It was developed in the 1970s, principally by American scientists. It is one of the greatest revolutions that have ever taken place in science in general and in the biological sciences in particular.

In essence, genetic engineering involves identifying a particular gene which confers a particular trait on a particular type of organism. That gene, that is, the section of a DNA molecule which constitutes the gene, is then extracted from that organism in the laboratory and transferred to the cells of another type of organism. The two types of organisms involved, the gene donor and the gene recipient, may be quite unrelated; it is quite possible to transfer a gene from an animal species to a plant species. The transferred gene then becomes part of the genotype of the recipient organism and confers on the recipient the trait which that gene controls. The recipient organism may then grow and reproduce quite normally and the new gene will be transferred to its offspring which will also have the new trait. The techniques involved in genetic engineering are highly complex and will not be discussed further here. Texts such as Atwell et al. (1999), Primrose et al. (2002), Curtis (2005) or Srivastava and Narula (2005) describe them.

From the point of view of tree breeding, or plant breeding in general, genetic engineering is restricted at present to the manipulation of qualitative traits only, that is, traits which are controlled by single alleles (Sect. 12.1.2) or at least very few alleles. The techniques of genetic engineering are not yet sufficiently sophisticated to allow identification and transfer of many genes at once, as would be necessary to transfer a quantitative trait from one type of organism to another. In fact, it may never be possible to transfer quantitative traits through genetic engineering. Some of the genes involved in determining one quantitative trait will be involved also in determining a quite different quantitative trait. It may prove impossible to select and transfer the genes controlling one quantitative trait without, at the same time, affecting another quantitative trait, perhaps with disastrous consequences for the organism concerned. Conventional plant breeding (Sect. 12.2) will continue to be the major breeding strategy used in forest plantations as long as it is quantitative traits which remain of principal interest to tree breeders.

Van Frankenhuyzen and Beardmore (2004) have reviewed the limited progress made to date in genetic engineering of forest trees. Much of the work remains experimental and has not yet developed to the stage where large-scale plantations of genetically engineered trees have been established. However, success has been achieved with engineering for tolerance to herbicides in species of poplar, eucalypt, larch and pine, resistance to insect defoliation in species of poplar and spruce, and reduction of lignin content of wood in species of poplar. Research continues to develop trees genetically engineered to display other desirable traits.

The potential advantages of using genetic engineering to produce trees with highly desirable traits are obvious. However, there are risks also (Walter et al. 1998; Coyle at al. 2002; van Frankenhuyzen and Beardmore 2004). For example, growing large areas of a genetically modified tree which produces an insect toxin may simply lead to rapid evolution of a variety of the insect resistant to the toxin. Ultimately, this may lead to very severe damage to the general plantation programme as a whole. Instances have been reported where an insect toxin aimed at controlling a pest insect has also affected other insects in the ecosystem which are not harmful to the plants. Of course, both these problems may exist equally where insecticides are sprayed extensively on plantations; these are important reasons for deploying integrated pest management strategies in plantation forestry (Sect. 10.1.2).

There is danger too that seed from a genetically modified tree might be spread from the plantation into the native populations of the species concerned. Its genetic modification may give it such a competitive advantage that the genetically modified tree comes to dominate the native populations, with unpredictable consequences for the native forests. A possible solution to this problem is to introduce a gene for sterility, at the same time as other genes are being introduced, to prevent the genetically modified tree producing seed (Harcourt et al. 1995; Burdon and Richardson 2000).

So revolutionary are the concepts of genetic engineering of plants that the widespread use of modified plants in agriculture or forestry is politically highly controversial at present (Burley 2001; van Frankenhuyzen and Beardmore 2004). Of particular concern are possible effects on human health where a genetically modified food plant has developed some damaging trait in addition to the intended trait. Works such as Murray (2003) discuss these issues. Concerns about unanticipated impacts of genetically engineered trees on the ecosystems within which plantations grow are now leading to similar controversies in forestry (van Frankenhuyzen and Beardmore 2004).

13 Mixed-Species Plantations

There are various reasons for establishing plantations containing two or more tree species growing together in mixture. These include (Kelty 1992; Ball et al. 1995; Keenan et al. 1995):

- Providing a diverse range of products, where the different species produce different things (say different types of fruit) as well as wood
- Uncertainty as to which species will produce the most valuable products and so several are grown to improve the chance of obtaining at least some very valuable products at harvest
- Providing a more diverse ecosystem to encourage a wider range of animal species to live in the plantation
- Obtaining a greater total yield of products from the mixture than from a monoculture of any of the species
- Providing a 'nurse' crop, where one species protects another in some way from damaging agents such as full sunlight, wind, pests or disease
- Improving the aesthetic appearance of the landscape.

In the first three of these cases, the option exists for the grower to establish small areas of monocultures of each of the species, sometimes referred to as a coarse-grained mixture or a 'mosaic'. In the other cases, it is implicit that the plantation should contain two or more tree species intermingled, sometimes referred to as a fine-grained mixture or an 'intimate mixture'.

This chapter will be concerned principally with the growth and development of intimate mixtures of tree species. Often it is intended that all of the species planted will be harvested to yield one product or another; however, in some cases, one of the species may have been included in the mixture because it offers some benefit to the others and is not intended to yield a specific product.

13.1 Growth of Mixed-Species Plantations

Much of our knowledge of the growth behaviour of two or more species growing in intimate mixture has come from research with agricultural crops (Trenbath 1974; Park et al. 2003); however, there is little reason to think that the principles established from agricultural research should not apply equally to tree species in forest plantations. Kelty (1992) and Binkley et al. (1997) have reviewed the somewhat limited experience with tree mixtures (see also Garber and Maguire 2004).

Research has suggested that the important differences between intimate mixtures and monocultures are:

- Because the different species have different biological characteristics, the amounts required by each of one or other of the resources essential for its development (carbon dioxide, light, water, nutrients and equable temperature—Sect. 2.1.1) may differ.
- One of the species may be capable of obtaining some of those resources from parts of its surroundings which are inaccessible to the other species. For example, a species with a deeper root system may be able to obtain water and nutrients from deeper within the soil than a species with a shallower root system.

Where such differences apply, it is said that the species are occupying different niches within their environment. Those niches will usually overlap substantially, because all of the species will obtain at least some of the resource they require from the same parts of their surroundings. However, as long as the niches do not overlap completely, a set of species growing in intimate mixture may be able to use more, in total, of the resources available from the site than if any one of them was growing in monoculture. In turn, this means that the combined growth of the species mixture could be greater than the growth of any of the species in monoculture. Obviously this is of great interest to plantation forestry; it offers the possibility that total production from an intimate mixture of tree species might be higher than from monocultures of any of them.

Where examples have been found of intimate mixtures with incomplete niche overlap and greater production than monocultures, there seems to have been differences between the species in any one of the following (Kelty 1992):

- Their height, growth form or the efficiency with which their leaves are capable of using sunlight to produce food in photosynthesis. An example of this would be an intimate mixture containing a species, which has leaves capable of undertaking photosynthesis in shade, growing beneath the canopy of a species which requires relatively high sunlight intensity for photosynthesis. This has been observed in plantations grown in Europe with Scots pine (*Pinus sylvestris*), as the upper-canopy species, mixed with either Norway spruce (*Picea abies*) or beech (*Fagus sylvatica*) in the lower canopy.

- Their phenology, that is, the timing during the year of various of their processes, such as the production of foliage. In forest plantations, this might occur where a deciduous species makes an upper canopy in mixture with an evergreen species; the evergreen species may have a period each year when it can grow fully exposed to light, before the deciduous species develops its new canopy.
- Their root system structure, particularly their depth of rooting. It has been argued that deeper-rooted species should be able to obtain water and nutrients from deeper in the soil, where shallower-rooted species cannot reach. Neave and Florence (1994) have given an example from native forests in New South Wales, Australia, where the pattern of occurrence across the landscape of spotted gum (*Eucalyptus maculata*), blackbutt (*E. pilularis*) and *Acacia mabellae* was determined by differences in their rooting depths and, hence, their abilities to access water and nutrients from different depths within the soil. Dunbabin et al. (1994) have shown how plants with differently shaped root systems may differ in their ability to take up nutrients from the soil. Bauhus and Messier (1999) found that hardwood and softwood species, growing together in native forests in Canada, developed their root systems in rather different ways so that each could exploit the available soil resources differently.

Apart from incomplete niche overlap, a second reason why intimate mixtures may have greater total production than monocultures is that one of the species may be able to give an advantage in some way to another. The principal ways in which this occurs are (Kelty 1992):

- Where coniferous forests dominate the landscape in colder regions of the northern hemisphere, leaf litter that accumulates on the ground may decompose only slowly because of the low temperatures. This means that the nutrients in the litter are returned only slowly to the soil for reuse by the trees (Sects. 2.1.4, 2.3.3, 6.3) and the trees may suffer some nutrient deficiency; this is a problem particularly with nitrogen, which is required in larger amounts than other nutrients (Sect. 2.1.4). As well, conifer litter is rather higher than other forest species types in the ratio of carbon to nitrogen it contains; this too limits the activity of the microbes which decompose litter.

Under these circumstances, a benefit might be conferred on the coniferous species by growing it in intimate mixture with a hardwood. For example, in Germany, a hardwood birch (*Betula*) species is sometimes established in mixture with the softwood species spruce (*Picea abies*). The leaves of birch are rather higher in their nitrogen content than those of spruce. When they fall as litter, they reduce the ratio of carbon to nitrogen in the litter as a whole and the litter is decomposed more rapidly. This ensures an adequate supply of nitrogen is available to newly developing spruce leaves and fine roots (Brandtberg and Lundkvist 2004).

- By inclusion in the mixture of a species with a nitrogen-fixing capability. This may increase the amount of the nitrogen available to the other species in the mixture. Nitrogen is abundant in the environment; nitrogen gas makes up about four fifths by volume of the air. Unfortunately, gaseous nitrogen cannot be taken up by plant roots. However, there is a large group of nitrogen-fixing plants (over 600 of which are tree species), which have the capability of transforming nitrogen from the atmosphere into a chemical form which other plant roots can then take up from the soil. Most of the plants with this capability are from the legume family (the scientific name of which is the *Fabaceae*, a well-known member of which is the common garden pea), although species from some other plant families have this capability also. Nitrogen-fixing plants develop nodules on their roots which contain certain types of bacteria. The bacteria carry out the chemical transformation of nitrogen which has diffused into the soil from the air. Because of their special capability, many nitrogen-fixing plants are extremely important in agriculture, generally because they can make nitrogen available in soils which otherwise have a limited supply. Examples of plantation mixtures which include a nitrogen-fixing species are given in the next section.

Apart from the potential for intimate mixtures to have a greater product yield than monocultures, there may be other advantages to be gained from mixtures (Keenan et al. 1995). The presence of one species may protect another from pests or disease. A taller species which shades a smaller one may lead to less branching or an improved stem form in the smaller species, hence leading ultimately to an improvement in the quality of the wood harvested. A taller species may shade a smaller species which is otherwise unable to grow well in full sunlight in the early years of establishment of the plantation. In these cases, the species which protects the other is termed a 'nurse' species. Often the nurse species is removed relatively early in the life of the plantation, once its nursing function has been completed. Examples of these types of mixtures are given in the next section.

Whilst there are differences in growth behaviour between intimate mixtures and monocultures, there are similarities also. The processes of symmetric and asymmetric competition (Sect. 2.4) continue to occur, both between trees of the same species in the mixture and between trees of the different species. However, unlike monocultures, competition below ground for water and nutrients from the soil may be asymmetric. If one species is physiologically better able to take up water and nutrients from the soil, it will be able to obtain a disproportionately large proportion of those resources (Hutchings et al. 2003). Just as with asymmetric competition for light above ground (Sect. 2.4), that will lead eventually to trees of that species becoming the larger trees in the plantation; in mixtures, competitively more successful species are often termed the more 'aggressive' species.

13.2 Examples of Mixed-Species Plantations

A number of examples will be given in this section to illustrate a variety of ways in which intimate mixtures may have a role in plantation forestry and how different mixtures grow and develop.

13.2.1 Sydney Blue Gum–*Falcataria* in Hawaii

One of the most striking and widely quoted examples of a successful mixed-species plantation comes from experimental work done in Hawaii by DeBell et al. (1997) (see also Binkley and Giardina 1997). The experiment involved an intimate mixture of two species, Sydney blue gum (*E. saligna*) and *Falcataria moluccana* (named previously either *Albizia falcataria* or *Paraserianthes falcataria*). The experimenters were investigating the possibility of producing high-quality timber from plantations established on sites which had been degraded after many years of repeated cropping with sugar cane. The soils were particularly lacking in nitrogen and *Falcataria* was of interest because it is a nitrogen-fixing species (Sect. 13.1).

The experiment involved growing both species in experimental plots planted with 2,500 trees per hectare. The experimental plots contained either each species alone or both species planted in intimate mixture. In the mixtures, different plots had different proportions of the two species. Experiments of this type are used commonly in studies of competition between two or more species and are known as 'replacement' experiments (Kelty and Cameron 1995).

Figure 13.1 shows the stand oven-dry biomass of stem wood, measured at 10 years of age, in the experimental plots with differing proportions of the two species. The results show that blue gum was the more aggressive competitor of the two, because its growth in mixture was consistently closer to its growth in monoculture (that is, in plots with 100% blue gum) than was the case for *Falcataria*.

The results show also that the total stem wood yield of the two species in mixture was consistently higher than the yield of either in monoculture, reaching a maximum of 230 t/ha when 34% of the trees planted were blue gum and 66% were *Falcataria*. DeBell et al. showed that the higher yields of the mixtures were due to the benefits gained by the blue gum from the nitrogen contributed to the site by *Falcataria* through fixation of nitrogen.

Because blue gum was the more aggressive competitor, in the mixtures its trees soon outstripped the *Falcataria* trees in height. The blue gum trees were then able to shade the *Falcataria* trees and reduce their growth substantially through asymmetric competition. At 10 years of age in the plots planted with 50% of each species, the blue gum trees averaged 29 m in height, 20 cm in stem diameter at breast height and 0.33 m^3 in stem wood

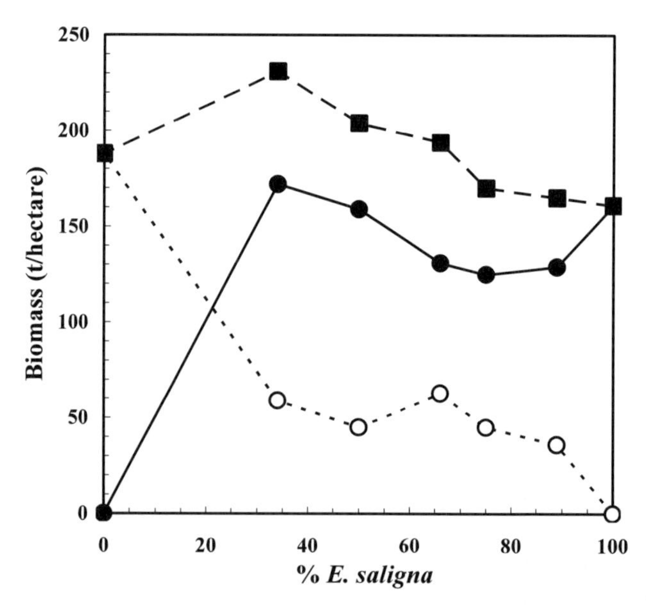

Fig. 13.1. The stand stem wood oven-dry biomass of Sydney blue gum (*Eucalyptus saligna*) (●), *Falcataria moluccana* (O) and the total of the two species (■) at 10 years of age in experimental plantations of the two species in mixture in Hawaii. The results are graphed against the percentage of *E. saligna* planted in different plots in the experiment (Source—DeBell et al. 1997)

volume. The *Falcataria* trees were much smaller, averaging 15 m in height, 13 cm in stem diameter and 0.12 m^3 in stem wood volume.

The largest blue gum trees in the experiment were found in the plots where total wood production was the highest, that is, where only 34% of the trees planted were blue gum. At 10 years of age, those trees averaged 31 m in height, 24 cm in diameter at breast height and 0.50 m^3 in stem wood volume. This was far more than the average 24 m in height, 15 cm in diameter and 0.16 m^3 in stem wood volume of the blue gum trees growing in the monoculture plots. In effect, the low initial stocking of the blue gum in that mixture (800 trees per hectare), coupled with their great competitive advantage over the *Falcataria* trees, ensured their much greater average size than blue gum growing in the much more heavily stocked (2,500 trees per hectare) blue gum monoculture. From a timber production point of view, the blue gum trees in that mixture would produce much higher quantities of more valuable, large logs than those in the blue gum monoculture. These results for blue gum are almost what might be expected if blue gum trees had been planted at a low stocking density and nitrogen fertiliser had been applied to them, rather than planting the *Falcataria* trees mixed with them to supply the nitrogen.

In the mixed-species plots, so competitively successful were the blue gum trees that DeBell et al. concluded the *Falcataria* trees were too small to produce any worthwhile amounts of timber. That did not concern them, because of the enormous gains they had achieved in timber production by the blue gum trees in the mixture. Figure 13.2 shows a recent photograph of a mixed-species stand in the experiment.

This experiment provides a classic example where a nitrogen-fixing, species planted in mixture with a more aggressive species can provide enormous benefits, both in timber production by the aggressive species and by avoiding the need to apply expensive nitrogen fertiliser. Some other experiments have illustrated also the potential advantages of plantation mixtures of eucalypts with a nitrogen-fixing species (e.g. Austin et al. 1997; Khanna 1997, Kumar et al. 1998; Bauhus et al. 2000, 2004; Forrester et al. 2004)

13.2.2 Black Alder–Poplar in Quebec

This example involves a replacement experiment (Côté and Camiré 1987) with a nitrogen-fixing species black alder (*Alnus glutinosa*), mixed with a hybrid poplar (*Populus nigra* × *Populus trichocharpa*) in plantations in Quebec, Canada. The experiment was established at an extraordinarily high stocking density (90,000 trees per hectare) and the trees were intended to be grown for only a few years as a bioenergy plantation (Sects.

Fig. 13.2. A photograph of a mixed Sydney blue gum (*E. saligna*)–*F. moluccana* stand, at 20 years of age, in experimental plantations of the two species in mixture in Hawaii (DeBell et al. 1997). The crowns of the *Falcataria* trees can be seen extending over the path and are well below the crowns of the much taller and larger, white-stemmed blue gums (Photo—J.B. Friday, University of Hawaii)

1.2, 7.2.4). Poplars are used for bioenergy plantations in Europe and North America because they grow rapidly and coppice (Sect. 5.5) readily. In this experiment, black alder was mixed with poplar to assess its contribution to nitrogen nutrition in the plantation.

Figure 13.3 shows the stand total biomass (aboveground and belowground parts of the trees), after 3 years of growth, of each species and the two combined. The results were similar to those for the Sydney blue gum–*Falcataria moluccana* mixtures (Sect. 13.2.1). The poplar was the more aggressive competitor. Even after only 1 year of growth, the poplar trees were more than twice as tall as the alder trees; a similar height difference was observed by Hansen and Dawson (1982) in another alder–poplar

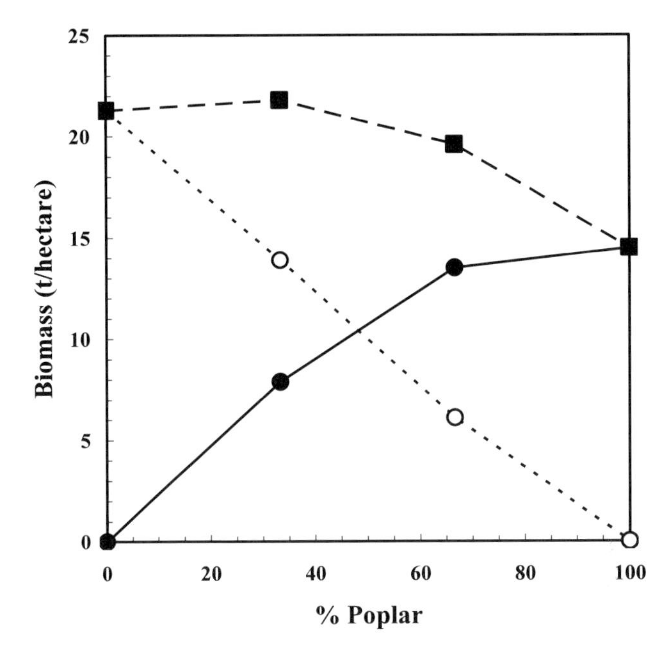

Fig. 13.3. The stand oven-dry biomass yields (shoots plus roots), after 3 years of growth, of hybrid poplar (*Populus nigra* × *P. trichocharpa*) (●), alder (*Alnus glutinosa*) (O) and the total of the two species (■) in a replacement experiment in a bioenergy plantation in Quebec, Canada. The results are graphed against the proportion of poplar planted in different parts of the experiment (data from Table 3 of Côté and Camiré 1987)

mixed planting. The biomass production was highest in the 33% poplar with 67% alder mixture, indicating an advantage was gained by poplar from nitrogen fixation by the alder.

The results suggest that inclusion of a nitrogen-fixing species in intimate mixture in a bioenergy plantation might increase total production. DeBell and Radwan (1979) found a similar effect in another alder–poplar mixed planting. If it was allowed to continue growing, the more aggressive poplar would be expected eventually to dominate the mixture, just as Sydney blue gum did in the example in Sect. 13.2.1. In natural stands, poplar has been found as the dominant species in stands where it is mixed with alder (Pezeshki and Oliver 1985).

13.2.3 *Cedrela–Cordia–Hyeronima* in Costa Rica

Menalled et al. (1998) have provided a very interesting example of a mixed-species plantation, where insect damage and incomplete niche overlap in all of canopy structure, phenology and rooting structure seemed to be determining tree growth behaviour in the mixture.

They considered three tropical tree species, native to Costa Rica and of interest to plantation forestry there, *Cedrela odorata*, *Cordia alliodora* and *Hyeronima alchorneoides*. They established an experimental plantation (planted with 2,887 trees per hectare) with plots containing monocultures of each species and an intimate mixture of the three with (in effect) one third of the trees in the mixture being of each species. Trees in the mixture grew rapidly and, by 4 years of age, the tallest trees were around 15-m tall and the canopy had closed.

Cedrela is a species susceptible to attack by larvae of an insect, the mahogany shoot borer (*Hypsipyla grandella*), which bore into the growing tips of trees and may damage them severely (Sect. 9.7.4). In the mixture in the experiment, damage to the *Cedrela* trees reduced their early growth so much that, by 4 years of age, they formed an understorey of small trees below the other two species. So severe was their initial disadvantage that it appeared that *Cedrela* would continue to be suppressed by the taller trees in the mixture, through asymmetric competition for light. Even though they suffered substantial insect damage too, *Cedrela* trees growing in monoculture continued to grow better than those in the mixture; mahogany shoot borer ceases to damage trees once they are more than about 5-m tall, because the adult insects do not fly much higher than that.

Cordia was the most aggressive competitor in the mixture and, by 4 years of age, its trees were tallest and formed an overstorey above the *Hyeronima* trees (with the small *Cedrela* trees below the *Hyeronima* trees). However, *Cordia* is a semideciduous species and maintains a rather open canopy which does not cast heavy shade. *Hyeronima* is a species well adapted to carry out photosynthesis in partial shade; hence, the *Hyeronima* trees were able to form a dense canopy below the *Cordia* trees and continued to grow well in the partial shade. Because of this niche differentiation in canopy positions between *Cordia* and *Hyeronima* trees (and the small *Cedrela* trees) the total leaf area index (Sect. 2.3.1) of the mixed-species plantation intercepted about the same amount of sunlight as the *Hyeronima* monoculture and appreciably more of the sunlight than the *Cordia* or *Cedrela* monocultures. That is to say, the mixed-species plantation is likely to show at least as much growth in total as any of the monocultures.

As well as this niche differentiation in canopy position, Menalled et al. argued that the semideciduous nature of *Cordia* would lead also to niche differentiation in phenology. For parts of the year when the canopy of *Cordia* had lost some of its leaves, the *Hyeronima* canopy would be better lit and be able to carry out more photosynthesis than when the *Cordia* canopy was fully developed.

In addition to canopy niche differentiation, it was argued there was likely to be niche differentiation between *Cordia* and *Hyeronima* in the ability of their root systems to obtain nitrogen from the soil. *Cordia* requires rather high amounts of nitrogen for its metabolic needs and *Hyeronima* is a rather deeper-rooting species than *Cordia*. Menalled et al. argued that in mixture, *Hyeronima* may be able to access nitrogen from deeper in the soil than *Cordia*. Thus, more nitrogen would be available to the trees in the mixture than would be available to either species in monoculture. This might be expected to produce greater production overall in the mixture than in the monoculture.

Menalled et al. were reporting results from the experiment at only 4 years of age, much too young to surmise how the tree sizes and wood yields would have developed by the time the trees were large enough to harvest for timber. However, their results illustrate that selection of the appropriate species to include in a mixed-species plantation can lead to substantial niche differentiation between the species. This means that overall better use can be made by the trees of the resources for growth available at a site, with consequent gains in production. Bauhus et al. (2004) have described an interesting case of an intimate mixture of a nitrogen-fixing species black wattle (*Acacia mearnsii*) with the eucalypt Tasmanian blue gum (*E. globulus*). They attributed the higher aboveground biomass of the mixture than of the monocultures to canopy niche differentiation rather than, or as well as, the nitrogen fixation.

13.2.4 Red Cedar–Silky Oak in Queensland

This example illustrates a case where a nurse species protected another species from damage by a pest, in this case an insect pest. It involved an experimental plantation of an intimate mixture of silky oak (*Grevillea robusta*), as the nurse species, and red cedar (*Toona ciliata*) in tropical north Queensland, Australia (Keenan et al. 1995). Both species are native to Australia and both produce desirable timber, but red cedar is prized particularly as a very high-quality timber for furniture making, both in Australia and in other parts of Asia where the species occurs naturally.

Red cedar is very susceptible to attack by larvae of an insect, the tip moth (*Hypsipyla robusta*), which bore into the growing tips of trees and may damage them severely (red cedar is a species related to Spanish red cedar, which is damaged similarly by a related insect—Sects. 9.7.4, 13.2.3). So severe is the damage to red cedar by the insect, that it has largely precluded it from consideration as a species for plantation monoculture in Australia.

The experimental plantation was established at 2,000 trees per hectare, with plots of either a monoculture of red cedar or an intimate mixture of equal numbers of red cedar with silky oak. As well, additional plots were established at 1,000 trees per hectare with monocultures of silky oak. Annually over the next 5 years, red cedar trees were planted at 1,000 trees per hectare in the silky oak monoculture plots. Eventually, this gave a sequence of intimate-mixture plots, with red cedar planted at successively later ages below the silky oak.

The protection from insect attack afforded to the red cedar by the silky oak was remarkable, provided the silky oak trees were at least 1 year old when the red cedar trees were mixed with them. Without the protection of silky oak, only about 30% of red cedar trees were still alive 10 years after planting. As well, more than 30% of those survivors had poor form, with multiple leading shoots sprouting from the damaged tree tips.

With protection from 1-year-old silky oak, about 75% of the red cedar trees survived to 10 years of age and only 3% of those had poor form. The proportion of red cedar trees surviving rose as the age of the silky oaks below which they were planted rose, reaching about 95% with protection from silky oaks which were 5-years-old when the red cedar trees were planted; the proportion of surviving red cedar trees with poor form remained at about 3%.

As yet, it has not been determined exactly how the presence of the silky oak protects red cedar from attack by *Hypsipyla*. Shading, reduced temperature and reduced wind below the protective canopy have all been identified as factors reducing *Hypsipyla* activity (Mitchell 1971; Mo and Tanton 1996; Campbell 1998). Recent work on the problem has been attempting to determine which red cedar properties favour attack by *Hypsipyla* (Bygrave and Bygrave 1998, 2001; Cunningham and Floyd 2004).

Keenan et al. concluded that the most appropriate management regime for red cedar trees would be to plant them below 2-year-old silky oak trees. Competition from the silky oak trees would then be minimised by removing them from the plantation at about 10 years of age. The red cedar trees would then be thinned twice, to a final stocking density of 150 trees per hectare, with a view to clear-felling for timber at about 50 years of age.

14 Conclusion

Mead (2005) has published a very thoughtful review where he attempted to identify which silvicultural practices have proved to be most effective, generally around the world, in increasing plantation productivity (in biomass or wood volume growth rate). Table 14.1 summarises his conclusions and shows also his assessment of the relative costs incurred with various practices. On the basis of these results, he suggested that, in general, 'the priorities for improving productivity should be:

1. First, ensure that the correct species and provenance for the site are being used.
2. Treat major nutrient deficiencies and ensure that there is good rooting depth by draining the site, etc. These will lead to long term improvements in [productivity] and the results can often be spectacular.
3. Use good planting stock, planting methods, weed control, and the optimum initial stocking level/rotation length. On droughty sites, weed control is critical.
4. A longer term option is tree breeding, while using [nitrogen] fertilizer on established stands produces rapid productivity gains.'

Irrigation is one silvicultural practice mentioned in Mead's table, but which has been considered little in this book. On drier sites, irrigation can lead to spectacular increases in productivity. However, the availability of water and the costs of large scale irrigation generally preclude it as a silvicultural practice expect in specialised cases, such as plantations being used for disposal of sewage waste (Sect. 1.2). Baker et al. (2005) have discussed some examples of irrigated plantations in Australia. Mead did not include pest or disease control (Chaps. 10, 11) in his list of silvicultural treatments. Of course, these are protective measures to prevent loss of productivity, rather than to increase it; where they are necessary, failure to implement them can lead to disastrous losses. Nor did he include thinning or pruning (Chaps. 8, 9). Both of these are concerned principally with improvement of the quality of the wood produced, through increased sizes of logs or production of wood clear of knots. Nor did he consider in any detail the effects of silvicultural practices on the quality of the wood produced, hence on its value.

Table 14.1. A summary of the gains in wood production achieved, over a short (up to 12–15 years) or long (more than 15 years) rotation, through the application of various silvicultural practices in commercial plantation forests around the world, together with the relative cost of each practice (adapted from Table 2 of Mead 2005)

Silvicultural practice	Gains (%) in wood yields over a rotation		Relative cost
	Short rotation	Long rotation	
Selection of species or provenance	25 to >75	25 to >75	Low
Correction of major mineral nutrient deficiencies on site	25–75	Up to >75	Moderate to high
Provision of adequate rooting depth by drainage or breaking up compacted layers in soil	>75	25–50	Moderate to high
Site cultivation	>75	10–25	High
Quality of seedlings and care in planting	25–50	<10	Moderate
Stocking density at planting and choice of appropriate rotation length	10–75	10–25	Moderate
Weed control	25 to >75	<10 to >75	Moderate
Applying fertiliser at planting	10–25	<10 or 10–25	Low to moderate
Irrigation	Up to 75	Up to 50	High
Application of fertiliser after canopy closure	25–50	<10 or 10–25	High
Tree breeding	10–75	10–50	Very high

Ultimately, the decision as to what silvicultural practices are employed in any particular plantation will be determined by economic factors. If the financial gains achieved through some practices are insufficient to more than offset the costs involved, then obviously those practices will be avoided. Nevertheless, productivity is often the most important factor determining the viability of a plantation enterprise and Mead has given some useful guidelines as to what issues are generally most important. Of course, any particular plantation will have to be considered on its merits to decide which factors are most crucial to ensure its viability.

Superimposed on consideration of the silvicultural technology which can be used today in plantation forestry must always be consideration of the sustainability of the plantation enterprise. That is, plantations must be managed in such a way that their productivity can be maintained in the long term and that sites are not damaged in any way which will preclude their use for other purposes by later generations. From thousands of years of agricultural experience, it is well known that repeated cropping and manipulation of sites can lead to site degradation and, ultimately, crop failure. The book by Lindenmayer and Franklin (2003) gives much information on the developments in forestry that have occurred in recent times to promote sustainable management practices.

Intensive plantation forestry has not been practised for a sufficiently long time to assess properly whether it really is being done in a sustainable fashion. Study of plantation yields from the same sites over several rotations has found that productivity has often increased in later rotations, owing to continuing improvements in silvicultural practice (Evans 1999; Powers 1999). Whether or not these increases are hiding long-term, deleterious effects on sites remains to be determined.

Considerable effort is being made in forestry in general, not just plantation forestry, to develop criteria by which long-term sustainability of forestry practices might be judged and indicators (that is, specific measurements) to assess whether or not those criteria are being met (Grayson 1995; Hickey et al. 2005). Powers (1999) is of the opinion that determination of the effects of plantation forestry on soil air porosity and the organic matter in the soil will be two of the most important indicators of any long-term deleterious effects of plantation forestry; air porosity determines the ability of roots to grow and develop (Sect. 5.1) and organic matter is associated intimately with the cycling of nutrients in the plantation and the availability of nutrients from the soil (Sects. 2.3.3, 6.1.2, 6.3, 8.1). Both Turner et al. (1999) and Fox (2000) have suggested that site-specific management will be essential to ensure sustainability. That is, it will be necessary to have '...detailed knowledge of soils as they occur on the landscape and their physical, chemical and biological properties that affect productivity...Understanding the processes and properties of a specific soil...will enable foresters to develop management regimes tailored to each soil' (Fox 2000). From the discussion in this book, it should be evident how complex the development of site-specific management practices for plantation forests can be.

Plantation forestry is being hailed today as a potential saviour of the remaining native forests of the world, forests which have been cleared and exploited ruthlessly in the past. However, much work remains to be done to ensure plantation forestry is a long-term, sustainable supplier of wood and other benefits and does not become, ultimately, an environmental problem itself.

References

ABARE-Jaakko Pöyry Consulting (1999) Global outlook for plantations. Research report 99.9. Australian Bureau of Agricultural and Resource Economics, Canberra

Achim A, Ruel J-C, Gardiner BA, Laflamme G, Meunier S (2005) Modelling the vulnerability of balsam fir forests to wind damage. For Ecol Manage 204:35–50

Adams PR, Beadle CL, Mendham NJ, Smethurst PJ (2003) The impact of timing and duration of grass control on growth of a young *Eucalyptus globulus* Labill. plantation. New For 26:147–165

Adler A, Verwijst T, Aronsson P (2005) Estimation and relevance of bark proportion in a willow stand. Biomass Bioenergy 29:102–113

Akashi N, Terazawa K (2005) Bark stripping damage to conifer plantations in relation to the abundance of sika deer in Hokkaido, Japan. For Ecol Manage 208:77–83

Albert DJ, Fry G, Poole BR (1980) An industrial company's view of nursery stock quality. N Z J For Sci 10:2–11

Allard RW (1960) Principles of plant breeding. Wiley, New York

Allen HL, Wentworth TR (1993) Vegetation control and site preparation affect patterns of shoot elongation for 3-year-old loblolly pine. Can J For Res 23:2110–2115

Allen HL, Dougherty PM, Campbell RG (1990) Manipulation of water and nutrients—practice and opportunity in southern U.S. pine forests. For Ecol Manage 30:437–453

Almeida AC, Landsberg JJ, Sands PJ, Ambrogi MS, Fonseca S, Barddal SM, Bertolucci, FL (2004) Needs and opportunities for using a process-based productivity model as a practical tool in *Eucalyptus* plantations. For Ecol Managet 193:167–177

Amateis RL, Radtke PJ, Burkhart HE (1996) Growth and yield of thinned and unthinned plantations. J For 94(12):19–23

Amponsah IG, Lieffers VJ, Comeau PG, Brockley RP (2004) Growth response and sapwood hydraulic properties of young lodgepole pine following repeated fertilization. Tree Physiol 24:1099–1108

Ancelin P, Courbaud B, Fourcaud T (2004) Development of an individual tree-based mechanical model to predict wind damage within forest stands. For Ecol Manage 203:101–121

Andersen L (2004) Field performance of *Quercus petraea* seedlings grown under competitive conditions: influence of prior undercutting in the seedbed. New For 28:37–47

Andersen RS, Towers W, Smith P (2005) Assessing the potential for biomass energy to contribute to Scotland's renewable energy needs. Biomass Bioenergy 29:73–82

Annapurna D, Rathore TS, Joshi G (2004) Effect of container type and size on the growth and quality of seedlings of Indian sandalwood (*Santalum album* L.). Aust For 67:82–87

Annapurna D, Rathore TS, Joshi G (2005) Refinement of potting medium ingredients for production of high quality seedlings of sandalwood (*Santalum album* L.). Aust For 68:44–49

Anta MB, González JGA (2005) Development of a stand density management diagram for even-aged pedunculate oak stands and its use in designing thinning schedules. Forestry 78:209–216

Anttonen S, Manninen A-M, Saranpää P, Kainulainen P, Linder S, Vapaavuori E (2002) Effects of long-term nutrient optimisation on stem wood chemistry in *Picea abies*. Trees 16:386–394

Aphalo PJ, Ballaré CL, Scopel AL (1999) Plant-plant signalling, the shade-avoidance response and competition. J Exp Bot 50:1629–1634

Applegate GB, Bragg AL (1989) Improved growth rates of red cedar (*Toona australis* (F. Muell.) Harms) seedlings in growtubes in north Queensland. Aust For 52:293–297

Ares A (1993) Application of multivariate analysis to site quality evaluation for coniferous plantations. S Afr For J 167:27–34

Ares A, St Louis D, Brauer,D. (2003) Trends in tree growth and understorey yield in silvopastoral practices with southern pines. Agrofor Syst 59:27–33

Arnold R, Bush D, Stackpoole D (2005) Genetic variation and tree improvement. In: Nambiar S, Ferguson I (eds) New forests. CSIRO, Melbourne, pp 25–49

Arthaud GJ, Klemperer WD (1988) Optimizing high and low thinning in loblolly pine with dynamic programming. Can J For Res 18:1118–1122

Araújo J, Almeida MH, Ramos A, Lemos L (1995) Influence of paclobutrazol in *Eucalyptus globulus* Labill. seed orchards. In: Potts BM, Borralho NMG, Reid JB, Cromer RN, Tibbits WN, Raymond CA (eds) Eucalypt plantations: improving fibre yield and quality. Proceedings CRCTHF-IUFRO conference, Hobart, 19–24 February. Cooperative Research Centre for Temperate Hardwood Forestry, Hobart, pp 311–312

Asaro C, Loeb SC, Hanula JL (2003) Cone consumption by southeastern fox squirrels: a potential basis for clonal preferences in a loblolly and slash pine seed orchard. For Ecol Manage 186:185–195

Assis T, Warburton P, Harwood C (2005) Artificially induced protogyny: an advance in the controlled pollination of *Eucalyptus*. Aust For 68:27–33

Attiwill PM, Adams MA (eds) (1996) Nutrition of eucalypts. CSIRO, Melbourne

Attiwill PM, Turvey ND, Adams MA (1985) Effects of mound cultivation (bedding) on concentration and conservation of nutrients in a sandy podzol. For Ecol Manage 11:97–110

Atwell BJ, Kriedemann PE, Turnbull CGN (eds) (1999) Plants in action. Macmillan, Melbourne

Augusto L, Ranger J, Ponette Q, Rapp M (2000) Relationships between forest tree species, stand production and stand nutrient amount. Ann For Sci 57:313–324

Aussenac G, Granier A (1988) Effects of thinning on water stress and growth in Douglas-fir. Can J For Res 18:100–105

Aussenac G, Granier A, Naud R (1982) Influence d'une éclaircie sur la croissance et le bilan hydrique d'un jeune peuplement de Douglas (*Pseudotsuga menziesii* (Mirb.) Franco). Can J For Res 12:222–231

Austin MT, Brewbaker JL, Wheeler R, Fownes JH (1997) Short-rotation biomass trial of mixed and pure stands of nitrogen-fixing trees and *Eucalyptus grandis*. Aust For 60:161–168

Avery TE, Burkhart HE (2002) Forest measurements, 5th edn. McGraw-Hill, Boston

Bachelard EP (1969) Studies on the formation of epicormic shoots on eucalypt stem segments. Aust J Biol Sci 2:1291–1296

Bacon GJ, Bachelard EP (1978) The influence of nursery conditioning treatments on some physiological responses of recently transplanted seedlings of *Pinus caribea* Mor. var. *hondurensis* B. & G. Aust For Res 8:171–183

Bacon GJ, Hawkins PJ (1977) Studies on the establishment of open root Caribbean pine planting stock in southern Queensland. Aust For 40:173–191

Baddeley JA, Watson CA (2004) Seasonal patterns of fine-root production and mortality in *Prunus avium* in Scotland. Can J For Res 34:1534–1537

Bailey JD, Harjanto NA (2005) Teak (*Tectona grandis* L.) tree growth, stem quality and health in coppiced plantations in Java, Indonesia. New For 30:55–65
Baird IA, Ive JR (1989) Using the LUPLAN land-use planning package to implement the Recreation Opportunity System Spectrum approach to park management planning. J Environ Manage 27:249–262
Baker T, Duncan M, Stackpole D (2005) Growth and silvicultural management of irrigated plantations. In: Nambiar S, Ferguson I (eds) New forests. CSIRO, Melbourne, pp 113–156
Balandier P, Dupraz C (1999) Growth of widely spaced trees. A case study from young agroforestry plantations in France. Agrofor Syst 43:151–167
Baldwin VC, Feduccia DP, Haywood JD (1989) Postthinning growth and yield of row-thinned and selectively thinned loblolly and slash pine plantations. Can J For Res 19:247–256
Baldwin VC, Peterson KD, Clark A, Ferguson RB, Strub MR, Bower DR (2000) The effects of spacing and thinning on stand and tree characteristics of 38-year-old loblolly pine. For Ecol Manage 137:91–102
Ball JB, Wormald TJ, Russo L (1995) Experience with mixed and single species plantations. Commonw For Rev 74:301–305
Balneaves JM (1988) Packaging and cool-storage effects on growth of *Cupressus macrocarpa* seedlings. N Z J For Sci 18:297–303
Balneaves JM, De La Mare PJ (1989) Root patterns of *Pinus radiata* on five ripping treatments in a Canterbury forest. N Z J For Sci 19:29–40
Bamber RK, Humphreys FR (1965) Variations in sapwood starch levels in some Australian forest species. Aust For 29:15–23
Barbour R, Bailey RE, Cook JA (1992) Evaluation of relative density, diameter growth, and stem form in a red spruce (*Picea rubens*) following thinning. Can J For Res 22:229–238
Bargali SS, Singh RP, Singh SP (1992) Structure and function of an age series of eucalypt plantations in central Himalaya. II. Nutrient dynamics. Ann Bot 69:413–421
Barnett JL, How RA, Humphreys WF (1977) Possum damage to pine plantations in north-eastern New South Wales. Aust For Res 7:185–195
Barry KM, Hall MF, Mohammed CL (2005) The effect of time and site on incidence and spread of pruning-related decay in plantation-grown *Eucalyptus nitens*. Can J For Res 35:495–502
Bartelink HH, Kramer K, Mohren GMJ (1997) Applicability of the radiation-use efficiency concept for simulating growth of forest stands. For Ecol Manage 88:169–179
Bassett OD, White G (2001) Review of the impact of retained overwood trees on stand productivity. Aust For 64:57–63
Battaglia M, Sands PJ (1997) Modelling site productivity of *Eucalyptus globulus* in response to climatic and site factors. Aust J Plant Physiol 24:831–850
Battaglia M, Sands PJ (1998) Process-based forest productivity models and their application in forest management. For Ecol Manage 102:13–32
Battaglia M, Mummery D, Smith A (2002) Economic analysis of site survey and productivity modelling for the selection of plantation areas. For Ecol Manage. 162:185–195
Battaglia M, Sands P, White D, Mummery D (2004) CABALA: a linked carbon, water and nitrogen model of forest growth for silvicultural decision support. For Ecol Manage 193:251–282
Bauhus J, Messier C (1999) Soil exploitation strategies of fine roots in different tree species of the southern boreal forests of eastern Canada. Can J For Res 29:260–273
Bauhus J, Khanna PK, Menden N (2000) Aboveground and belowground interactions in mixed plantations of *Eucalyptus globulus* and *Acacia mearnsii*. Can J For Res 30:1886–1894

Bauhus J, van Winden AP, Nicotra AB (2004) Aboveground interactions and productivity in mixed-species plantations of *Acacia mearnsii* and *Eucalyptus globulus*. Can J For Res 34:686–694

Beadle CL (1997) Dynamics of leaf and canopy development. In: Nambiar EKS, Brown AG (eds) Management of soil, nutrients and water in tropical plantation forests. Monograph no 43. Australian Centre for International Agricultural Research, Canberra, pp 169–212

Beadle CL, Turnbull CRA (1986) Leaf and branch development in two contrasting species of *Eucalyptus* in relation to early growth and biomass production. In: Fujimori T, Whitehead D (eds) Crown and canopy structure in relation to productivity. Forestry and Forest Products Research Institute, Ibaraki, Japan, pp 263–283

Beadle CL, McLeod DE, Turnbull CRA, Ratkowsky DA, McLeod R (1989) Juvenile/total foliage ratios in *Eucalyptus nitens* and growth of stands and individual trees. Trees 3:117–124

Beadle CL, Honeysett JL, Turnbull CRA, White DA (1995) Site limits to achieving genetic potential. In: Potts BM, Borralho NMG, Reid JB, Cromer RN, Tibbits WN, Raymond CA (eds) Eucalypt plantations: improving fibre yield and quality. Proceedings CRCTHF-IUFRO conference, Hobart, 19–24 February. Cooperative Research Centre for Temperate Hardwood Forestry, Hobart, pp 325–330

Beets PN, Madgwick HAI (1988) Above-ground dry matter and nutrient content of *Pinus radiata* as affected by lupin, fertiliser, thinning, and stand age. N Z J For Sci 18:43–64

Beets PN, Pollock DS (1987) Uptake and accumulation of nitrogen in *Pinus radiata* stands as related to age and thinning. N Z J For Sci 17:353–371

Bekunda MA, Smethurst PJ, Khanna PK, Willett IR (1990) Effects of post-harvest residue management on labile soil phosphorus in a *Pinus radiata* plantation. For Ecol Manage 38:13–25

Benson AD, Shepherd KR (1976) Effect of nursery practice on *Pinus radiata* seedling characteristics and field performance: I. Nursery seedbed density. N Z J For Sci 6:19–26

Benson AD, Shepherd KR (1977) Effects of nursery practice on *Pinus radiata* seedling characteristics and field performance: II. Nursery root wrenching. N Z J For Sci 7:68–76

Bergeron J-M, Tardif J (1988) Winter browsing preferences of snowshoe hares for coniferous seedlings and its implication in large-scale reforestation programs. Can J For Res 18:280–282

Bergkvist P, Ledin S (1998) Stem biomass yields at different planting designs and spacings in willow coppice system. Biomass Bioenergy 14:149–156

Bernier PY (1993) Comparing natural and planted black spruce seedlings. I. Water relations and growth. Can J For Res 23:2427–2434

Betters DR, Steinkamp EA, Turner MT (1991) Singular path solutions and optimal rates for thinning even-aged forest stands. For Sci 37:1632–1640

Bi H, Turvey ND (1994) Effects of *Eucalyptus obliqua* (L'Herit) density on young stands of even-aged *Pinus radiata* (D.Don). New For 8:25–42

Binkley D (2004) A hypothesis about the interaction of tree dominance and stand production through stand development. For Ecol Manage 190:265–271

Binkley D, Giardina C (1997) Nitrogen fixation in tropical forest plantations. In: Nambiar EKS, Brown AG (eds) Management of soil, nutrients and water in tropical plantation forests. Monograph no 43. Australian Centre for International Agricultural Research, Canberra, pp 297–337

Binkley D, O'Connell AM, Sankaran KV (1997) Stand development and productivity. In: Nambiar EKS, Brown AG (eds) Management of soil, nutrients and water in tropical plantation forests. Monograph no 43. Australian Centre for International Agricultural Research, Canberra, pp 419–442

Birk EM (1993) Biomass and nutrient distribution in radiata pine in relation to previous land use. II Nutrient accumulation, distribution and removal. Aust For 56:148–156

Birk EM, Turner J (1992) Response of flooded gum (*E. grandis*) to intensive cultural treatment: biomass and nutrient content of eucalypt plantations and native forests. For Ecol Manage 47:1–28

Bishaw B, DeBell DS, Harrington CA (2003) Patterns of survival, damage, and growth for western white pine in a 16-year-old spacing trial in western Washington. West J Appl For 18:35–43

Black TA, Tan CS, Nnyamah JU (1980) Transpiration rates of Douglas fir trees in thinned and unthinned stands. Can J Soil Sci 60:625–631

Blake TJ (1983) Coppice systems for short-rotation intensive forestry: the influence of cultural, seasonal and plant factors. Aust For Res 13:279–291

Blakemore P (2004) Density and shrinkage of four low-rainfall plantation-grown eucalypts. Aust For 67:152–155

Blazier MA, Hennessey TC, Deng S (2005) Effects of fertilization and vegetation control on microbial biomass carbon and dehydrogenase activity in a juvenile loblolly pine plantation. For Sci 51:449–459

Blumfield TJ, Xu ZH, Chen C (2005) Mineral nitrogen dynamics following soil compaction and cultivation during hoop pine plantation establishment. For Ecol Manage 204:129–135

Bogeat-Triboulot,M-B. Bartoli F, Garbaye J, Marmeisse R, Tagu D (2004) Fungal ectomycorrhizal community and drought affect root hydraulic properties and soil adherence to roots of *Pinus pinaster* seedlings. Plant Soil 267:213–223

Bonham KJ, Mesibov R, Bashford R (2002) Diversity and abundance of some ground-dwelling invertebrates in plantation vs. native forests in Tasmania, Australia. For Ecol Manage 158:237–247

Boomsma DB, Cellier KM, McGuire DO, Nethercott K, Sedgley J, Nambiar EKS (1997) Omission of nitrogen fertilizer on second rotation sites: effects on *Pinus radiata* growth in southern Australia. Aust For 60:240–250

Booth T (2005) Environment, species selection and productivity prediction. In: Nambiar S, Ferguson I (eds) New forests. CSIRO, Melbourne, pp 5–23

Booth TH (1985) A new method for assisting species selection. Commonw For Rev 64:241–250

Booth TH (1991) Where in the world? New climatic analysis methods to assist species and provenance selection for trials. Unasylva 165:51–57

Booth TH, Jones PG (1998) Identifying climatically suitable areas for growing particular trees in Latin America. For Ecol Manage 108:167–173

Booth TH, Stein JA, Nix HA, Hutchinson MF (1989) Mapping regions climatically suitable for particular species: an example using Africa. For Ecol Manage 28:19–31

Booth TH, Jovanovic T, New M (2002) A new world climatic mapping program to assist species selection. For Ecol Manage 163:111–117

Bootle KR (1983) Wood in Australia. McGraw-Hill, Sydney

Bouriaud O, Bréda N, Le Moguédec G, Nepveu G (2004) Modelling variability of wood density in beech as affected by ring age, radial growth and climate. Trees 18:264–276

Bowyer JL, Shmulsky R, Haygreen JC (2003) Forest products and wood science: an introduction, 4th edn. Iowa State University Press, Ames

Boyle TJB, Cossalter C, Griffin AR (1997) Genetic resources for plantation forestry. In: Nambiar EKS, Brown AG (eds) Management of soil, nutrients and water in tropical plantation forests. Monograph no 43. Australian Centre for International Agricultural Research, Canberra, pp 25–63

Bradshaw RE (2004) Dothistroma (red-band) needle blight of pines and the dothistromin toxin: a review. For Pathol 34:163–185

Bradshaw RE, Ganley RJ, Jones WT, Dyer PS (2000) High levels of dothistromin toxin produced by the forest pathogen *Dothistroma pini*. Mycol Res 104:325–332

Bradshaw RE, Bhatnagar D, Ganley RJ, Gillman CJ, Monahan BJ, Seconi JM (2002) *Dothistroma pini*, a forest pathogen, contains homologs of aflatoxin biosynthetic pathway genes. Appl Environ Microbiol 68:2885–2892

Bradstock R (1981) Biomass in an age series of *Eucalyptus grandis* plantations. Aust For Res 11:111–127

Brady NC, Weil RR (2001) The nature and properties of soil, 13th edn. MacMillan, New York

Brandenburg W (1993) The coppice. The most productive, sustainable and attractive of all tree management methods. Growing Today June:28–37

Brandtberg P-E, Lundkvist H (2004) Does an admixture of Betula species in *Picea abies* stands increase organic matter quality and nitrogen release? Scand J For Res 19:127–141

Bravo-Oviedo A, Montero G (2005) Site index in relation to edaphic variables in stone pine (*Pinus pinea* L.) stands in south west Spain. Ann For Sci 62:61–72

Bredenkamp BV (1984) Row thinnings do not adversely affect yields or form of the final crop in improved *Pinus taeda*. S Afr For J 131:28–33

Briggs DG (1995) Pruning in relation to forest inventory, wood quality, and products. In: Hanley DP, Oliver CD, Maguire DA, Briggs DG, Fight RD (eds) Forest pruning and wood quality. Contribution no 77, Institute of Forest Resources, University of Washington, Seattle, pp 21–35

Brix H, Mitchell AK (1986) Thinning and nitrogen fertilization effects on soil and tree water stress in a Douglas-fir stand. Can J For Res 16:1334–1338

Brown BN (2000) Management of disease during eucalypt propagation. In: Keane PJ, Kile GA, Podger FD, Brown BN (eds) Diseases and pathogens of eucalypts. CSIRO, Melbourne, pp 487–517

Bruijnzeel LA (1997) Hydrology of plantation forests in the tropics. In: Nambiar EKS, Brown AG (eds) Management of soil, nutrients and water in tropical plantation forests. Monograph no 43. Australian Centre for International Agricultural Research, Canberra, pp 125–167

Brundett M, Malajczuk N, Mingqin G, Daping X, Snelling S, Dell B (2005) Nursery inoculation of *Eucalyptus* seedlings in Western Australia and southern China using spores and mycelial inoculum of diverse ectomycorrhizal fungi from different climatic regions. For Ecol Manage 209:193–205

Bubb KA, Xu ZH, Simpson JA, Saffigna PG (1998) Some nutrient dynamics associated with litterfall and litter decomposition in hoop pine plantations of southeast Queensland, Australia. For Ecol Manage 110:343–352

Bubb KA, Yu B, Cakurs U, Costantini A (2000) Impacts of site preparation techniques on runoff, soil and nitrogen losses during the establishment phase in hoop pine plantations of southeast Queensland. Aust For 63:241–247

Bubb KA, Frayne PF, Wittmer TR, Grimmett JL (2003) Efficacy of atrazine and simazine applications over harvest residue in Queensland's subtropical softwood plantations. Aust For 66:102–107

Bulinski J, McArthur C (2000) Spatial distribution of browsing damage and mammalian herbivores in Tasmanian eucalypt plantations. Aust For 63:27–33

Bulinski J, McArthur C (2003) Identifying factors related to the severity of mammalian browsing damage in eucalypt plantations. For Ecol Manage 183:239–247

Bullard SH, Sherali HD, Klemperer WD (1985) Estimating optimal thinning and rotation for mixed-species timber stands using a random search algorithm. For Sci 31:303–315

Bunce HWF, McLean JA (1990) Hurricane Gilbert's impact on the natural forests and *Pinus caribaea* plantations of Jamaica. Commonw For Rev 69:147–155

Burdett AN (1979) New methods for measuring root growth capacity: their value in assessing lodgepole pine stock quality. Can J For Res 9:63–67

Burdett AN (1983) Quality control in the production of forest planting stock. For Chron 59:132–138

Burdett AN (1990) Physiological processes in plantation establishment and the development of specifications for forest planting stock. Can J For Res 20:415–27

Burdett AN, Herring LJ, Thompson CF (1984) Early growth of planted spruce. Can J For Res 14:644–651

Burdon R, Richardson T (2000) Seeing the wood from the genes. Aust Sci 21(3):23–25

Burdon RD, Kibblewhite RP, Walker JCF, Megraw RA, Evans R, Cown DJ (2004) Juvenile versus mature wood: a new concept, orthogonal to corewood versus outerwood, with special reference to *Pinus radiata* and *P. taeda*. For Sci 50:399–415

Burgess D (1991) Western hemlock and Douglas-fir seedling development with exponential rates of nutrient addition. For Sci 37:54–67

Burley J (2001) Genetics in sustainable forestry: the challenges for forest genetics and tree breeding in the new millennium. Can J For Res 31:561–565

Burley J, Kanowski PJ (2005) Breeding strategies for temperate hardwoods. Forestry 78:199–208

Bygrave FL, Bygrave PL (1998) *Cedrela* species are attacked by the tip moth *Hypsipila robusta* when grafted on to red cedar *Toona ciliata*. Aust For 61:45–47

Bygrave FL, Bygrave PL (2001) Host preference of the Meliaceae shootborer *Hypsipila*: further information from grafting *Cedrela odorata* and *Cedrela fissilis* on *Toona ciliata* (Australian red cedar). Aust For 64:216–219

Byington EK (1990) Agroforestry in the temperate zone. In: MacDicken KG Vergara NT (eds) Agroforestry: classification and management. Wiley, New York, pp 228–289

Cameron AD (2002) Importance of early selective thinning in the development of long-term stand stability and improved log quality: a review. Forestry 75:25–35

Campbell KG (1998) Observations on red cedar and tip moth. Aust For 61:40–44

Campbell RA, Howard CA (1993) Priorities for forestry herbicide application technology research. Can J For Res 23:2204–2212

Candy SG, Elliott HJ, Bashford R, Greener A (1992) Modelling the impact of defoliation by the leaf beetle, *Chrysophtharta bimaculata* (Coleoptera: Chrysomelidae), on height growth of *Eucalyptus regnans*. For Ecol Manage 54:69–87

Cañellas I, Del Río M, Roig S, Montero G (2004) Growth response to thinning in *Quercus pyrenaica* Willd. coppice stands in Spanish central mountain. Ann For Sci 61:243–250

Cannell MGR, Smith RI (1980) Yields of minirotation closely spaced hardwoods in temperate regions: review and appraisal. For Sci 26:415–428

Cao QV, Dean TJ, Baldwin VC (2000) Modeling the size-density relationship in direct-seeded slash pine stands. For Sci 46:317–321

Carlson CE (1994) Germination and early growth of western larch (*Larix occidentalis*), alpine larch (*Larix lyallii*), and their reciprocal hybrids. Can J For Res 24:911–916

Carlson WC, Preisig CL, Promnitz LC (1980) Comparative root system morphologies of seeded-in-place, bareroot, and container-cultured Sitka spruce seedlings after outplanting. Can J For Res 10:250–256

Carlyle JC (1995) Nutrient management in a *Pinus radiata* plantation after thinning: the effect of nitrogen fertilizer on soil nitrogen fluxes and tree growth. Can J For Res 25:1673–1683

Carlyle JC (1998) Relationships between nitrogen uptake, leaf area, water status and growth in an 11-year-old *Pinus radiata* plantation in response to thinning, thinning residue, and nitrogen fertiliser. For Ecol Manage 108:41–55

Carlyle JC, Nambiar EKS (2001) Relationships between net nitrogen mineralization, properties of the forest floor and mineral soil, and wood production in *Pinus radiata* plantations. Can J For Res 31:889–898

Carlyle JC, Bligh MW, Nambiar EKS (1998) Woody residue management to reduce nitrogen and phosphorus leaching from sand soil after clear-felling *Pinus radiata* plantations. Can J For Res 28:1222–1232

Carson MJ, Inglis CS (1988) Genotype and location effects on internode length of *Pinus radiata* in New Zealand. N Z J For Sci 18:267–279

Carson SD (1989) Selecting *Pinus radiata* for resistance to Dothistroma needle blight. N Z J For Sci 19:3–21

Carter R, Klinka K (1992) Use of ecological site classification in the prediction of forest productivity and response to fertilisation. S Afr For J 160:19–23

Carter WG (1974) Growing and harvesting eucalypts on short rotations for pulping. Aust For 36:214–225

Caulfield JP, Schönau APG, Donald DGM (1993) Making eucalyptus species-site selection decisions using stochastic dominance analysis. For Ecol Manage 56:147–162

Charman PEV, Murphy BW (eds) (2001) Soils: their properties and management, 2nd edn. Oxford University Press, Melbourne

Chauhan SS, Walker J (2004) Relationship between longitudinal growth strain and some wood properties in *Eucalyptus nitens*. Aust For 67:254–260

Chen CM, Rose DW, Leary RA (1980) Derivation of optimal stand density over time—a discrete stage, continuous state dynamic programming solution. For Sci 26:217–227

Chen CR, Xu ZH (2005) Soil carbon and nitrogen pools and microbial properties in a 6-year-old slash pine plantation of subtropical Australia: impacts of harvest residue management. For Ecol Manage 206:237–247

Clarke B (1975) Establishment of eucalypt plantations Coffs Harbour, N.S.W. Sci Technol 12:10–13

Clason TR (1993) Hardwood competition reduces loblolly pine plantation productivity. Can J For Res 23:2133–2140

Clay DV, Dixon FL, Willoughby I (2005) Natural products as herbicides for tree establishment. Forestry 78:1–9

Clearwater MJ, Meinzer FC (2001) Relationships between hydraulic architecture and leaf photosynthetic capacity in nitrogen-fertilized *Eucalyptus grandis* trees. Tree Physiol 21:683–690

Close DC, Beadle CL (2004) Total, and chemical fractions, of nitrogen and phosphorus in *Eucalyptus* seedling leaves: effects of species, nursery fertiliser management and transplanting. Plant Soil 259:85–95

Close DC, Beadle CL, Holz GK, Ravenwood IC (1999) A photobleaching event at the North Forest Products' Somerset nursery reduces growth of *Eucalyptus globulus* seedlings. Tasforests 11:59–67

Close DC, Beadle CL, Brown PH, Holz GK (2000) Cold-induced photoinhibition affects establishment of *Eucalyptus nitens* (Deane and Maiden) Maiden and *Eucalyptus globulus* Labill. Trees 15:32–41

Close DC, Bail I, Beadle CL, Clasen QC (2003) Physical and nutritional characteristics and performance after planting of *Eucalyptus globulus* Labill. seedlings from ten nurseries: implications for seedling specifications. Aust For 66:145–152

Close DC, Beadle CL, Brown PH (2005) The physiological basis of containerised tree seedling 'transplant shock': a review. Aust For 68:112–120

Coates W (2005) Tree species selection for a mine tailings bioremediation project in Peru. Biomass Bioenergy 28: 418–423

Cole EC, Newton M (1987) Fifth-year response of Douglas-fir to crowding and nonconiferous competition. Can J For Res 17:181–186

Connell MJ, Raison RJ, Jenkins P (2004) Effects of thinning and coppice control on stand productivity and structure in a silvertop ash (*Eucalyptus sieberi* L. Johnson) forest. Aust For 67:30–38

Coops N, Stanford M, Old K, Dudzinski M, Culvenor D, Stone C (2003) Assessment of dothistroma needle blight of *Pinus radiata* using airborne hyperspectral imagery. Phytopathology 93:1524–1532

Cornelius JP, Watt AD (2003) Genetic variation in a *Hypsipyla*-attacked clonal trial of *Cedrela odorata* under two pruning regimes. For Ecol Manage 183:341–349

Corona P, Scotti R, Tarchiani N (1998) Relationship between environmental factors and site index in Douglas-fir plantations in central Italy. For Ecol Manage 110:195–207

Costantini A, Doley D (2001a) Management of compaction during the harvest of *Pinus* plantations in Queensland: I. Policy considerations for controlling machine activity. Aust For 64:181–185

Costantini A, Doley D (2001b) Management of compaction during the harvest of *Pinus* plantations in Queensland: II. Preliminary evaluation of compaction effects on productivity. Aust For 64:186–192

Costantini A, Doley D (2001c) Management of compaction during the harvest of *Pinus* plantations in Queensland: III. Preliminary investigation of the potential for selected soil parameters to predict rut compaction. Aust For 64:193–198

Costantini A, Dunn GM, Grimmett JL (1997a) Towards sustainable management of forest plantations in south-east Queensland. II: Protecting soil and water values during second rotation *Pinus* plantation management. Aust For 60:226–232

Costantini A, Grimmett JL, Dunn GM (1997b) Towards sustainable management of forest plantations in south-east Queensland. I: Logging and understorey residue management between rotations in steep country *Araucaria cunninghamii* plantations. Aus For 60:218–225

Côté B, Camiré C (1987) Tree growth and nutrient cycling in dense plantings of hybrid poplar and black alder. Can J For Res 17:516–523

Cotterill PP, Dean CA (1990) Successful tree breeding with index selection. CSIRO, Melbourne

Coyle DR, McMillin JD, Hall RB, Hart E (2002) Deployment of tree resistance to insects in short rotation *Populus* plantations. In: Wagner MR, Clancy KM, Lieutier F, Paine TD (eds) Mechanisms and deployment of resistance in trees to insects. Kluwer, Dordrecht, pp 189–215

Crane WJB (1981) Growth following fertilisation of thinned *Pinus radiata* stands near Canberra in south-eastern Australia. Aust For 44:14–25

Cregg BM, Dougherty PM, Hennessey TC (1988) Growth and wood quality of young loblolly pine trees in relation to stand density and climatic factors. Can J For Res 18:851–858

Cregg BM, Hennessey TC, Dougherty PM (1990) Water relations of loblolly pine trees in southeastern Oklahoma following precommercial thinning. Can J For Res 20:1508–1513

Cremer KW (1973) Ability of *Eucalyptus regnans* and associated evergreen hardwoods to recover from cutting or complete defoliation in different seasons. Aust For Res 6:9–22

Cremer KW, Meredith EM (1976) Growth of radiata pine after row thinning compared with selection thinning. Aust For 39:193–200

Cremer KW, Myers BJ, van der Duys F, Craig IE (1977) Silvicultural lessons from the 1974 windthrow in radiata pine plantations near Canberra. Aust For 40:274–292

Cremer KW, Borough CJ, McKinnell FH, Carter PR (1982) Effects of stocking and thinning on wind damage in plantations. N Z J For Sci 12:244–268

Cremer KW, Cromer RN, Florence RG (1984) Stand establishment. In: Hillis WE, Brown AG (eds) Eucalypts for wood production. CSIRO, Academic, New York, pp 81–135

Cromer RN (1996) Silviculture of eucalypt plantations in Australia. In: Attiwill PM, Adams MA (eds) Nutrition of eucalypts, CSIRO, Melbourne, pp 259–273

Cromer RN, Jarvis PG (1990) Growth and biomass partitioning in *Eucalyptus grandis* seedlings in response to nitrogen supply. Aust J Plant Physiol 17:503–515

Cromer RN, Cameron DM, Rance SJ, Ryan PA, Brown M (1993a) Response to nutrients in *Eucalyptus grandis*. 1. Biomass accumulation. For Ecol Manage 62:211–230

Cromer RN, Cameron DM, Rance SJ, Ryan PA, Brown M (1993b) Response to nutrients in *Eucalyptus grandis*. 2. Nitrogen accumulation. For Ecol Manage 62:231–243

Cucchi V, Bert D (2003) Wind-firmness in *Pinus pinaster* Ait. stands in southwest France: influence of stand density, fertilisation and breeding in two experimental stands damaged during the 1999 storm. Ann For Sci 60:209–226

Cucchi V, Meredieu C, Stokes A, de Coligny F, Suarez J, Gardiner BA (2005) Modelling the windthrow risk for simulated forest stands of Maritime pine (*Pinus pinaster* Ait.). For Ecol Manage 213:184–196

Cunningham SA, Floyd RB (2004) Leaf compositional differences predict variation in *Hypsipila robusta* damage to *Toona ciliata* in field trials. Can J For Res 34:642–648

Cunningham SA, Floyd RB, Griffiths MW, Wylie FR (2005) Patterns of host use by the shoot-borer *Hypsipyla robusta* (Pyralidae:Lepidoptera) comparing five Meliaceae tree species in Asia and Australia. For Ecol Manage 205:351–357

Curtis IS (ed) (2005) Transgenic crops of the world. Springer, Berlin Heidelberg New York

Curtis RO, Marshall DD, Bell JF (1997) LOGS: a pioneering example of silvicultural research in coast Douglas-fir. J For 95(7):19–25

Davis AS, Jacobs DF (2005) Quantifying root system quality of nursery seedlings and relationship to outplanting performance. New For 30:295–311

Davis LS, Johnson KN, Bettinger PS, Howard TE (2001) Forest management, 4th edn. McGraw Hill, Boston

Deal RL, Barbour RJ, McClellan MH, Parry DL (2003) Development of epicormic sprouts in Sitka spruce following thinning and pruning in south-east Alaska. Forestry 76:401–412

Dean TJ (2004) Basal area increment and growth efficiency as functions of canopy dynamics and stem mechanics. For Sci 50:106–116

Dean TJ, Jokela EJ (1992) A density-management diagram for slash pine plantations in the lower coastal plain. South J Appl For 16:178–185

Dean TJ, Roberts SD, Gilmore DW, Maguire DA, Long JN, O'Hara KL, Seymour RS (2002) An evaluation of the uniform stress hypothesis based on stem geometry in selected North American conifers. Trees 16:559–568

DeBell DS, Radwan MA (1979) Growth and nitrogen relations of coppiced black cottonwood and red alder in pure and mixed plantings. Bot Gaz 140:S97–S101

DeBell DS, Harrington CA (2002) Density and rectangularity of planting influence 20-year growth and development of red alder. Can J For Res 32:1244–1253

DeBell DS, Cole TG, Whitesell CD (1997) Growth, development, and yield in pure and mixed stands of *Eucalyptus* and *Albizia*. For Sci 43:286–298

Dell B (1996) Diagnosis of nutrient deficiencies in eucalypts. In: Attiwill PM, Adams MA (eds) Nutrition of eucalypts. CSIRO, Melbourne, pp 417–440

Desch HE, Dinwoodie JM (1996) Timber structure, properties, conservation and use, 7th edn. Macmillan, London

Dick AMP (1989) Control of Dothistroma needle blight in the *Pinus radiata* stands of Kinleith forest. N Z J For Sci 19:171–179

Dighton J, Harrison AF (1990) Changes in phosphate status of Sitka-spruce plantations of increasing age, as determined by root bioassy. For Ecol Manage 31:35–44

Dighton J, Jones HE (1992) The use of roots to test N, P and K deficiencies in *Eucalyptus* nutrition. S Afr For J 160:33–37

Dighton J, Jones HE, Poskitt JM (1993) The use of nutrient bioassays to assess the response of *Eucalyptus grandis* to fertilizer application. 1. Interactions between nitrogen, phosphorus, and potassium in seedling nutrition. Can J For Res 23:1–6

Doley D (1978) Effects of shade on gas exchange and growth in seedlings of *Eucalyptus grandis* Hill in response to nitrogen supply. Aust J Plant Physiol 5:723–738

Donald DGM (1987) The application of fertilisers to pines following second thinning. S Afr For J 142:13–16

Donald DGM, Lange PW, Schutz CJ, Morris AR (1987) The application of fertilizers to pines in southern Africa. S Afr For J 141:53–62

Donner BL, Running SW (1986) Water stress response after thinning *Pinus contorta* stands in Montana. For Sci 32:614–625

Downes GM, Hudson IL, Raymond CA, Dean GH, Michell AJ, Schimleck LR, Evans R, Muneri A (1997) Sampling plantation eucalypts for wood and fibre properties. CSIRO, Melbourne

Draper NR, Smith H (1988) Applied regression analysis, 3rd edn. Wiley, New York

Drew TJ, Flewelling JW (1977) Some recent Japanese theories of yield-density relationships and their application to Monterey pine plantations. For Sci 23:517–534

Drew TJ, Flewelling JW (1979) Stand density management: an alternative approach and its application to Douglas-fir plantations. For Sci 25:518–532

Dunbabin V, Rengel Z, Diggle AJ (2004) Simulating form and function of root systems: efficiency of nitrate uptake is dependent on root system architecture and the spatial and temporal variability of nitrate supply. Funct Ecol 18:204–211

Dungey HS, Potts BM (2002) Susceptibility of some *Eucalyptus* species and their hybrids to possum damage. Aust For 65:23–30

Eckersten H, Slapokas T (1990) Modelling nitrogen turnover and production in an irrigated short-rotation forest. For Ecol Manage 50:99–123

Egnell G, Örlander G (1993) Using infrared thermography to assess viability of *Pinus sylvestris* and *Picea abies* seedlings before planting. Can J For Res 23:1737–1743

Eldridge K, Davidson J, Harwood C, van Wyk G (1994) Eucalypt domestication and breeding. Clarendon, Oxford

Eldridge RH, Turner J, Lambert MJ (1981) *Dothistroma* needle blight in a New South Wales *Pinus radiata* plantation in relation to soil types. Aust For 44:42–45

Elliott GS, Mason RW, Ferry DG, Edwards IR (1989) Dothistromin risk assessment for forestry workers. N Z J For Sci 19:163–170

Elliott HJ, Bashford R, Greener A, Candy SG (1992) Integrated pest management of the Tasmanian *Eucalyptus* leaf beetle, *Chrysophtharta bimaculata*, (Olivier) (Coleoptera: Chyrosomelidae). For Ecol Manage 53:29–38

Elliott KJ, Vose JM (1993) Site preparation and burning to improve southern Appalachian pine-hardwood stands: photosynthesis, water relations, and growth of planted *Pinus strobus* during establishment. Can J For Res 23:2278–2285

Encyclopædia Britannica (2004) Deluxe edition 2004 CD. Encyclopædia Britannica, Sydney

Ericsson T, Ingestad T (1988) Nutrition and growth of birch seedlings at varied relative phosphorus addition rates. Physiol Plant 72:227–235

Ericsson T, Kähr M (1993) Growth and nutrition of birch seedlings in relation to potassium supply rate. Trees 7:78–85

Ericsson T, Rytter L, Linder S (1992) Nutritional dynamics and requirements of short rotation forests. In: Mitchell CP, Ford-Robertson JB, Hinckley T, Sennerby-Forsse L (eds) Ecophysiology of short rotation forest crops. Elsevier, London, pp 35–65

Estes BL, Enebak SA, Chappelka AH (2004) Loblolly pine seedling growth after inoculation with plant growth-promoting rhizobacteria and ozone exposure. Can J For Res 34:1410–1416

Evans J (1979) The effects of leaf position and leaf age in foliar analysis of *Gmelina arborea*. Plant Soil 52:547–552

Evans J (1997) Bioenergy plantations—experience and prospects. Biomass Bioenergy 13:189–191

Evans J (1999) Sustainability of forest plantations: a review of evidence and future prospects. Int For Rev 1:153–162

Evans J, Turnbull JW (2004) Plantation forestry in the tropics, 3rd edn. Oxford University Press, Oxford.

Evans R, Stringer S, Kibblewhite RP (2000) Variation of microfibril angle, density and fibre orientation in twenty-nine *Eucalyptus nitens* trees. Appita J 53:450–457

Everham EM (1995) A comparison of methods for quantifying catastrophic wind damage to forests. In: Coutts MP, Grace J (eds) (1995) Wind and trees. Cambridge University Press, Cambridge, pp 340–357

Eyles A, Davies NW, Mohammed C (2003) Wound formation in *Eucalyptus globulus* and *Eucalyptus nitens*: anatomy and chemistry. Can J For Res 33:2331–2339

Fahey TD, Willits SA (1995) Volume and quality of clear wood from pruned trees. In: Hanley DP, Oliver CD, Maguire DA, Briggs DG, Fight RD (eds) Forest pruning and wood quality. Contribution no 77. Institute of Forest Resources, University of Washington, Seattle, pp 115–126

Fahey TJ, Hughes JW (1994) Fine root dynamics in a northern hardwood forest ecosystem, Hubbark Brook Experimental Forest, NH. J Ecol 82:533–548

Falconer DS, Mackay TFC (1996) Introduction to quantitative genetics. Longman, Harlow

FAO (2001) State of the world's forests 2001. Food and Agricultural Organisation of the United Nations, Rome

Fleck S, Niinemets Ü, Cescatti A, Tenhunen JD (2003) Three-dimensional lamina architecture alters light harvesting efficiency in *Fagus*: a leaf-scale analysis. Tree Physiol 23:577–589

Flinn DW, Aeberli BC (1982) Establishment techniques for radiata pine on poorly drained soils deficient in phosphorus. Aust For 45:164–173

Flinn DW, Hopmans P, Farrell PW, James JM (1979) Nutrient losses from the burning of *Pinus radiata* logging residues. Aust For Res 9:17–23

Florence RG (1996) Ecology and silviculture of eucalypt forests. CSIRO, Melbourne

Fölster H, Khanna PK (1997) Dynamics of nutrient supply in plantation soils. In: Nambiar EKS, Brown AG (eds) Management of soil, nutrients and water in tropical plantation forests. Monograph no 43. Australian Centre for International Agricultural Research, Canberra, pp 339–378

Forrester DI, Bauhus J, Khanna PK (2004) Growth dynamics in a mixed-species plantationn of *Eucalyptus globulus* and *Acacia mearnsii*. For Ecol Manage 193:81–95

Foster DR, Boose ER (1995) Hurricane disturbance regimes in temperate and tropical forest ecosystems. In: Coutts MP, Grace J (eds) Wind and trees. Cambridge University Press, Cambridge, pp 305–339

Foster PG, Costantini A (1991a) *Pinus* plantation establishment in Queensland: I. Field surveys for site preparation planning and site design. Aust For 54:75–82

Foster PG, Costantini A (1991b) *Pinus* plantation establishment in Queensland: II. Site preparation classes. Aust For 54:83–89

Foster PG, Costantini A (1991c) *Pinus* plantation establishment in Queensland: III. Site preparation design. Aust For 54:90–94

Fourcaud T, Lac P (2003) Numerical modelling of shape regulation and growth stresses in trees I. An incremental static finite element formulation. Trees 17:23–30

Fourcaud T, Blaise F, Lac P, Castera P, de Reffye P (2003) Numerical modelling of shape regulation and growth stresses in trees II. Implementation in the AMAPpara software and simulation of tree growth. Trees 17:31–39

Fox TR (2000) Sustained productivity in intensively managed forest plantations. For Ecol Manage 138:187–202

Francis PJ, Bacon GJ, Gordon P (1984) Effect of ripping and nitrogen fertiliser on the growth and windfirmness of slash pine on a shallow soil on the Queensland coastal lowlands. Aust For 47:90–94

Franich RA (1988) Chemistry of weathering and solubilisation of copper fungicide and the effect of copper on germination, growth, metabolism, and reproduction of *Dothistroma pini*. N Z J For Sci 18:318–328

Frederick DJ, Madgwick HAI, Jurgensen MF, Oliver GR (1985a) Dry matter, energy, and nutrient contents of 8-year-old stands of *Eucalyptus regnans*, *Acacia dealbata* and *Pinus radiata* in New Zealand. N Z J For Sci 15:142–157

Frederick DJ, Madgwick HAI, Jurgensen MF, Oliver GR (1985b) Dry matter content and nutrient distribution in an age series of *Eucalyptus regnans* plantations in New Zealand. N Z J For Sci 15:158–179

Fredericksen TS, Zedaker SM, Smith DW, Seiler JR, Kreh RE (1993) Interference interactions in experimental pine-hardwood stands. Can J For Res 23:2032–2043

Friend GR (1982) Mammal populations in exotic pine plantations and indigenous eucalypt forests in Gippsland, Victoria. Aust For 45:3–18

Fuwape JA, Akindele SO (1997) Biomass yield and energy value of some fast-growing multipurpose trees in Nigeria. Biomass Bioenergy 12:101–106

Gadgil PD, Bawden AD (1981) Infection of wounds in *Eucalyptus delegatensis*. N Z J For Sci 11:262–270

Gadgil PD, Harris JM (eds) (1980) Planting stock quality. Special issue. N Z J For Sci 10:1–303

Ganjegunte GK, Condron LM, Clinton PW, Davis MR, Mahieu N (2004) Decomposition and nutrient release from radiata pine (*Pinus radiata*) coarse woody debris. For Ecol Manage 187:197–211

Ganley RJ, Bradshaw RE (2001) Rapid identification of polymorphic microsatellite loci in a forest pathogen, *Dothistroma pini*, using anchored PCR. Mycol Res 105:1075–1078

Gapare WJ, Barnes RD, Gwaze DP, Nyoka BI (2003) Genetic improvement of *Eucalyptus grandis* using breeding seedling orchards and the multiple population breeding strategy in Zimbabwe. S Afr For J 197:13–19

Garber SM, Maguire DA (2004) Stand productivity and development in two mixed-species spacing trials in the central Oregon Cascades. For Sci 50:92–105

Gardiner B, Peltola H, Kellomäki S (2000) Comparison of two models for predicting the critical wind speeds required to damage coniferous trees. Ecol Model 129:1–23

Gardiner BA, Quine CP (2000) Management of forests to reduce the risk of abiotic damage—a review with particular reference to the effects of strong winds. For Ecol Manage 135:261–267

Gasana JK, Loewenstein H (1984) Site classification for Maiden's gum, *Eucalyptus globulus* subsp. *maidenii*, in Rwanda. For Ecol Manage 8:107–116

Gerrand A, Keenan RJ, Kanowski P, Stanton,R (2003) Australian forest plantations: an overview of industry, environmental and community issues and benefits. Aust For 66:1–8

Gerrand AM, Neilsen WA (2000) Comparing square and rectangular spacings in *Eucalyptus nitens* using a Scotch plaid design. For Ecol Manage 129:1–6

Gerrand AM, Neilsen WA, Medhurst JL (1997) Thinning and pruning eucalypt plantations for sawlog production in Tasmania. Tasforests 9:15–34

Gholz HL, Fisher RF, Pritchett WL (1985a) Nutrient dynamics in slash pine plantation ecosystems. Ecology 66:647–659

Gholz HL, Perry CS, Cropper WP, Hendry LC (1985b) Litterfall, decomposition, and nitrogen and phosphorus dynamics in a chronosequence of slash pine (*Pinus elliottii*) plantations. For Sci 31:463–478

Ginn SE, Seiler JR, Cazell BH, Kreh RE (1991) Physiological and growth responses of eight-year-old loblolly pine stands to thinning. For Sci 37:1030–1040

Glover GR, Zutter BR (1993) Loblolly pine and mixed hardwood stand dynamics for 27 years following chemical, mechanical, and manual site preparation. Can J For Res 23:2126–2132

Gonçalves JLM, Barros NF, Nambiar EKS, Novais RF (1997) Soil and stand management for short-rotation plantations. In: Nambiar EKS, Brown AG (eds) Management of soil, nutrients and water in tropical plantation forests. Monograph no 43. Australian Centre for International Agricultural Research, Canberra, pp 379–417

Gorham E (1979) Shoot height, weight and standing crop in relation to density of monospecific stands. Nature 279:148–150

Gower ST, McMurtrie RE, Murty D (1996) Above-ground net primary production decline with stand age: potential causes. Trends Ecol Evol 11:378–382

Grayson AJ (ed) (1995) The world's forests: international initiatives since Rio. Commonwealth Forestry Association, Oxford

Greaves BL, Borralho NMG, Raymond CA (1997) Breeding objectives for plantation eucalypts grown for production of kraft pulp. For Sci 43:265–272

Green S (2003) A review of the potential for the use of bioherbicides to control forest weeds in the UK. Forestry 76:285–298

Gregg BM, Dougherty PM, Hennessey TC (1988) Growth and wood quality of young loblolly pine trees in relation to stand density and climatic factors. Can J For Res 18:851–858

Grossnickle SC, Major JE (1994a) Interior spruce seedlings compared with emblings produced from somatic embryogenesis. II. Stock quality assessment prior to planting. Can J For Res 24:1385–1396

Grossnickle SC, Major JE (1994b) Interior spruce seedlings compared with emblings produced from somatic embryogenesis. III. Physiological response and morphological development on a reforestation site. Can J For Res 24:1397–1407

Grossnickle SC, Major JE, Folk RS (1994) Interior spruce seedlings compared with emblings produced from somatic embryogenesis. I. Nursery development, fall acclimation, and over-winter storage. Can J For Res 24:1376–1384

Grove TS, Thomson BD, Malajczuk N (1996) Nutritional physiology of eucalypts: uptake, distribution and utilization. In: Attiwill PM, Adams MA (eds) Nutrition of eucalypts. CSIRO, Melbourne, pp 77–108

Gullan PJ, Cranston PS (1999) The insects. An outline of entomolgy, 2nd edn. Blackwell, Oxford

Guo D, Mou P, Jones RH, Mitchell RJ (2004) Spatio-temporal patterns of soil available nutrients following experimental disturbance in a pine forest. Oecologia 138:613–621

Gupta SD, Ibaraki Y (eds) (2005) Plant tissue culture engineering. Springer, Berlin Heidelberg New York

Hackett C (1991) Mobilising environmental information about lesser-known plants: the value of two neglected levels of description. Agrofor Syst 14:131–143

Hackett C, Vanclay JK (1998) Mobilizing expert knowledge of tree growth with the PLANTGRO and INFER systems. Ecol Model 106:233–246

Hale SE, Levy PE, Gardiner BA (2004) Trade-offs between seedling growth, thinning and stand stability in Sitka spruce stands: a modelling analysis. For Ecol Manage 187:105–115

Hall JE (1997) Canada's Model Forest Program: a participatory approach to sustainable forest management in Canada. Commonw For Rev 76:261–267

Hamelin RC, Hunt RS, Geils BW, Jensen GD, Jacobi V, Lecours N (2000) Barrier to gene flow between eastern and western populations of *Cronartium ribicola* in North America. Phytopathology 90:1073–1078

Hanley TA, Taber RD (1980) Selective plant species inhibition by elk and deer in three conifer communities in western Washington. For Sci 26:97–107

Hansen E, Heilman P, Strobl S (1992) Clonal testing and selection for field plantations. In: Mitchell CP, Ford-Robertson JB, Hinckley T, Sennerby-Forsse L (eds) Ecophysiology of short rotation forest crops. Elsevier, London, pp 124–145

Hansen EA, Dawson JO (1982) Effect of *Alnus glutinosa* on hybrid *Populus* height growth in a short-rotation intensively cultured plantation. For Sci 28:49–59

Hansen LW, Ravn HP, Geldmann J (2005) Within- and between-stand distribution of attacks by pine weevil [*Hylobius abietis* (L.)] Scand J For Res 20:122–129

Hansson L (1985) Damage by wildlife, especially small rodents, to North American *Pinus contorta* provenances introduced into Sweden. Can J For Res 15:1167–1171

Hara T (1986a) Growth of individuals in plant populations. Ann Bot 57:55–68

Hara T (1986b) Effects of density and extinction coefficient on size variability in plant populations. Ann Bot 57:885–892

Harcourt R, Kyozuka J, Zhu X, Southerton S, Llewellyn D, Dennis E, Peacock J (1995) Genetic engineering for sterility in temperate plantation eucalypts. In: Potts BM, Borralho NMG, Reid JB, Cromer RN, Tibbits WN, Raymond CA (eds) Eucalypt plantations: improving fibre yield and quality. Proceedings CRCTHF-IUFRO conference, Hobart, 19–24 February. Cooperative Research Centre for Temperate Hardwood Forestry, Hobart, pp 403–405

Harlow BA, Duursma RA, Marshall JD (2005) Leaf longevity of western red cedar (*Thuja plicata*) increases with depth in the canopy. Tree Physiol 25:557–562

Harrington CA, Brissette C, Carlson WC (1989) Root system structure in planted and seeded loblolly and shortleaf pine. For Sci 35:469–480

Harrington TB, Edwards MB (1999) Understorey vegetation, resource availability, and litterfall responses to pine thinning and woody vegetation control in longleaf pine plantations. Can J For Res 29:1055–1064

Harrison AF, Dighton J, Jones HJ (1992) Application of root bioassays to detect nutrient deficiencies in fast-growing trees and agroforestry crops. In: Calder IR, Hall RL, Adlard PG (eds) Growth and water use of forest plantations. Wiley, Chichester, pp 161–163

Hartmann HT, Kester DE, Davies FT, Geneve RL (2001) Plant propagation: principles and practices, 7th edn. Prentice Hall, Englewood Cliffs

Hasselgren K (1998) Use of municipal waste products in energy forestry: highlights from 15 years of experience. Biomass Bioenergy 15:71–74

Hawke MF (1991) Pasture production and animal performance under pine agroforestry in New Zealand. For Ecol Manage 45:109–118

Hawkins BJ, Burgess D, Mitchell AK (2005) Growth and nutrient dynamics of western hemlock with conventional or exponential greenhouse fertilization and planting in different fertility conditions. Can J For Res 35:1002–1016

Hawkins RJ (1992) The response of *Chamaecyparis nootkatensis* stecklings to seven nutrient regimes. Can J For Res 22:647–653

Haywood JD (2005) Effects of herbaceous and woody plant control on *Pinus palustris* growth and foliar nutrients through six growing seasons. For Ecol Manage 214:384–397

Heilman PE, Ekuan G, Fogle D (1994) Above- and below-ground biomass and fine roots of 4-year-old hybrids of *Populus trichocarpa* × *Populus deltoides* and parental species in short-rotation culture. Can J For Res 24:1186–1192

Helenius P (2005) Effect of thawing regime on growth and mortality of frozen-stored Norway spruce container seedlings planted in cold and warm soil. New For 29:33–41

Helenius P, Luoranen J, Rikala R (2005) Effect of preplanting drought on survival, growth and xylem water potential of actively growing *Picea abies* container seedlings. Scand J For Res 20:103–109

Hennessey TC, Dougherty PM, Lynch TB, Wittwer RF, Lorenzi EM (2004) Long-term growth and ecophysiological responses of a southeastern Oklahoma loblolly pine plantation to early rotation thinning. For Ecol Manage 192:97–116

Henskens FL, Battaglia M, Cherry ML, Beadle CL (2001) Physiological basis of spacing effects on tree growth and form in *Eucalyptus globulus*. Trees 15:365–377

Herbert MA (1996) Fertilizers and eucalypt plantations in South Africa. In: Attiwill PM, Adams MA (eds) Nutrition of eucalypts. CSIRO, Melbourne, pp 303–325

Herbohn JL, Emtage NF, Harrison SR, Smorfitt DB (2005) Attitudes of landholders to farm forestry in tropical eastern Australia. Aust For 68:50–58

Herbohn KF, Harrison SR, Herbohn JL (2000) Lessons from small-scale forestry initiatives in Australia: the effective integration of environmental and commercial values. For Ecol Manage 128:227–240

Hertel H, Kaetzel R (1999) Susceptibility of Norway spruce clones (*Picea abies* (L.) Karst.) to insects and roe deer in relation to genotype and foliar phytochemistry. Phyton Ann REI Bot 39:65–72

Hessburg PF, Hansen EM (1986) Mechanism of intertree transmission of *Ceratocystis wageneri* in young Douglas-fir. Can J For Res 16:1250–1254

Hessburg PF, Hansen EM (2000) Infection of Douglas-fir by *Leptographium wageneri*. Can J Bot 78:1254–1261

Hickey GM, Innes JL, Kozak RA, Bull GQ, Vertinsky I (2005) Monitoring and reporting for sustainable forest management: an international multiple case study. For Ecol Manage 209:237–259

Hibbs DE (1987) The self-thinning rule and red alder management. For Ecol Manage 18:273–281

Higgins HG (1984) Pulp and paper. In: Hillis WE, Brown AG (eds) Eucalypts for wood production. CSIRO, Academic, New York, pp 290–316

Hillel D (1980) Applications of soil physics. Academic, New York

Hillis WE (1984) Wood quality and utilization. In: Hillis WE, Brown AG (eds) Eucalypts for wood production. CSIRO, Academic, New York, pp 259–289

Hobbs R, Catling PC, Wombey JC, Clayton M, Atkins L, Reid A (2003) Faunal use of bluegum (*Eucalyptus globulus*) plantations in southwestern Australia. Agrofor Syst 58:195–212

Hoffman D, Weih M (2005) Limitations and improvement of the potential utilisation of woody biomass for energy derived from short rotation woody crops in Sweden and Germany. Biomass Bioenergy 28:267–279

Hogg B, Nester M (1991) Productivity of direct thinning regimes in south-east Queensland hoop pine plantations. Commonw For Rev 70:37–45

Hökkä H, Ojansuu R (2004) Height development of Scots pine on peatlands: describing change in site productivity with a site index model. Can J For Res 34:1081–1092

Hood JV, Libby WJ (1980) A clonal study of intraspecific variability in radiata pine I. Cold and animal damage. Aust For Res 10:9–20

Huber A, Peredo HL (1988) Stem sunscald after thinning and pruning young *Pinus radiata* in the sandy soil region of Chile. N Z J For Sci 18:9–14

Hubert J, Lee, S (2005) A review of the relative roles of silviculture and tree breeding in tree improvement: the example of Sitka spruce in Britain and possible lessons for hardwood breeding. Forestry 78:109–120

Huggard DJ, Klenner W, Vyse A (1999) Windthrow following four harvest treatments in an Engelmann spruce-subalpine fir forest in southern interior British Columbia, Canada. Can J For Res 29:1547–1556

Hummel S (2000) Coppice sprouts in *Cordia alliodora*. J Trop For Sci 12:552–560
Hunt MA, Beadle CL (1998) Whole-tree transpiration and water-use partitioning between *Eucalyptus nitens* and *Acacia dealbata* weeds in a short rotation plantation in northeastern Tasmania. Tree Physiol 18:557–563
Hunt MA, Unwin GL, Beadle CL (1999) Effects of naturally regenerated *Acacia dealbata* on the productivity of a *Eucalyptus nitens* plantation in Tasmania, Australia. For Ecol Manage 117:75–85
Hunt RS (1998) Pruning western white pine in British Columbia to reduce white pine blister rust: 10-year results. West J Appl For 13:60–63
Hunt RS (2002) Can solid deer protectors prevent blister rust from attacking white pines? Can J Plant Pathol 24:74–76
Hutchings MJ, John EA (2004) The effects of environmental heterogeneity on root growth and root/shoot partitioning. Ann Bot 94:1–8
Hutchings MJ, John EA, Wijesinghe DK (2003) Toward understanding the consequences of soil heterogeneity for plant populations and communities. Ecology 84:2322–2334
Huxley P (1999) Tropical agroforestry. Blackwell, Oxford
Hyytiäinen K, Hari P, Kokkila T, Mäkelä A, Tahvonen O, Taipale J (2004) Connecting a process-based forest growth model to stand-level economic optimization. Can J For Res 34:2060–2073
Hyytiäinen K, Tahvonen O, Valsta L (2005) Optimum juvenile density, harvesting, and stand structure in even-aged Scots pine stands. For Sci 51:120–133
Ilic J (1999) Shrinkage-related degrade and its association with some physical properties in *Eucalyptus regnans* F. Muell. Wood Sci Technol 33:425–437
Ilstedt U, Malmer A, Nordgren A, Liau P (2004) Soil rehabilitation following tractor logging: early results on amendments and tilling in a second rotation *Acacia mangium* plantation in Sabah, Malaysia. For Ecol Manage 194:215–222
Ingestad T (1982) Relative addition rate and external concentration: driving variables used in plant nutrition research. Plant Cell Environ 5:443–453
Ingestad T, Ågren GI (1992) Theories and methods on plant nutrition and growth. Physiol Plant 84:177–184
Ingestad T, Kähr M (1985) Nutrition and growth of coniferous seedlings at varied relative nitrogen addition rate. Physiol Plant 65:109–116
Ingestad T, McDonald AJS (1989) Interaction between nitrogen and photon flux density in birch seedlings at steady-state nutrition. Physiol Plant 77:1–11
Ive JR, Cocks KD (1988) A decision support system for land planners and managers. In: Newton PW, Taylor MAP, Sharpe R (eds) Desktop planning: advanced microcomputer applications for physical and social infrastructure planning. Hargreen, Melbourne
Jaakkola T, Mäkinen H, Saranpää P(2005) Wood density in Norway spruce: changes with thinning intensity and age. Can J For Res 35:1767–1778
Jack SB, Long JN (1991) Analysis of stand density effects on canopy structure: a conceptual approach. Trees 5:44–49
Jack SB, Long JN (1996) Linkages between silviculture and ecology: an analysis of density management diagrams. For Ecol Manage 86:205–220
Jacobs DF, Timmer VR (2005) Fertilizer-induced changes in rhizosphere electrical conductivity: relation to forest tree seedling root system growth and function. New For 30:147–166
Jacobs DF, Salifu KF, Seifert JR (2005a) Relative contribution of initial root and shoot morphology in predicting field performance of seedlings. New For 30:235–251
Jacobs DF, Salifu KF, Seifert JR (2005b) Growth and nutritional response of hardwood seedlings to controlled-release fertilization at outplanting. For Ecol Manage 214:28–39
Jacobs MR (1955) Growth habits of the eucalypts. Forestry and Timber Bureau, Canberra
Jarvis PG (ed) (1991) Agroforestry: principles and practice. Elsevier, Amsterdam

Jayawickrama KJS, Carson MJ (2000) A breeding strategy for the New Zealand Radiata Pine Breeding Cooperative. Silvae Genet 49:82–90

Jiang Y, Zwiazek JJ, Macdonald SE (1994) Effects of prolonged cold storage on carbohydrate and protein content and field performance of white spruce bareroot seedlings. Can J For Res 24:1369–1375

Johnsen KH, Major JE (1995) Gas exchange of 20-year-old black spruce families displaying a genotype × environment interaction in growth rate. Can J For Res 25:430–439

Johnson AKL, Cramb RA, McAlpine JR (1994) Integrated land evaluation as an aid to land use planning in northern Australia. J Environ Manage 40:139–154

Johnson DW, Curtis PS (2001) Effects of forest management on soil C and N storage: meta analysis. For Ecol Manage 140:227–238

Johnson JM, Petruncio MD, Gara RI (1995) Impact of forest pest problems on intensive management practices such as pruning. In: Hanley DP, Oliver CD, Maguire DA, Briggs DG, Fight RD (eds) Forest pruning and wood quality. Contribution no 77. Institute of Forest Resources, University of Washington, Seattle, pp 245–251

Jokela EJ, Dougherty PM, Martin TA (2004) Production dynamics of intensively managed loblolly pine stands in the southern United States: a synthesis of seven long-term experiments. For Ecol Manage 192:117–130

Jones HE, Dighton J (1993) The use of nutrient bioassays to assess the response of *Eucalyptus grandis t*o fertilizer application 2. A field experiment. Can J For Res 23:7–13

Jordan L, Daniels RF, Clark A, He R (2005) Multilevel nonlinear mixed-effects models for the modeling of earlywood and latewood microfibril angle. For Sci 51:357–371

Jovanovic T, Arnold R, Booth T (2000) Determining the climatic suitability *of Eucalyptus dunnii* for plantations in Australia, China and Central and South America. New For 19:215–226

Jozsa LA, Brix H (1989) The effects of fertilization and thinning on wood quality of a 24-year-old Douglas-fir stand. Can J For Res 19:1137–1145

Judd TS (1996) Simulated nutrient losses due to timber harvesting in highly productive eucalypt forests and plantations. In: Attiwill PM, Adams MA (eds) Nutrition of eucalypts. CSIRO, Melbourne, pp 249–258

Kangas J, Kangas A (2005) Multiple criteria decision support in forest management—the approach, methods applied and experiences gained. For Ecol Manage 207:133–143

Kanninen M, Peréz D, Montero M, Víquez E (2004) Intensity and timing of the first thinning of *Tectona grandis* plantations in Costa Rica: results of a thinning trial. For Ecol Manage 203:89–99

Kanowski J, Catterall CP, Wardell-Johnson GW (2005) Consequences of broadscale timber plantations for biodiversity in cleared rainforest landscapes of tropical and subtropical Australia. For Ecol Manage 208:359–372

Kao C, Brodie JD (1980) Simultaneous optimization of thinnings and rotation with continuous stocking and entry intervals. For Sci 26:338–346

Karlsson K (2000) Stem form and taper changes after thinning and nitrogen fertilization in *Picea abies* and *Pinus sylvestris* stands. Scand J For Res 15:621–632

Kauman WG, Gerard J, Jiqing H, Huaijun W (1995) Processing of eucalypts. Commonw For Rev 74:147-154

Kaumi SYS (1983) Four rotations of a *Eucalyptus* fuel field trial. Commonw For Rev 62:19–24

Keane PJ, Kile GA, Podger FD, Brown BN (eds) (2000) Diseases and pathogens of eucalypts. CSIRO, Melbourne

Kearney D (1999) Characterisation of branching patterns, changes caused by variation in initial stocking and implications for silviculture for *E. grandis* and *E. pilularis* plantations in the North Coast region of NSW. BSc (forestry) honours thesis, Department of Forestry, Australian National University

Keenan R, Lamb D, Sexton G (1995) Experience with mixed species rainforest plantations in north Queensland. Commonw For Rev 74:315–321

Keenan R, Lamb D, Woldring O, Irvine T, Jensen R (1997) Restoration of plant biodiversity beneath tropical tree plantations in northern Australia. For Ecol Manage 99:117–131

Keenan RJ, Lamb D, Parotta J, Kikkawa J (1999) Ecosystem management in tropical timber plantations: satisfying economic, conservation and social objectives. J Sustain For 9:117–134

Keirle RM, Johnson IG, Magor VE (1983) Winter storage of *Pinus radiata* logs in forest and sawmill at Tumut, New South Wales. Aust For 46:210–215

Kellomäki S, Oker-Blom P, Valtonen E, Väisänen H (1989) Structural development of Scots pine stands with varying initial density: effect of pruning on branchiness of wood. For Ecol Manage 27:219–233

Kellomäki S, Ikonen V-P, Peltola H, Kolström T (1999) Modelling the structural growth of Scots pine with implications for wood quality. Ecol Model 122:117–134

Kelty MJ (1992) Comparative productivity of monocultures and mixed-species stands. In: Kelty MJ, Larson BC, Oliver CD (eds) The ecology and silviculture of mixed-species forests. Kluwer, Dordrecht, pp 125–141

Kelty MJ, Cameron IR (1995) Plot designs for the analysis of species interactions in mixed stands. Commonw For Rev 74:322–332

Kern CC, Friend AL, Johnson JM-F, Coleman MD (2004) Fine root dynamics in a developing *Populus deltoides* plantation. Tree Physiology 24, 651-660

Kerr G (2003) Effects of spacing on the early growth of planted *Fraxinus excelsior* L. Can J For Res 33:1196–1207

Kerr G, Cahalan C (2004) A review of site factors affecting the early growth of ash (*Fraxinus excelsior* L.). For Ecol Manage 188:225–234

Kerr RJ, Dieters MJ, Tier B (2004) Simulation of the competitive gains from four different hybrid tree breeding strategies. Can J For Res 34:209–220

Khanna PK (1997) Comparison of growth and nutrition of young monocultures and mixed stands of *Eucalyptus globulus* and *Acacia mearnsii*. For Ecol Manage 94:105–113

Kikuzawa K (1982) Yield-density diagram for natural deciduous broad-leaved forest stands. For Ecol Manage 4:341–358

Kim MS, Brunsfeld SJ, McDonald GI, Klopfenstein NB (2003) Effect of white pine blister rust (*Cronartium ribicola*) and rust-resistance breeding on genetic variation in western white pine (*Pinus monticola*). Theor Appl Genet 106:1004–1010

Kimmins JP, Comeau PG, Kurz W (1990) Modelling the interactions between moisture and nutrients in the control of forest growth. For Ecol Manage 30:361–379

King D, Loucks OL (1978) The theory of tree bole and branch form. Radiat Environ Biophys 15:141–165

King DA (1981) Tree dimensions: maximizing the rate of height growth in dense stands. Oecologia 51:351–356

King DA (1986) Tree form, height growth, and susceptibility to wind damage in *Acer saccharum*. Ecology 67:980–990

Kinloch BB (2003) White pine blister rust in North America: past and prognosis. Phytopathology 93:1044–1047

Knight PJ (1988) Seasonal fluctuations in foliar nutrient concentration in a young nitrogen-deficient stand of *Eucalyptus fastigata* with and without applied nitrogen. N Z J For Sci 18:15–32

Knight PJ, Nicholas ID (1996) Eucalypt nutrition: New Zealand experience. In: Attiwill PM, Adams MA (eds) Nutrition of eucalypts. CSIRO, Melbourne, pp 275–302

Knowles RL (1991) New Zealand experience with silvopastoral systems: a review. For Ecol Manage 45:251–267

Knowles RL (1995) New Zealand experience with pruning radiata pine. In: Hanley DP, Oliver CD, Maguire DA, Briggs DG, Fight RD (eds) Forest pruning and wood quality. Contribution no 77. Institute of Forest Resources, University of Washington, Seattle, pp 255–264

Knowles RL, Tahau F (1979) A repellent to protect radiata pine seedlings from browsing by sheep. N Z J For Sci 9:3-9

Koch JM, Ward SC (2005) Thirteen-year growth of jarrah (*Eucalyptus marginata*) on rehabilitated bauxite mines in south-western Australia. Aust For 68:176–186

Kopp RF, White EH, Abrahamson LP, Nowak CA, Zsuffa L, Burns KF (1993) Willow biomass trials in central New York State. Biomass Bioenergy 5:179–187

Kopp RF, Abrahamson LP, White EH, Nowak CA, Zsuffa L, Burns KF (1996) Woodgrass spacing and fertilization effects on wood biomass production by a willow clone. Biomass Bioenergy 11:451–457

Kopp RF, Abrahamson LP, White EH, Burns KF, Nowak CA (1997) Cutting cycle and spacing effects on biomass production by a willow clone in New York. Biomass Bioenergy 12:313–319

Koski V, Rousi M (2005) A review of the promises and constraints of breeding silver birch (*Betula pendula* Roth) in Finland. Forestry 78:187–198

Krasowski MJ (2003) Root system modifications by nursery culture reflect on post-planting growth and development of coniferous seedlings. For Chron 79:882–891

Kriedemann PE, Cromer RN (1996) The nutritional physiology of the eucalypts—nutrition and growth. In: Attiwill PM, Adams MA (eds) Nutrition of eucalypts. CSIRO, Melbourne, pp 109–121

Kubin E (1995) The effect of clear cutting, waste wood collecting and site preparation on the nutrient leaching to groundwater. In: Nilsson LO, Hüttl RF, Johansson UT (eds) Nutrient uptake and cycling in forest ecosystems. Kluwer, Dordrecht, pp 661–670

Kuiper LC, Sikkema R, Stolp JAN (1998) Establishment needs for short rotation forestry in the EU to meet the goals of the Commission's white paper on renewable energy (November 1997). Biomass Bioenergy 15:451–456

Kumar BM, Kumar SS, Fisher RF (1998) Intercropping teak with *Leucaena* increases tree growth and modifies soil characteristics. Agrofor Syst 42:81–89

Kumar BM, Long JN, Kumar P (1995) A density management diagram for teak plantations in peninsular India. For Ecol Manage 74:125–131

Kumar S, Garrick DJ (2001) Genetic responses to within-family selection using molecular markers in some radiata pine breeding schemes. Can J For Res 31:779–785

Kuuluvainen T (1988) Crown architecture and stemwood production in Norway spruce (*Picea abies* (*L.*) *Karst.*). Tree Physiol 4:337–346

Kuuluvainen T, Kanninen M (1992) Patterns in aboveground carbon allocation and tree architecture that favour stem growth in young Scots pine from high latitudes. Tree Physiol 10:69–80

Labrecque M, Teodorescu TI, Daigle S (1997) Biomass productivity and wood energy of *Salix* species after 2 years growth in SRIC fertilized with wastewater sludge. Biomass Bioenergy 12:409–417

Lacey ST, Brennan PD, Parekh J (2001) Deep may not be meaningful: cost and effectiveness of various ripping tine configurations in a plantation cultivation trial in eastern Australia. New For 21:231–248

Laclau J-P, Deleporte P, Ranger J, Bouillet J-P, Kazotti G (2003) Nutrient dynamics throughout the rotation of *Eucalyptus* clonal stands in Congo. Ann Bot 91:879–892

Laclau J-P, Ranger J, Deleporte P, Nouvellon Y, Saint-André L, Marlet S, Bouillet, J-P (2005) Nutrient cycling in a clonal stand of *Eucalyptus* and an adjacent savanna ecosystem in Congo 3. Input-output budgets and consequences for the sustainability of the plantations. For Ecol Manage 210:375–391

Ladrach WE (2004) Harvesting and comparative thinning alternatives in *Gmelina arborea*. New Forests 28:255-268

Laffan M (2000) A rapid field method for assessing site suitability for plantations in Tasmania's state forests. Tasforests 12:83–104

Laffan MD (1994) A methodology for assessing and classifying site productivity and land suitability for eucalypt plantations in Tasmania. Tasforests 6:61–67

Lal R (1997) Soils of the tropics and their management for plantation forestry. In: Nambiar EKS, Brown AG (eds) Management of soil, nutrients and water in tropical plantation forests. Monograph no 43. Australian Centre for International Agricultural Research, Canberra, pp 97–123

Lambert M, Turner J (2000) Commercial forest plantations on saline lands. CSIRO, Melbourne

Lambert MJ (1986) Sulphur and nitrogen nutrition and their interactive effects on *Dothistroma* infection in *Pinus radiata*. Can J For Res 16:1055–1062

Lambert MJ, Turner J (1988) Interpretation of nutrient concentrations in *Pinus radiata* foliage at Belanglo State Forest. Plant Soil 108:237–244

Landsberg JJ, Gower ST (1997) Applications of physiological ecology to forest management. Academic, San Diego

Larson P (1965) Stem form of young *Larix* as influenced by wind and pruning. For Sci 11:412-424

Lasserre J-P, Mason EG, Watt MS (2005) The effects of genotype and spacing on *Pinus radiata* [D. Don] corewood stiffness in an 11-year experiment. For Ecol Manage 205:375–383

Lauer DK, Glover GR, Gjerstad DH (1993) Comparison of duration and method of herbaceous weed control on loblolly pine response through mid rotation. Can J For Res 23:2116–2125

Lawley IR, Foley WJ (1999) Swamp wallabies and Tasmanian pademelons show intraspecific preferences for foliage. Aust For 62:17–20

le Mar K, McArthur C (2001) Changes in marsupial herbivore densities in relation to a forestry 1080-poisoning operation. Aust For 64:175–180

Le Roux X, Lacointe A, Escobar-Gutiérrez A, Le Dizès S (2001) Carbon-based models of individual tree growth: a critical appraisal. Ann For Sci 58:469–506

Leon A (1989) The Tasmanian *Eucalyptus* leaf beetle, *Chrysophtharta bimaculata*: an overview of the problem and current control methods. Tasforests 1:33–37

Leslie AD (1991) Agroforestry practices in Somalia. In: Jarvis PG (ed) Agroforestry: principles and practice. Elsevier, Amsterdam, pp 293–308

Lewis NB, Keeves A, Leech JW (1976) Yield regulation in South Australian Pinus radiata plantations. Bulletin no 23, Woods and Forests Department, South Australia

Li B, McKeand S, Weir R (1999) Tree improvement and sustainable forestry—impact of two cycles of loblolly pine breeding in the U.S.A. For Genet 6:229–234

Lieffers VJ, Mugasha AG, MacDonald SE (1993) Ecophysiology of shade needles of *Picea glauca* saplings in relation to removal of competing hardwoods and degree of prior shading. Tree Physiol 12:271–280

Liegel LH (1984) Assessment of hurricane rain/wind damage in *Pinus caribaea* and *Pinus oocarpa* provenance trials in Puerto Rico. Commonw For Rev 63:47–53

Liesebach M, von Wuehlisch G, Muhs H-J (1999) Aspen for short-rotation coppice plantations on agricultural sites in Germany: effects of spacing and rotation time on growth and biomass production of aspen progenies. For Ecol Manage 121:25–39

Lindenmayer DB, Franklin JF (eds) (2003) Towards forest sustainability. CSIRO, Melbourne

Lindenmayer DB, Hobbs RJ, Salt D (2003) Plantation forests and biodiversity conservation. Aust For 66:62–66

Linder S (1985) Potential and actual production in Australian forest stands. In: Landsberg JJ, Parsons W (eds) Research for forest management. CSIRO, Melbourne, pp 11–35

Lindgren D, Cui J, Son SG, Sonesson J (2004) Balancing seed yield and breeding value in clonal seed orchards. New For 28:11–22

Little KM, du Toit B (2003) Management of *Eucalyptus grandis* coppice regeneration of seedling parent stock in Zululand, South Africa. Aust For 66:108–112

Little KM, Gardner RAW (2003) Coppicing ability of 20 *Eucalyptus* species grown at two high altitude sites in South Africa. Can J For Res 33:181–189

Little KM, Smith CW, Norris CH (2000) The influence of various methods of plantation residue management on replanted *Acacia mearnsii* growth. Aust For 63:226–234

Little KM, van Staden J, Clarke GPY (2003a) *Eucalytpus grandis* × *E. camaldulensis* variability and intra-genotypic competition as a function of different vegetation management treatments. New For 25:227–242

Little KM, van Staden J, Clarke GPY (2003b) The relationship between vegetation management and the wood and pulping properties of a Eucalyptus hybrid clone. Ann For Sci 60:673–680

Loch AD (2005) Mortality and recovery of eucalypt beetle pest and beneficial arthropod populations after commercial application of the insecticide alpha-cypermethrin. For Ecol Manage 217:225–265

Lockie S (2003) Conditions for building social capital and community well-being through plantation forestry. Aust For 66:24–29

Long JN, Dean TJ, Roberts SD (2004) Linkages between silviculture and ecology: examination of several important conceptual models. For Ecol Manage 200:249–261

Lörz H, Wenzel G (2005) Molecular marker systems in plant breeding and crop improvement. Springer, Berlin Heidelberg New York

Louw JH, Scholes M (2002) Forest site classification and evaluation: a South African perspective. For Ecol Manage 171:153–168

Lowery RF, Lambeth CC, Endo M, Kane M (1993) Vegetation management in tropical forest plantations. Can J For Res 23:2006–2014

Lu P, Sinclair RW, Boult TJ, Blake SG (2005) Seedling survival of *Pinus strobus* and its interspecific hybrids after artificial inoculation of *Cronartium ribicola*. For Ecol Manage 214:344–357

Luckert MK, Williamson T (2005) Should sustained yield be part of sustainable forest management? Can J For Res 35:356–364

Luoga EJ, Witkowski ETF, Balkwill K (2004) Regeneration by coppicing (resprouting) of miombo (African savanna) trees in relation to land use. For Ecol Manage 189:23–35

Luostarinen K, Kauppi A (2005) Effects of coppicing on the root and stump carbohydrate dynamics in birches. New For 29:289–303

Mabvurira D, Miina J (2002) Individual-tree growth and mortality models for *Eucalyptus grandis* (Hill) Maiden plantations in Zimbabwe. For Ecol Manage 61:231–245

MacDicken KG, Vergara NT (1990) Agroforestry: classification and management. Wiley, New York

MacDonald E, Hubert J (2002) A review of the effects of silviculture on timber quality of Sitka spruce. Forestry 75:107–138

Macmillan WP (1984) Reconstituted wood products. In: Hillis WE, Brown AG (eds) Eucalypts for wood production. CSIRO, Academic, New York, pp 317–321

Maguire DA, Petruncio MD (1995) Pruning and growth of western Cascade species: Douglas-fir, western hemlock, Sitka spruce. In: Hanley DP, Oliver CD, Maguire DA, Briggs DG, Fight RD (eds) Forest pruning and wood quality. Contribution no 77. Institute of Forest Resources, University of Washington, Seattle, pp 179–215

Mahendrappa MK, Foster NW, Weetman GF, Krause HH (1986) Nutrient cycling and availability in forest soils. Can J Soil Sci 66:547–572

Maillard P, Garriou D, Deléens E, Gross P, Guehl J-M (2004) The effects of lifting on mobilisation and net asimilation of C and N during regrowth of transported Corsican pine seedlings. A dual ^{13}C and ^{15}N labelling approach. Ann For Sci 61:795–805
Mäkelä A (1997) A carbon balance model of growth and self-pruning. For Sci 43:7–24
Mäkelä A (2002) Derivation of stem taper from the pipe theory in a carbon balance framework. Tree Physiol 22:891–905
Mäkelä A (2003) Process-based modelling of tree and stand growth: towards a hierarchical treatment of multiscale processes. Can J For Res 33:398–409
Mäkelä A, Vanninen P (2000) Estimation of fine root mortality and growth from simple measurements: a method based on system dynamics. Trees 14:316–323
Mäkelä A, Landsberg J, Ek AR, Burk TE, Ter-Mikaelian M, Ågren GI, Oliver CD, Puttonen P (2000) Process-based models for ecosystem management: current state of the art and challenges for practical implementation. Tree Physiol 20:289–98
Mäkinen H (1999) Growth, suppression, death, and self-pruning of branches of Scots pine in southern and central Finland. Can J For Res 29:585–594
Mäkinen H (2002) Effect of stand density on the branch development of silver birch (*Betula pendula* Roth) in central Finland. Trees 16:346–353
Mäkinen H, Colin F (1998) Predicting branch angle and branch diameter of Scots pine from usual tree measurements and stand structural information. Can J For Res 28:1686–1696
Mäkinen H, Isomäki A (2004a) Thinning intensity and long-term changes in increment and stem form of Norway spruce trees. For Ecol Manage 201:295–309
Mäkinen H, Isomäki A (2004b) Thinning intensity and growth of Scots pine stands in Finland. For Ecol Manage 201:311–325
Mäkinen H, Isomäki A (2004c) Thinning intensity and growth of Norway spruce stands in Finland. Forestry 77:349–364
Mäkinen H, Isomäki A (2004d) Thinning intensity and long-term changes in increment and stem form of Scots pine trees. For Ecol Manage 203:21–34
Mäkinen H, Saranpää P, Linder S (2001) Effect of nutrient optimization on branch characteristics in *Picea abies* (L.) Karst. Scand J For Res 16:354–362
Mäkinen H, Nöjd P, Isomäki A (2002a) Radial, height and volume increment variation in *Picea abies* (L.) Karst. stands with varying thinning intensities. Scand J For Res 17:304–316
Mäkinen H, Saranpää P, Linder S (2002b) Wood-density variation of Norway spruce in relation to nutrient optimization and fibre dimensions. Can J For Res 32:185–194
Mäkinen H, Saranpää P, Linder S (2002c) Effect of growth rate on fibre characteristics in Norway spruce (*Picea abies* (L.) Karst.). Holzforschung 56:449–60
Mäkinen H, Ojansuu R, Niemistö P (2003a) Predicting external branch characteristics of planted silver birch (*Betula pendula* Roth.) on the basis of routine stand and tree measurements. For Sci 49:301–317
Mäkinen H, Ojansuu R, Sairanen P, Yli-Kojola H (2003b) Predicting branch characteristics of Norway spruce (*Picea abies* (L.) Karst.), from simple stand and tree measurements. Forestry 76:525–546
Malan FS, Hoon M (1992) Effect of initial spacing and thinning on some wood properties of *Eucalyptus grandis*. S Afr For J 163:13–20
Malimbwi RE, Persson A, Iddi S, Chamshama SAO, Mwihomeke ST (1992) Effects of spacing on yield and some wood properties of *Pinus patula* at Rongai, northern Tanzania. For Ecol Manage 53:297–306
Manion PD (1991) Tree disease concepts, 2nd edn. Prentice Hall, Englewood Cliffs
Marcar N, Morris J (2005) Plantation productivity in saline landscapes. In: Nambiar S, Ferguson I (eds) New forests. CSIRO, Melbourne, pp 51–74
Margolis HA, Brand DG (1990) An ecophysiological basis for understanding plantation establishment. Can J For Res 20:375–390

Margolis HA, Waring RH (1986) Carbon and nitrogen allocation patterns of Douglas-fir seedlings fertilized with nitrogen in autumn. II. Field performance. Can J For Res 16:903–909

Markin GP, Gardner DE (1993) Status of biological control in vegetation management in forestry. Can J For Res 23:2023–2031

Marks GC, Smith IW (1987) Effects of canopy closure and pruning on *Dothistroma septospora* needle blight of *Pinus radiata* D. Don. Aust For Res 17:145–150

Marks GC, Incoll WD, Long IR (1986) Effects of crown development, branch shed and competition on wood defect in *Eucalyptus regnans* and *E. sieberi*. Aust For Res 16:117–129

Marks GC, Smith IW, Cook IO (1989) Spread of *Dothistroma septospora* in plantations of *Pinus radiata* in Victoria between 1979 and 1988. Aust For 52:10–19

Martin JG, Kloeppel BD, Schaefer TL, Kimbler DL, McNulty SG (1998) Aboveground biomass and nitrogen allocation of ten deciduous southern Appalachian tree species. Can J For Res 28:1648–1659

Martin TA, Jokela EJ (2004) Stand development and production dynamics of loblolly pine under a range of cultural treatments in north-central Florida USA. For Ecol Manage 192:39–58

Mason EG (1985) Causes of juvenile instability of *Pinus radiata* in New Zealand. N Z J For Sci 15:263–280

Mason EG, Cullen AWJ (1986) Growth of *Pinus radiata* on ripped and unripped Taupo pumice soils. N Z J For Sci 16:3–18

Mason EG, Cullen AWJ, Rijkse WC (1988) Growth of two *Pinus radiata* stock types on ripped and ripped/bedded plots at Karioi forest. N Z J For Sci 18:287–296

Matala J, Hynynen J, Miina J, Ojansuu R, Peltola H, Sievänen R, Väisänen H, Kellomäki S (2003) Comparison of a physiological model and a statistical model for prediction of growth and yield in boreal forests. Ecol Model 161:95–116

Matheson AC, Raymond CA (1984) The impact of genotype × environment interactions on Australian *Pinus radiata* breeding programs. Aust For Res 14:11–25

Mayo O (1987) The theory of plant breeding, 2nd edn. Oxford University Press, Oxford

Mazanec RA, Mason ML, Vellios C (2003) Performance of spotted gum provenances for timber production on former bauxite mines in Western Australia. Aust For 66:129–136

McArthur C, Appleton R (2004) Effect of seedling characteristics at planting on browsing of *Eucalyptus globulus* by rabbits. Aust For 67:25–29

McArthur C, Marsh R, Close DC, Walsh A, Paterson S, Fitzgerald H, Davies NW (2003) Nursery conditions affect seedling chemistry, morphology and herbivore preferences for *Eucalyptus nitens*. For Ecol Manage 176:585–594

McCarter JB, Long JN (1986) A lodgepole pine density management diagram. West J Appl For 1:6–11

McDonald AJS, Lohammar T, Ericsson A (1986) Uptake of carbon and nitrogen at decreased nutrient availability in small birch (*Betula pendula* Roth.) plants. Tree Physiol 2:61–71

McDonald MA, Malcolm DC, Harrison AF (1991a) The use of a ^{32}P root bioassay to indicate the phosphorus status of forest trees. 1. Seasonal variation. Can J For Res 21:1180–1184

McDonald MA, Malcolm DC, Harrison AF (1991b) The use of a ^{32}P root bioassay to indicate the phosphorus status of forest trees. 2. Spatial variation. Can J For Res 21:1185–1193

McDonald PM, Fiddler GO (1993) Feasibility of alternatives to herbicides in young conifer plantations in California. Can J For Res 23:2015–2022

McGrath JF, Dumbrell IC, Hingston RA, Copeland B (2003) Nitrogen and phosphorus increase growth of thinned late-rotation *Pinus radiata* on coastal sands in Western Australia. Aust For 66:217–222

McJannett D, Vertessy R (2001) Effects of thinning on wood production, leaf area index, transpiration and canopy interception of a plantation subject to drought. Tree Physiol 21:1001–1008

McKeand SE (1988) Optimum age for family selection for growth in genetic tests of loblolly pine. For Sci 34:400–411

McKenzie H, Hay E (1996) Performance of *Eucalyptus saligna* in New Zealand. Bulletin 51. New Zealand Forest Research Institute, Rotorua

McKimm RJ, Waugh G, Northway RL (1988) Utilisation potential of plantation-grown *Eucalyptus nitens*. Aust For 51:63–71

McMurtrie RE, Medlyn BE, Dewar RC (2001) Increased understanding of nutrient immobilization in soil organic matter is critical for predicting the carbon sink strength of forest ecosystems over the next 100 years. Tree Physiol 21:831–839

McNally J (1955) Damage to Victorian exotic pine plantations by native animals. Aust For 19:87–99

Mead DJ (2005) Opportunities for improving plantation productivity. How much? How quickly? How realistic? Biomass Bioenergy 28:249–266

Mead DJ, Will GM (1976) Seasonal and between-tree variation in the nutrient levels in *Pinus radiata* foliage. N Z J For Sci 6:3–13

Mead DJ, Draper D, Madgwick HAI (1984) Dry matter production of a young stand of *Pinus radiata*: some effects of nitrogen fertiliser and thinning. N Z J For Sci 14:97–108

Medhurst JL, Beadle CL (2001) Crown structure and leaf area index development in thinned and unthinned *Eucalyptus nitens* plantations. Tree Physiol 21:989–999

Medhurst JL, Beadle CL (2005) Photosynthetic capacity and foliar nitrogen distribution in *Eucalyptus nitens* is altered by high-intensity thinning. Tree Physiology 25:981–991

Medhurst JL, Beadle CL, Neilsen WA (2001) Early-age and later-age thinning affects growth, dominance, and intraspecific competition in *Eucalyptus nitens* plantations. Can J For Res 31:187–197

Medhurst JL, Battaglia M, Beadle CL (2002) Measured and predicted changes in tree and stand water use following high-intensity thinning of an 8-year-old *Eucalyptus nitens* plantation. Tree Physiol 22:775–784

Menalled FD, Kelty MJ, Ewel JJ (1998) Canopy development in tropical tree plantations: a comparison of species mixtures and monocultures. For Ecol Manage 104:249–263

Mena-Petite A, Estavillo JM, Duñabeitia M, González-Moro B, Muñoz-Rueda A, Lacuesta M (2004) Effect of storage conditions on post planting water status and performance of *Pinus radiata* D. Don stock-types. Ann For Sci 61:695–704

Mendham DS, Smethurst PJ, Holz GK, Menary RC, Grove TS, Weston C, Baker T (2002) Soil analyses as indicators of phosphorus response in young eucalypt plantations. Soil Sci Soc Am J 66:959–968

Mendham DS, O'Connell AM, Grove TS, Rance SJ (2003) Residue management effects on soil carbon and nutrient contents and growth of second rotation eucalypts. For Ecol Manage 181:357–372

Mendoza GA, Prabhu R (2005) Combining participatory modeling and multi-criteria analysis for community-based forest management. For Ecol Manage 207:145–156

Mercuri AM, Duggin JA, Grant CD (2005) The use of saline mine water and municipal wastes to establish plantations on rehabilitated open-cut coal mines, Upper Hunter Valley NSW, Australia. For Ecol Manage 204:195–207

Merino A, López AR, Brañas J, Rodríguez-Soalleiro R (2003) Nutrition and growth in newly established plantations of *Eucalyptus globulus* in northwestern Spain. Ann For Sci 60:509–517

Merino A, Balboa MA, Soalleiro R, González JGA (2005) Nutrient exports under different harvesting regimes in fast-growing forest plantations in southern Europe. For Ecol Manage 207:325–339

Messina MG (1992) Response of *Eucalyptus regnans* F.Muell. to thinning and urea fertilization in New Zealand. For Ecol Manage 51:269–283

Miller HG (1981) Forest fertilization: some guiding concepts. Forestry 54:157–167

Miller HG (1995) The influence of stand development on nutrient demand, growth and allocation. In: Nilsson LO, Hüttl RF, Johansson UT (eds) Nutrient uptake and cycling in forest ecosystems. Kluwer, Dordrecht, pp 225–232

Miller HG, Cooper JM, Miller JD, Pauline OJL (1979) Nutrient cycles in pine and their adaptation to poor soils. Can J For Res 9:19–26

Mitchell AL (1971) Planting trials with red cedar. Aust For 35:8–16

Mitchell SJ (1995a) A synopsis of windthrow in British Columbia: occurrence, implications, assessment and management. In: Coutts MP, Grace J (eds) (1995) Wind and trees. Cambridge University Press, Cambridge, pp 448–459

Mitchell SJ (1995b) The windthrow triangle: a relative windthrow hazard assessment procedure for forest managers. For Chron 71:446–450

Mitchell SJ (1998) A diagnostic framework for windthrow risk estimation. For Chron 74:100–105

Mitchell SJ (2000) Stem growth responses in Douglas-fir and Sitka spruce following thinning: implications for assessing wind-firmness. For Ecol Manage 135:105–114

Mo J, Tanton MT (1996) Diel activity patterns and the effects of wind on the mating success of red cedar tip moth, *Hypsipila robusta* Moore (Lepidoptera: Pyralidae). Aust For 59:42–45

Möller CM (1954) The influence of thinning on volume increment. 1. Results of investigations. In: Heiberg SO (ed) Thinning problems and practice in Denmark. World forestry series, bulletin 1. College of Forestry Syracuse, State University of New York, Syracuse, pp 5–32

Moncur MW, Rasmussen GF, Hasan O (1994) Effect of paclobutrazol on flower-bud production in *Eucalyptus nitens* espalier seed orchards. Can J For Res 24:46–49

Montagu KD, Kearney DE, Smith RGB (2003) The biology and silviculture of pruning planted eucalypts for clear wood production—a review. For Ecol Manage 179:1–13

Montague TL (1993) An assessment of the ability of tree guards to prevent browsing damage using captive swamp wallabies (*Wallabia bicolor*). Aust For 56:145–147

Montague TL (1994) Wallaby browsing and seedling palatability. Aust For 57:171–175

Montague TL (1996) The extent, timing and economics of browsing damage in eucalypt and pine plantations of Gippsland, Victoria. Aust For 59:120–129

Moore JR (2000) Differences in maximum resistive bending moments of *Pinus radiata* trees grown on a range of soil types. For Ecol Manage 135:63–71

Moore JR, Maguire DA (2004) Natural sway frequencies and damping ratios of trees: concepts, review and synthesis of previous studies. Trees 18:195–203

Moore JR, Maguire DA (2005) Natural sway frequencies and damping ratios of trees: influence of crown structure. Trees 19:363–373

Morikawa Y, Hattori S, Kiyono Y (1986) Transpiration of a 31-year-old *Chamaecyparis obtusa* Endl. stand before and after thinning. Tree Physiol 2:105–114

Mörling T (2002) Evaluation of annual ring width and ring density development following fertilisation and thinning of Scots pine. Ann For Sci 59:29–40

Moroni MT, Smethurst PJ, Holz GK (2004) Indices of soil nitrogen availability in five Tasmanian *Eucalyptus nitens* plantations. Aust J Soil Res 42:719–725

Morris J, Benyon R (2005) Plantation water use. In: Nambiar S, Ferguson I (eds) New forests. CSIRO, Melbourne, pp 75–104

Mósena M, Dillenburg LR (2004) Early growth of Brazilian pine (*Araucaria angustifolia* [Bertol.] Kuntze) in response to soil compaction and drought. Plant Soil 258:293–306

Muhammed N, Koike M, Sajjaduzzaman M, Sophanarith K (2005) Reckoning social forestry in Bangladesh: policy and plan versus implementation. Forestry 78:373–383

Mummery D, Battaglia M (2001) Applying ProMod spatially across Tasmania with sensitivity analysis to screen for prospective *Eucalyptus globulus* plantation sites. For Ecol Manage 140:51–63

Mummery D, Battaglia M (2002) Data input quality and resolution effects on regional and local scale Eucalyptus globulus productivity predictions in north-east Tasmania. Ecol Model 156:13–25

Mummery D, Battaglia M, Beadle CL, Turnbull CRA, McLeod R (1999) An application of terrain and environmental modelling in a large-scale forestry experiment. For Ecol Manage 118:149–159

Munson AD, Bernier PY (1993) Comparing natural and planted black spruce seedlings. II. Nutrient uptake and efficiency of use. Can J For Res 23:2435–2442

Murray DR (2003) Seeds of concern. The genetic manipulation of plants. University of New South Wales Press, Sydney, and CABI, Oxford

Murty D, McMurtrie RE (2000) The decline of forest productivity as stands age: a model-based method for analysing causes for the decline. Ecol Model 134:185–205

Muukkonen P (2005) Needle biomass turnover rates of Scots pine (*Pinus sylvestris* L.) derived from the needle-shed dynamics. Trees 19:273–279

Muukkonen P, Lehtonen A (2004) Needle and branch biomass turnover rates of Norway spruce (*Picea abies*). Can J For Res 34:2517–2527

Myers BJ, Theiveyanathan S, O'Brien ND, Bond WJ (1996) Growth and water use of *Eucalyptus grandis* and *Pinus radiata* plantations irrigated with effluent. Tree Physiol 16:211–219

Myers BJ, Benyon RG, Bond WJ, Falkiner RA, O'Brien ND, Polglase PJ, Smith CJ, Snow VO, Theiveyanathan S (1998) Water, salt, nutrients and growth in effluent-irrigated plantations. Paper presented at the WaterTECH conference, Australian Water and Wastewater Association, Brisbane, April 1998

Nair PKR (1991) State-of-the-art of agroforestry systems. For Ecol Manage 45:5–29

Nair PKR (1993) An introduction to agroforestry. Kluwer, Dordrecht

Nambiar EKS (1980) Root configuration and root regeneration in *Pinus radiata* seedlings. N Z J For Sci 10:249–263

Nambiar EKS (1990) Interplay between nutrients, water, root growth and productivity in young plantations. For Ecol Manage 30:213–232

Nambiar EKS (1990/1991) Management of forests under nutrient and water stress. Water Air Soil Pollut 54:209–230

Nambiar EKS, Sands R (1993) Competition for water and nutrients in forests. Can J For Res 23:1955–1968

Nambiar EKS, Zed PG (1980) Influence of weeds on the water potential, nutrient content and growth of young radiata pine. Aust For Res 10:279–288

Neave IA, Florence RG (1994) Effect of root configuration on the relative competitive ability of *Eucalyptus maculata* Hook. regrowth following clearfelling. Aust For 57:49–58

Negi JDS, Sharma SC (1996) Mineral nutrition and resource conservation in *Eucalyptus* plantations and other forest covers in India. In: Attiwill PM, Adams MA (eds) Nutrition of eucalypts. CSIRO, Melbourne, pp 399–416

Neil PE, Barrance AJ (1987) Cyclone damage in Vanuatu. Commonw For Rev 66:255–264

Neilsen WA, Lynch T (1998) Implications of pre- and post-fertilizing changes in growth and nitrogen pools following multiple applications of nitrogen fertilizer to a *Pinus radiata* stand over 12 years. Plant Soil 202:295–307

Neilsen WA, Gerrand AM (1999) Growth and branching habit of *Eucalyptus nitens* at different spacing and the effect on final crop selection. For Ecol Manage 123:217–229

Neilsen WA, Pinkard EA (2003) Effects of green pruning on growth of *Pinus radiata*. Can J For Res 33:2067–2073

Neilsen WA, Ringrose C (2001) Effect of initial herbicide treatment and planting material on woody weed development and the growth of *Eucalyptus nitens* and *Eucalyptus regnans*. Weed Res 41:301–309

Nelder J (1962) New kinds of systematic designs for spacing experiments. Biometrics 18:283–307

Neumann FG, Marks GC (1990) Status and management of insect pests and diseases in Victorian softwood plantations. Aust For 53:131–144

Neumann FG, Collett NG, Smith IW (1993) The Sirex wasp and its biological control in plantations of radiata pine variably defoliated by *Dothistroma septospora* in north-eastern Victoria. Aust For 56:129–139

New D (1989) Forest health—an industry perspective of the risks to New Zealand's plantations. N Z J For Sci 19:155–158

Newnham RM (1965) Stem form and the variation of taper with age and thinning regime. Forestry 38:218–224

Newton AC, Baker P, Ramnarine S, Mesén JF, Leakey RRB (1993) The mahogany shoot borer: prospects for control. For Ecol Manage 57:301–328

Newton PF (1997a) Algorithmic versions of black spruce stand density management diagrams. For Chron 73:257–265

Newton PF (1997b) Stand density management diagrams: review of their development and utility in stand-level management planning. For Ecol Manage 98:251–265

Newton PF, Weetman GF (1993) Stand density management diagrams and their development and utility in black spruce management. For Chron 69:421–430

Newton PF, Weetman GF (1994) Stand density management diagram for managed black spruce stands. For Chron 70:65–74

Newton PF, Lei Y, Zhang SY (2004) A parameter recovery model for estimating black spruce diameter distributions within the context of a stand density management diagram. For Chron 80:349–358

Newton PF, Lei Y, Zhang SY (2005) Stand-level diameter distribution yield model for black spruce plantations. For Ecol Manage 209:181–192

Nicholas ID (1993) Growing firewood–management considerations. In: Shula RG, Hay AE, Tarlton GL (eds) The firewood venture. New Zealand Forest Research Institute, Rotorua, pp 36–50

Niemistö P (1995) Influence of initial spacing and row-to-row distance on the crown and branch properties and taper of silver birch (*Betula pendula*). Scand J For Res 10:235–244

Niinemets Ü, Afas NA, Cescatti A, Pellis A, Ceulemans R (2004) Petiole length and biomass investment in support modify light interception efficiency in dense poplar plantations. Tree Physiol 24:141–154

Niinemets U, Sparrow A, Cescatti A (2005) Light capture efficiency decreases with increasing tree age and size in the southern hemisphere gymnosperm *Agathis australis*. Trees 19:177–190

Nilsson LO, Hüttl RF, Johansson UT (eds) (1995) Nutrient uptake and cycling in forest ecosystems. Kluwer, Dordrecht

Nilsson U, Örlander G (1995) Effects of regeneration methods on drought damage to planted Norway spruce seedlings. Can J For Res 25:790–802

Nordborg F, Nilsson U (2003) Growth, damage and net nitrogen uptake in *Picea abies* (L.) Karst. seedlings, effects of site preparation and fertilisation. Ann For Sci 60:657–666

Nyland RD (1996) Silviculture concepts and applications. McGraw-Hill, New York

O'Connell AM, Sankaran KV (1997) Organic mater accretion, decomposition and mineralisation. In: Nambiar EKS, Brown AG (eds) Management of soil, nutrients and water in tropical plantation forests. Monograph no 43. Australian Centre for International Agricultural Research, Canberra, pp 443–480

Ogawa K (2005) Time-trajectory of mean phytomass and density during a course of self-thinning in a sugi (*Cryptomeria japonica* D.Don) plantation. For Ecol Manage 214:104–110

Ogawa K, Hagihara A (2003) Self-thinning and size variation in a sugi (*Cryptomeria japonica* D. Don) plantation. For Ecol Manage 174:413–421

O'Hara KL, Parent DR, Hagle SK (1995) Pruning eastern Cascade and northern Rocky Mountain species: biological opportunities. In: Hanley DP, Oliver CD, Maguire DA, Briggs DG, Fight RD (eds) Forest pruning and wood quality. Contribution no 77. Institute of Forest Resources, University of Washington, Seattle, pp 216–237

Ohmart CP (1982) Destructive insects of native and planted *Pinus radiata* in California, and their relevance to Australian forestry. Aust For Res 12:151–161

Old KM, Davison EM (2000) Canker diseases of eucalypts. In: Keane PJ, Kile GA, Podger FD, Brown BN (eds) Diseases and pathogens of eucalypts. CSIRO, Melbourne, pp 241–257

Oliver AR (2000) Advances in drying plantation-grown eucalypt timber: an overview of Tasmanian research. Aust For 63:248–251

Oliver GR (1991) Growth and biomass of a *Eucalyptus nitens* coppice trial in New Zealand. In: Menzies MI, Parrott GE, Whitehouse LJ (eds) Efficiency of stand establishment operations. Bulletin 156. New Zealand Forest Research Institute, Rotorua, pp 261–268

Opie JE, Curtin RA, Incoll WD (1984) Stand management. In: Hillis WE, Brown AG (eds) Eucalypts for wood production. CSIRO, Academic, New York, pp 179–197

Oppenheimer MJ, Shiver BD, Rheney JW (1989) Ten-year growth response of midrotation slash pine plantations to control of competing vegetation. Can J For Res 19:329–334

O'Reilly JM, McArthur C (2000) Damage to and intake of plantation seedlings by captive European rabbits (*Oryctolagus cuniculus*). Aust For 63:1–6

O'Reilly-Wapstra JM, McArthur C, Potts BM (2002) Genetic variation in resistance of *Eucalyptus globulus* to marsupial browsers. Oecologia 130:289–296

O'Reilly-Wapstra JM, McArthur C, Potts BM (2004) Linking plant genotype, plant defensive chemistry and mammal browsing in a *Eucalyptus* species. Funct Ecol 18:677–684

O'Reilly-Wapstra JM, Potts BM, McArthur C, Davies NW (2005) Effects of nutrient availability on the genetic-based resistance of *Eucalyptus globulus* to a mammalian herbivore and on plant defensive chemistry. Oecologia 142:597–605

Örlander G, Due K (1986) Location of hydraulic resistance in the soil-plant pathway in seedlings of *Pinus sylvestris* L. grown in peat. Can J For Res 16:115–123

Örlander G, Nordlander G (2003) Effects of field vegetation control on pine weevil (*Hylobius abietis*) damage to newly planted Norway spruce seedlings. Ann For Sci 60:667–671

Osler GHR, West PW, Laffan MD (1996a) Test of a system to predict productivity of eucalypt plantations in Tasmania. Aust For 59:57–63

Osler GHR, West PW, Downes GM (1996b) Effects of bending stress on taper and growth of stems of young *Eucalyptus regnans* trees. Trees 10:239–246

Page MW (1984) Production of sawn wood from small eucalypt logs. In: Hillis WE, Brown AG (eds) Eucalypts for wood production. CSIRO, Academic, New York, pp 322–327

Pandey D (1983) Growth and yield of plantation species in the tropics. Forest Resources Division, Food and Agricultural Organisation, Rome

Pape R (1999a) Influence of thinning and tree diameter class on the development of basic density and annual ring width in *Picea abies*. Scand J For Res 14:27–37

Pape R (1999b) Effects of thinning regimes on the wood properties and stem quality of *Picea abies*. Scand J For Res 14:38–50
Papesch AJG, Moore JR, Hawke AE (1997) Mechanical stability of *Pinus radiata* trees at Eyrewell Forest investigated using static tests. N Z J For Sci 27:188–204
Paredes GL, Brodie JD (1987) Efficient specification and solution of the even-aged rotation and thinning problem. For Sci 33:14–29
Park SE, Benjamin LR, Watkinson AR (2003) The theory and application of plant competition models: an agronomic perspective. Ann Bot 92:741–748
Parladé J, Luque J, Pera J, Rincón AM (2004) Field performance of *Pinus pinea* and *P. halepensis* seedlings inoculated with *Rhizopogon* spp. and outplanted in formerly arable land. Ann of For Sci 61:507–514
Parrotta JA (1999) Productivity, nutrient cycling, and succession in single- and mixed-species plantations of *Casuarina equisetifolia*, *Eucalyptus robusta* and *Leucaena leucocephala* in Puerto Rico. For Ecol Manage 124:45–77
Paul KI, Polglase PJ, O'Connell AM, Carlyle CJ, Smethurst PJ, Khanna PK (2002) Soil nitrogen availability predictor (SNAP): a simple model for predicting mineralisation of nitrogen in forest soils. Aust J Soil Res 40:1011–1026
Pawlick T (1989) Coppice with care. Agrofor Today 1(3):15–17
Pederick LA, Eldridge KG (1983) Characteristics of radiata pine achievable by breeding. Aust For 46:287–293
Peng C (2000a) Growth and yield models for uneven-aged stands: past, present and future. For Ecol Manage 132:259–79
Peng C (2000b) Understanding the role of forest simulation models in sustainable forest management. Environ Impact Assess Rev 20:481–501
Pennanen T, Heiskanen J, Korkama T (2005) Dynamics of ectomycorrhizal fungi and growth of Norway spruce seedlings after planting on a mounded forest clearcut. For Ecol Manage 213:243–252
Pérez D, Kanninen M (2005) Stand growth scenarios for *Tectona grandis* plantations in Costa Rica. For Ecol Manage 210:425–441
Pérez LD, Viquez E, Kanninen M (2003) Preliminary pruning programme for *Tectona grandis* plantations in Costa Rica. J Trop For Sci 15:557–569
Perry MA, Mitchell RJ, Zutter BR, Glover GR, Gjerstad DH (1993) Competitive responses of loblolly pine to gradients in loblolly pine, sweetgum and broomsedge densities. Can J For Res 23:2049–2058
Peterson JA, Seiler JR, Nowak J, Ginn SE, Kreh RE (1997) Growth and physiological responses of young loblolly pine stands to thinning. For Sci 43:529–534
Petersson M, Örlander G, Nordlander G (2005) Soil features affecting damage to conifer seedlings by the pine weevil *Hylobius abietis*. Forestry 78:83–92
Petruncio M, McFadden G (1995) Developments in forest pruning equipment. In: Hanley DP, Oliver CD, Maguire DA, Briggs DG, Fight RD (eds) Forest pruning and wood quality. Contribution no 77. Institute of Forest Resources, University of Washington, Seattle, pp 317–25
Petruncio M, Briggs D, Barbour RJ (1997) Predicting pruned branch stub occlusion in young, coastal Douglas-fir. Can J For Res 27:1074–1082
Pezeshki SR, Oliver CD (1985) Early growth patterns of red alder and black cottonwood in mixed species plantations. For Sci 31:190–200
Philipson JJ (1995) Effects of cultural treatments and gibberellin $A_{4/7}$ on flowering of container-grown European and Japanese larch. Can J For Res 25:184–192
Phillips JD (2001) The relative importance of intrinsic and extrinsic factors in pedodiversity. Ann Assoc Am Geogr 91:609–621
Phillips JD, Marion DA (2004) Pedological memory in forest soil development. For Ecol Manage 188:363–380

Pietrzykowski E, McArthur C, Fitzgerald H, Goodwin AN (2003) Influence of patch characteristics on browsing of tree seedlings by mammalian herbivores. J Appl Ecol 40:458–469

Piirto DD, Valkonen S (2005) Structure and development of pitch canker infected Monterey pine stands at Año Nuevo, California. For Ecol Manage 213:160–174

Pike CC, Robison DJ, Maynard CA, Abrahamson LP (2003) Evaluating growth and resistance of eastern and western white pine to white pine weevil and blister rust in the northeast. North J Appl For 20:19–26

Pinkard EA (2002) Effects of pattern and severity of pruning on growth and branch development of pre-canopy closure *Eucalyptus nitens*. For Ecol Manage 157:27–30

Pinkard EA, Battaglia M (2001) Using hybrid models to develop silvicultural prescriptions for *Eucalyptus nitens*. For Ecol Manage 154: 337–345

Pinkard EA, Beadle CL (1998) Aboveground biomass partitioning and crown architecture of *Eucalyptus nitens* following green pruning. Can J For Res 28:1419–1428

Pinkard EA, Beadle CL (2000) A physiological approach to pruning. Int For Rev 2:295–305

Pinkard EA, Battaglia M, Beadle CL, Sands PJ (1999) Modeling the effect of physiological responses to green pruning on net biomass production of *Eucalyptus nitens*. Tree Physiol 19:1–12

Pinkard EA, Mohammed C, Beadle CL, Hall MF, Worledge D, Mollon A (2004) Growth responses, physiology and decay associated with pruning plantation-grown *Eucalyptus globulus* Labill. and *E. nitens* (Deane and Maiden) Maiden. For Ecol Manage 200:263–277

Plauborg KU (2004) Analysis of radial growth response to changes in stand density for four tree species. For Ecol Manage 188:65–75

Pontailler JY, Ceulemans R, Guittet J (1999) Biomass yield of poplar after five 2-year coppice rotations. Forestry 72:157–163.

Porté A, Bartelink HH (2002) Modelling mixed forest growth: a review of models for forest management. Ecol Model 150:141–188

Potts BM (2004) Genetic improvement of eucalypts. In: Encyclopedia of forest science. Elsevier, Oxford, pp 104–114

Potts BM, Dungey HS (2004) Interspecific hybridization of *Eucalyptus*: key issues for breeders and geneticists. New For 27:115–138

Potts BM, Reid JB (2003) Tasmania's eucalypts: their place in science. Pap Proc R Soc Tasmania 137:21–37

Power AB, Dodd RS (1984) Early differential susceptibility of juvenile seedlings and more mature stecklings of *Pinus radiata* to *Dothistroma pini*. N Z J For Sci 14:223–228

Powers RF (1999) On the sustainable productivity of planted forests. New For 17:263–306

Prado JA, Toro JA (1996) Silviculture of eucalypt plantations in Chile. In: Attiwill PM, Adams MA (eds) Nutrition of eucalypts. CSIRO, Melbourne, pp 357–369

Prasolova NV, Xu ZH, Saffigna PG, Dieters MJ (2000) Spatial-temporal variability of soil moisture, nitrogen availability indices and other chemical properties in hoop pine (*Araucaria cunninghamii*) plantations of subtropical Australia. For Ecol Manage 136:1–10

Primrose SB, Twyman RM, Old RW (2002) Principles of gene manipulation, 6th edn. Blackwell, Oxford

Pukkala T, Miina J (1997) A method for stochastic multiobjective optimization of stand management. For Ecol Manage 98:189–203

Pukkala T, Miina J, Kellomäki S (1998) Response to different thinning intensities in young *Pinus sylvestris*. Scand J For Res 13:141–150

Quine CP (1995) Assessing the risk of wind damage to forests: practice and pitfalls. In: Coutts MP, Grace J (eds) (1995) Wind and trees. Cambridge University Press, Cambridge, pp 379–403

Quine CP (2000) Estimation of mean wind climate and probability of strong winds for wind risk assessment. Forestry 73:247–258

Quine CP, Bell PD (1998) Monitoring of windthrow occurrence and progression in spruce forests in Britain. Forestry 71:87–97

Race D, Curtis A (1997) Socio-economic considerations for regional farm forestry development. Aust For 60:233–239

Radvanyi A (1980) Control of small mammal damage in the Alberta oil sands reclamation and afforestation program. For Sci 26:687–702

Rangen SA, Hawley AWL, Hudson RJ (1994) Relationship of snowshoe hare feeding preferences to nutrient and tannin content of four conifers. Can J For Res 24:240–245

Raymond CA (2002) Genetics of *Eucalyptus* wood properties. Ann For Sci 59:525–531

Raymond CA, Banham P, MacDonald AC (1998) Within tree variation and genetic control of basic density, fibre length and coarseness in *Eucalyptus regnans* in Tasmania. Appita J 51:299–305

Raymond CA, Volker PW, Williams ER (1997) Provenance variation, genotype by environment interactions and age-age correlations for *Eucalyptus regnans* on nine sites in south eastern Australia. For Genet 4:235–251

Raymond CA, Kube PD, Pinkard L, Savage L, Bradley AD (2004) Evaluation of non-destructive methods of measuring growth stress in *Eucalyptus globulus*: relationships between strain, wood properties and stress. Fo Ecol Manage 190:187–200

Read DJ, Lewis DH, Fitter AH, Alexander IJ (eds) (1992) Mycorrhizas in ecosystems. CABI, Oxford

Reich PB, Walters MB, Krause SC, Vanderklein DW, Raffa KF, Tabone T (1993) Growth, nutrition and gas exchange of *Pinus resinosa* following artificial defoliation. Trees 7:67–77

Reid DEB, Lieffers VJ, Silins U (2004) Growth and crown efficiency of height repressed lodgepole pine; are suppressed trees more efficient? Trees 18:390–398

Reid JB, Hasan O, Moncur MW, Hetherington S (1995) Paclobutrazol as a management tool for tree breeders to promote early and abundant seed production. In: Potts BM, Borralho NMG, Reid JB, Cromer RN, Tibbits WN, Raymond CA (eds) Eucalypt plantations: improving fibre yield and quality. Proceedings CRCTHF-IUFRO conference, Hobart, 19–24 February. Cooperative Research Centre for Temperate Hardwood Forestry, Hobart, pp 293–298

Retuerto R, Fernandez-Lema B, Rodriguez-Roiloa, Obeso JR (2004) Increased photosynthetic performance in holly trees infested by scale insects. Funct Ecol 18:664–669

Reutebuch SE, Hartsough BR (1995) Early pruning of Douglas-fir: production rates and product implications In: Hanley DP, Oliver CD, Maguire DA, Briggs DG, Fight RD (eds) Forest pruning and wood quality. Contribution no 77. Institute of Forest Resources, University of Washington, Seattle, pp 370–385

Richards BN, Bevege DI (1972) Principles and practice of foliar analysis as a basis for crop-logging in pine plantations. Plant Soil 36:109–119

Richardson B (1993) Vegetation management practices in plantation forests of Australia and New Zealand. Can J For Res 23:1989–2005

Riitters K, Brodie JD (1984) Implementing optimal thinning strategies. For Sci 30:82–85

Rikala R, Heiskanen J, Lahti M (2004) Autumn fertilization in the nursery affects growth of *Picea abies* container seedlings after transplanting. Scand J For Res 19:400–414

Risenhoover KL, Maass SA (1987) The influence of moose on the composition and structure of Isle Royale forests. Can J For Res 17:357–364

Ritchie GA, Dunlap JR (1980) Root growth potential: its development and expression in forest tree seedlings. N Z J For Sci 10:218–248

Roberts ER, McCormack RJ (1991) Thinning technologies. In: Kerruish CM, Rawlins WHM (eds) The young eucalypt report. CSIRO, Melbourne. pp 50–106

Robinson KM, Karp A, Taylor G (2004) Defining leaf traits linked to yield in short-rotation coppice *Salix*. Biomass Bioenergy 26:417–431
Robison DJ (2002) Deploying pest resistance in genetically-limited forest plantations: developing ecologically-based strategies for managing risk. In: Wagner MR, Clancy KM, Lieutier F, Paine TD (eds) Mechanisms and deployment of resistance in trees to insects. Kluwer, Dordrecht, pp 169–188
Rodgers AR, Williams D, Sinclair ADE, Sullivan TP, Andersen RJ (1993) Does nursery production reduce antiherbivore defences of white spruce? Evidence from feeding experiments with snowshoe hares. Can J For Res 23:2358–2361
Rodríguez R, Sánchez F, Gorgoso J, Castedo F, López C, v Gadow K (2002) Evaluating standard treatment options for *Pinus radiata* plantations in Galicia (north-western Spain). Forestry 75:273–284
Roig,S, del Río,M, Cañellas I, Montero G (2005) Litter fall in Mediterranean *Pinus pinaster* Ait. stands under different thinning regimes. For Ecol Manage 206:179–190
Romanyà J, Vallejo VR (2004) Productivity of *Pinus radiata* plantations in Spain in response to climate and soil. For Ecol Manage 195:177–189
Rook DA (1971) Effect of undercutting and wrenching on growth of *Pinus radiata* D.Don seedlings. J Appl Ecol 8:477–490
Rose R, Rosner L (2005) Eight-year response of Douglas-fir seedlings to area of weed control and herbaceous versus woody weed control. Ann For Sci 62:481–492
Rosenqvist H, Dawson M (2005a) Economics of using wastewater irrigation of willow in Northern Ireland. Biomass Bioenergy 29:83–92
Rosenqvist H, Dawson M (2005b) Economics of willow growing in Northern Ireland. Biomass Bioenergy 28:7–14
Rosenqvist H, Aronsson P, Hasselgren K, Perttu K (1997) Economics of using municipal wastewater irrigation of willow coppice crops. Biomass Bioenergy 12:1–8
Roy V, Plamondon AP, Bernier PY (2004) Persistence of early growth of planted *Picea mariana* seedlings following clear-cutting and drainage in Quebec wetlands. Can J For Res 34:1157–1160
Ruark GA, Bockheim JG (1988) Biomass, net primary production, and nutrient distribution for an age sequence of *Populus tremuloides* ecosystems. Can J For Res 18:435–443
Ruel J-C, Larouche C, Achim A (2003) Changes in root morphology after precommercial thinning in balsam fir stands. Can J For Res 33:2452–2459
Running SW, Coughlan JC (1988) FOREST-BGC, a general model of forest ecosystem processes for regional applications. I. Hydrological balance, canopy gas exchange and primary production processes. Ecol Model 42:125–154
Running SW, Gower ST (1991) FOREST-BGC, a general model of forest ecosystem processes for regional applications. II. Dynamic carbon allocation and nitrogen budgets. Tree Physiol 9:147–160
Russell K (1995) Sealing of pruning wounds and associated diseases, and prevention of disease by pruning. In: Hanley DP, Oliver CD, Maguire DA, Briggs DG, Fight RD (eds) Forest pruning and wood quality. Contribution no 77. Institute of Forest Resources, University of Washington, Seattle, pp 238–244
Ryan MG, Binkley D, Fownes JH (1997) Age-related decline in forest productivity: pattern and process. Adv Ecol Res 27:213–262
Ryan PJ, Harper RJ, Laffan M, Booth TH, McKenzie NJ (2002) Site assessment for farm forestry in Australia and its relationship to scale, productivity and sustainability. For Ecol Manage 171:133–152
Rytter L, Stener L-G (2005) Productivity and thinning effects in hybrid aspen (*Populus tremula* L. × *P. tremuloides* Michx.) stands in southern Sweden. Forestry 78:285–295
Rytter R-M, Rytter L (1998) Growth, decay, and turnover rates of fine roots of basket willows. Can J For Res 28:893–902

Sakai A, Sakai S, Akiyama F (1997) Do sprouting tree species on erosion-prone sites carry large reserves of resources? Ann Bot 79:625–630

Salminen H, Varmola M (1993) Influence of initial spacing and planting design on the development of young Scots pine (*Pinus sylvestris* L.) stands. Silva Fenn 27:21–28

Sampson DA, Albaugh TJ, Johnsen KH, Allen HL, Zarnoch SJ (2003) Monthly leaf area index estimates from point-in-time measurements and needle phenology for *Pinus taeda*. Can J For Res 33:2477–2490

Sands PJ, Battaglia M, Mummery D (2000) Application of process-based models to forest management: experience with ProMod, a simple plantation productivity model. Tree Physiol 20:383–392

Sands R (1984) Transplanting stress in radiata pine. Aust For Res 14:67–72

Sands R, Nambiar EKS (1984) Water relations of *Pinus radiata* in competition with weeds Can J For Res 14:233–237

Sands R, Greacen EL, Gerard CJ (1979) Compaction of sandy soils in radiata pine forests. I. A penetrometer study. Aust J Soil Res 17:101–113

Sands R, Nugroho PB, Leung DWM, Sun OJ, Clinton PW (2000) Changes in soil CO_2 and O_2 concentrations when radiata pine is grown in competition with pasture or weeds and possible feedbacks with radiata pine root growth and respiration. Plant Soil 225:213–225

Santantonio D, Santantonio E (1987) Effect of thinning on production and mortality of fine roots in a *Pinus radiata* plantation on a fertile site in New Zealand. Can J For Res 17:919–928

Savill PS, Fennessy J, Samuel CJA (2005) Approaches in Great Britain and Ireland to the genetic improvement of broadleaved trees. Forestry 78:63–73

Schaetzl RJ, Johnson DL, Burns SF, Small TW (1989) Tree uprooting: a review of terminology, process, and environmental implications. Can J For Res 19:1–11

Schewe AM, Stewart JM (1986) Twig weight-diameter relationships for selected browse species on the Duck Mountain Forest Reserve, Manitoba. Can J For Res 16:675–680

Schirmer J, Tonts M (2003) Plantations and sustainable rural communities. Aust For 66:67–74

Scholes H (1998) Can energy crops become a realistic CO_2 mitigation option in south west England? Biomass Bioenergy 15:333–344

Schönau APG, Coetzee J (1989) Initial spacing, stand density and thinning in eucalypt plantations. For Ecol Manage 29:245–266

Schönau APG, Gardener RAW (1991) Eucalypts for colder areas in southern Africa. In: Proceedings of IUFRO symposium on intensive forestry: the role of eucalypts, Durban, South Africa, pp 467–479

Schowalter TD (2000) Insect ecology. An ecosystem approach. Academic, San Diego

Schulze E-D, Beck E, Müller-Hohenstein K (2005) Plant ecology. Springer, Berlin Heidelberg New York

Schwinning S, Weiner J (1998) Mechanisms determining the degree of size asymmetry in competition among plants. Oecologia 113:447–455

Scott DA, Burger JA, Kaczmarek DJ, Kane MB (2004a) Nitrogen supply and demand in short-rotation sweetgum plantations. For Ecol Manage 189:331–343

Scott DA, Burger JA, Kaczmarek DJ, Kane MB (2004b) Growth and nutrition response of young sweetgum plantations to repeated nitrogen fertilization on two site types. Biomass Bioenergy 27:313–325

Sequeira W, Gholz HL (1991) Canopy structure, light penetration and tree growth in a slash pine (*Pinus elliottii*) silvo-pastoral system at different stand configurations in Florida. For Chron 67:263–267

Seymour RS, Smith DM (1987) A new stocking guide formulation applied to eastern white pine. For Sci 33:469–484

Shao G, Shugart GF (1997) A compatible growth-density stand model derived from a distance-dependent individual tree model. For Sci 43:443–446
Sharma M, Burkhart HE, Amateis RL (2002) Spacing rectangularity effect on the growth of loblolly pine plantations. Can J For Res 32:1451–1459
Shelbourne CJA (1969) Tree breeding methods. Forest Research Institute, technical paper no 55. New Zealand Forest Service, Wellington
Shelbourne CJA, Carson MJ, Wilcox MD (1989) New techniques in the genetic improvement of radiata pine. Commonw For Rev 68:191–201
Sheppard SRJ (2005) Participatory decision support for sustainable forest management: a framework for planning with local communities at the landscape level in Canada. Can J For Res 35:1515–1526
Sheppard SRJ, Meitner M (2005) Using multi-criteria analysis and visualisation for sustainable forest management planning with stakeholder groups. For Ecol Manage 207:171–187
Shoulders E, Tiarks AE (1980) Predicting height and relative performance of major southern pines from rainfall, slope, and available soil moisture. For Sci 26:437–447
Sierra-Lucero V, McKeand SE, Huber DA, Rockwood DL, White TL (2002) Performance differences and genetic parameters for four coastal provenances of loblolly pine in the southeastern United States. For Sci 48:732–742
Siipilehto J, Heikkilä R (2005) The effects of moose browsing on the height structure of Scots pine saplings in a mixed stand. For Ecol Manage 205:117–126
Simpson JA, Ades PK (1990) Screening *Pinus radiata* families and clones for disease and pest insect resistance. Aust For 53:194–199
Sims REH, Senelwa K, Maiava T, Bullock BT (1999) *Eucalyptus* species for biomass energy in New Zealand—part II: coppice performance. Biomass Bioenergy 17:333–343
Sjolte-Jorgensen J (1967) The influence of spacing on the growth and development of coniferous plantations. In: Romberger JA, Mikola P (eds) International review of forestry research, vol 2, Academic, New York, pp 43–94
Slodičák M (1995) Thinning regime in stands of Norway spruce subjected to wind and snow damage. In: Coutts MP, Grace J (eds) (1995) Wind and trees. Cambridge University Press, Cambridge, pp 436–447
Smethurst P, Holz G, Moroni M, Baillie C (2004) Nitrogen management in *Eucalyptus nitens* plantations. For Ecol Manage 193:63–80
Smethurst PJ (2000) Soil solution and other soil analyses as indicators of nutrient supply: a review. For Ecol Manage 138:397–411
Smethurst PJ, Nambiar EKS (1989) Role of weeds in the management of nitrogen in a young *Pinus radiata* plantation. New For 3:203–224
Smethurst PJ, Nambiar EKS (1990) Effects of slash and litter management on fluxes of nitrogen and tree growth in a young *Pinus radiata* plantation. Can J For Res 20:1498–1507
Smith CW, Little KM, Norris CH (2001) The effect of land preparation at re-establishment on the productivity of fast growing hardwoods in South Africa. Aust For 64:165–174
Smith FW, Long JN (1989) The influence of canopy architecture on stemwood production and growth efficiency of *Pinus contorta* var. *latifolia*. J App Ecol 26:681–691
Smith FW, Long JN (2001) Age-related decline in forest growth: an emergent property. For Ecol Manage 144:175–181
Smith JP, Hoffman JT (2001) Site and stand characteristics related to white pine blister rust in high-elevation forests of southern Idaho and western Wyoming. West North Am Nat 61:409–416
Smith NJ (1986) A model of stand allometry and biomass allocation during the self-thinning process. Can J For Res 16:990–995

Smith NJ (1989) A stand-density control diagram for western red cedar, *Thuja plicata*. For Ecol Manage 27:235–244

Smith NJ, Hann DW (1986) A growth model based on the self-thinning rule. Can J For Res 16:330–334

Smith SE, Read DJ (1997) Mycorrhizal symbiosis, 2nd edn. Academic, San Diego

Sniezko RA, Mullin LJ (1987) Taxonomic implication of bush pig damage and basal shoots in *Pinus tecunumanii*. Commonw For Rev 66:313–316

Snowdon P (2002) Modeling type 1 and type 2 growth responses in plantations after fertilizer or other silvicultural treatments. For Ecol Manage 163:229–244

Solberg S, Andreassen K, Clarke N, Tørseth K, Tveito OE, Strand GH, Tomter S (2004) The possible influence of nitrogen and acid deposition on forest growth in Norway. For Ecol Manage 192:241–249

Somerville A (1979) Root anchorage and root morphology of *Pinus radiata* on a range of ripping treatments. N Z J For Sci 9:294–315

Somerville A (1981) Wind-damage profiles in a *Pinus radiata* stand. N Z J For Sci 11:75–78

South DB, Mitchell RJ, Zutter BR, Balneaves JM, Barber BL, Nelson DG, Zwolinski JB (1993a) Integration of nursery practices and vegetation management: economic and biological potential for improving regeneration. Can J For Res 23:2083–2092

South DB, Zwolinski JB, Donald DGM (1993b) Interaction among seedling diameter grade, weed control and soil cultivation for *Pinus radiata* in South Africa. Can J For Res 23:2078–2082

South DB, Harris SW, Barnett JP, Hainds MJ, Gjerstad DH (2005) Effects of container type and seedling size on survival and early height growth of *Pinus palustris* seedlings in Alabama, U.S.A. For Ecol Manage 204:385–98

Spencer RD, Jellinek LO (1995) Public concerns about pine plantations in Victoria. Aust For 58:99–106

Spencer RD, Bugg AL, Frakes IA (2003) Evaluating tradeoffs in regional forest planning. Aust For 66:120–128

Spiecker H, Mielikäinen K, Köhl M, Skovsgaard J (eds) (1996) Growth trends in European forests. Springer, Berlin Heidelberg New York

Srivastava S, Narula A (eds) (2005) Plant biotechnology and molecular markers. Springer, Berlin Heidelberg New York

Stape JL, Gonçalves JLM, Gonçalves AN (2001) Relationship between nursery practices and field performance for *Eucalyptus* plantations in Brazil. New For 22:19–41

Stape JL, Binkley D, Ryan MG (2004) Eucalyptus production and the supply, use and efficiency of use of water, light and nitrogen across a geographic gradient in Brazil. For Ecol Manage 193:17–31

Stegemoeller KA, Chappell HN (1990) Growth response of unthinned and thinned Douglas-fir stands to single and multiple applications of nitrogen. Can J For Res 20:343–349

Stener L-G, Jansson G (2005) Improvement of *Betula pendula* by clonal and progeny testing of phenotypically selected trees. Scand J For Res 20:292–303

Sterba H, Amateis RL (1998) Crown efficiency in a loblolly pine (*Pinus taeda*) spacing experiment. Can J For Res 28:1344–1351

Sterba H, Monserud RA (1993) The maximum density concept applied to uneven-aged mixed-species stands. For Sci 39:432–452

Stogsdill WR, Whittwer RF, Hennessey TC, Dougherty PM (1992) Water use in thinned loblolly pine plantations. For Ecol Manage 50:233–245

Stokes A, Fitter AH, Coutts MP (1995) Responses of young trees to wind: effects on root growth. In: Coutts MP, Grace J (eds) (1995) Wind and trees. Cambridge University Press, Cambridge, pp 264–275

Stone C, Birk E (2001) Benefits of weed control and fertiliser application to young *Eucalyptus dunnii* stressed from waterlogging and insect damage. Aust For 64:151–158

Stone C, Simpson JA (1987) Influence of *Ips grandicollis* on the incidence and spread of bluestain fungi in *Pinus elliottii* billets in north-eastern New South Wales. Aust For 50:86–94

Stone C, Chisholm LA, McDonald S (2003) Spectral reflectance characteristics of *Pinus radiata* needles affected by dothistroma needle blight. Can J Bot 81:560–569

Studholme WP (1995) The experience of and management strategy adopted by the Selwyn Plantation Board, New Zealand. In: Coutts MP, Grace J (eds) (1995) Wind and trees. Cambridge University Press, Cambridge, pp 468–476

Sullivan TP (1993) Feeding damage by bears in managed forests of western hemlock-western red cedar in midcoastal British Columbia. Can J For Res 23:49–54

Sullivan TP, Sullivan DS (1982) Barking damage by snowshoe hares and red squirrels in lodgepole pine stands in central British Columbia. Can J For Res 12:443–448

Sullivan TP, Sullivan DS (1986) Impact of feeding damage by snowshoe hares on growth rates of juvenile lodgepole pine in central British Columbia. Can J For Res 16:1145–1149

Sullivan TP, Vyse A (1987) Impact of red squirrel feeding damage on spaced stands of lodgepole pine in the Cariboo region of British Columbia. Can J For Res 17:666–674

Sullivan TP, Jackson WT, Pojar J, Banner A (1986) Impact of feeding damage by the porcupine on western hemlock-Sitka spruce forests of north-coastal British Columbia. Can J For Res 16:642–647

Sullivan TP, Coates H, Jozsa LA, Diggle PK (1993) Influence of feeding damage by small mammals on tree growth and wood quality in young lodgepole pine. Can J For Res 23:799–809

Sullivan TP, Krebs JA, Diggle PK (1994) Prediction of stand susceptibility to feeding damage by red squirrels in young lodgepole pine. Can J For Res 24:14–20

Sutton RF (1980) Planting stock quality, root growth capacity, and field performance of three boreal conifers. N Z J For Sci 10:54–71

Swenson JJ, Waring RH, Fan W, Coops N (2005) Predicting site index with a physiologically based growth model across Oregon, USA. Can J For Res 35:1697–1707

Szujecki A (1987) Ecology of forest insect pests. Junk, Dordrecht

Tahvanainen L, Rytkönen V-M (1999) Biomass production of *Salix viminalis* in southern Finland and the effect of soil properties and climate conditions on its production and survival. Biomass Bioenergy 16:103–117

Tait DE, Cieszewski CJ, Bella IE (1988) The stand dynamics of lodgepole pine. Can J For Res 18:1255–1260

Tang Z, Chambers JL, Guddanti S, Barnett JP (1999) Thinning, fertilization, and crown position interact to control physiological responses of loblolly pine. Tree Physiol 19:87–94

Tani N, Takahashi T, Ujino-Ihara T, Iwata H, Yoshimura K, Tsumura Y (2004) Development and characteristics of microsatellite markers for sugi (*Cryptomeria japonica* D. Don) derived from microsatellite-enriched libraries. Ann For Sci 61:569–575

Tasissa G, Burkhart HE (1998) An application of mixed effects analysis to modeling thinning effects on stem profile of loblolly pine. For Ecol Manage 103:87–101

Taylor JS, Blake TJ, Pharis RP (1982) The role of plant hormones and carbohydrates in the growth and survival of coppiced *Eucalyptus* seedlings. Physiol Plant 55:421–430

Teeter L, Bliss JC, Henry WA (1993) Adoption of herbaceous weed control by southern forest industry. Can J For Res 23:2312–2316

Terlesk CJ, McConchie M (1988) Stand reorganisation: results from the trials at Hautu Forest, New Zealand. N Z J For Sci 18:329–344

Teste FP, Schmidt MG, Berch SM, Bulmer C, Egger KN (2004) Effects of ectomycorrhizal inoculants on survival and growth of interior Douglas-fir seedlings on reforestation sites and partially rehabilitated landings. Can J For Res 34:2074–2088

Thakur ML (2000) Forest entomology. Sai, Dehra Dun

Theiveyanathan S, Benyon RG, Marcar NE, Myers BJ, Polglase PJ, Falkiner RA (2004) An irrigation-scheduling model for application of saline water to tree plantations. For Ecol Manage 193:97–112

Theodorou C, Cameron JN, Bowen GD (1991) Root characteristics of several *Pinus radiata* genotypes growing on different sites in Gippsland. Aust For 54:40–51

Thiffault N, Jobidon R, Munson AD (2003) Performance and physiology of large containerized and bare-root spruce seedlings in relation to scarification and competition in Québec (Canada). Ann For Sci 60:645–655

Thiffault N, Titus BD, Munson AD (2005) Silvicultural options to promote seedling establishment on *Kalmia-Vaccinium*-dominated sites. Scand J For Res 20:110–121

Thomas RC, Mead DJ (1992a) Uptake of nitrogen by *Pinus radiata* and retention within the soil after applying ^{15}N-labelled urea at different frequencies. 1. Growth response and nitrogen budgets. For Ecol Manage 53:131–151

Thomas RC, Mead DJ (1992b) Uptake of nitrogen by *Pinus radiata* and retention within the soil after applying ^{15}N-labelled urea at different frequencies. 2. Seasonal trends and processes. For Ecol Manage 53:153–174

Thomson AJ (1986) Trend surface analysis of spatial patterns of tree size, microsite effects, and competitive stress. Can J For Res 16:279–282

Thwaites R (2002) Spatial terrain analysis for matching native tree species to sites: a methodology. New For 24:81–95

Thwaites RN, Slater BK (2000) Soil-landscape resource assessment for plantations—a conceptual framework towards an explicit multi-scale approach. For Ecol Manage 138:123–138

Tibbits W, Hodge G (1998) Genetic parameters and breeding value predictions for *Eucalyptus nitens* wood fiber production traits. For Sci 44:587–598

Timmer VR, Armstrong G (1987) Growth and nutrition of containerized *Pinus resinosa* at exponentially increasing nutrient additions. Can J For Res 17:644–647

Timmer VR, Armstrong G, Miller BD (1991) Steady-state nutrient preconditioning and early outplanting performance of containerized black spruce seedlings. Can J For Res 21:585–594

Tingey DT, Phillips DL, Johnson MG, Rygiewicz PT, Beedlow PA, Hogsett WE (2005) Estimates of Douglas-fir fine root production and mortality from minirhizotrons. For Ecol Manage 204:359–370

Toivonen RT, Tahvanainen LJ (1998) Profitability of willow cultivation for energy production in Finland. Biomass Bioenergy 15:27–37

Toro J, Gessel SP (1999) Radiata pine plantations in Chile. New For 18:33–44

Trenbath BR (1974) Biomass productivity of mixtures. Adv Agron 26:177–210

Tschaplinski TJ, Blake TJ (1989a) Photosynthetic reinvigoration of leaves following shoot decapitation and accelerated growth of coppice shoots. Physiol Plant 75:157–165

Tschaplinski TJ, Blake TJ (1989b) The role of sink demand in carbon partitioning and photosynthetic reinvigoration following shoot decapitation. Physiol Plant 75:166–173

Tschaplinski TJ, Blake TJ (1994) Carbohydrate mobilization following shoot defoliation and decapitation in hybrid poplar. Tree Physiol 14:141–151

Tschaplinski TJ, Blake TJ (1995) Growth and carbohydrate status of coppice shoots of hybrid poplar following shoot pruning. Tree Physiol 15:333–338

Turjaman M, Tamai Y, Segah H, Limin SH, Cha JY, Osaki M, Tawaraya K (2005) Inoculation with the ectomycorrhizal fungi *Pisolithus arhizus* and *Scleroderma* sp. improves early growth of *Shorea pingana* nursery seedlings. New For 30:67–73

Turnbull CRA, Beadle CL, McLeod R, Cherry ML (1997) Clearing with excavators and nitrogen fertiliser increases the yield of *Eucalyptus nitens* in plantations established on a native forest site in southern Tasmania. Aust For 64:109–115

Turner J, Knott JH, Lambert M (1996) Fertilization of *Pinus radiata* plantations after thinning. I Productivity gains. Aust For 59:7–21

Turner J, Gessel SP, Lambert MJ (1999) Sustainable management of native and exotic plantations in Australia. New For 18:17–32

Turner J, Lambert MJ, Hopmans P, McGrath J (2001) Site variation in *Pinus radiata* plantations and implictations for site specific management. New For 21:249–282

Turvey ND, Cameron JN (1986a) Site preparation for a second rotation of radiata pine: soil and foliage chemistry, and effect on tree growth. Aust For Res 16:9–19

Turvey ND, Cameron JN (1986b) Site preparation for a second rotation of radiata pine: growing costs and production of wood and kraft pulp. Aust For 49:160–165

Turvey ND, Booth TH, Ryan PJ (1990) A soil technical classification system for *Pinus radiata* (D.Don) plantations. II. A basis for estimation of crop yield. Aust J Soil Res 28:813–824

Udawatta RP, Nygren P, Garrett HE (2005) Growth of three oak species during establishment of an agrofresry practice for watershed protection. Can J For Res 35:602–609

Updegraff K, Baughman MJ, Taff SJ (2004) Environmental benefits of cropland conversion to hybrid poplar: economic and policy considerations. Biomass Bioenergy 27:411–428

Urban ST, Lieffers VJ, Macdonald SE (1994) Release in the trunk and structural roots of white spruce as measured by dendrochronology. Can J For Res 24:1550–1556

Uzoh FCC (2001) A height increment equation for young ponderosa pine plantations using precipitation and soil factors. For Ecol Manage 142:193–203

Valentine HT, Mäkela A (2005) Bridging process-based and empirical approaches to modeling tree growth. Tree Physiol 25:769–779

Valinger E (1992) Effects of thinning and nitrogen fertilization on stem growth and form of *Pinus sylvestris* trees. Scand J For Res 7:219–228

Valinger E, Elfving B, Mörling T (2000) Twelve-year growth response of Scots pine to thinning and nitrogen fertilisation. For Ecol Manage 134:45–53

van Buijtenen JP (2001) Genomics and quantitative genetics. Can J For Res 31:617–622

van den Broek R, van den Burg T, van Wijk A, Turkenburg W (2000) Electricity generation from eucalyptus and bagasse by sugar mills in Nicaragua: a comparison with fuel oil electricity generation on the basis of costs, macro-economic impacts and environmental emissions. Biomass Bioenergy 19:311–335

van den Driessche R (1982) Relation between spacing and nitrogen fertilization of seedlings in the nursery, seedling size, and outplanting performance. Can J For Res 12:865–875

van den Driessche R (1983) Growth, survival, and physiology of Douglas-fir seedlings following root wrenching and fertilization. Can J For Res 13:270–278

van den Driessche R (1984) Relation between spacing and nitrogen fertilization of seedlings in the nursery, seedling mineral nutrition, and outplanting performance. Can J For Res 14:431–436

van den Driessche R (1985) Late-season fertilization, mineral nutrient reserves, and retranslocation in planted Douglas fir (*Pseudotsuga menziesii* (Mirb.) Franco) seedlings. For Sci 31:485–496

van den Driessche R (1992) Changes in drought resistance and root growth capacity of container seedlings in response to nursery drought, nitrogen, and potassium treatments. Can J For Res 22:740–749

van der Pas JB (1981) Reduced early growth rates of *Pinus radiata* caused by *Dothistroma pini*. N Z J For Sci 11:210–220 van der Pas JB, Bulman L, Horgan GP (1984) Disease control by aerial spraying of *Dothistroma pini* in tended stands of *Pinus radiata* in New Zealand. N Z J For Sci 14:23–40

van Frankenhuyzen K, Beardmore T (2004) Current status and environmental impact of transgenic forest trees. Can J For Res 34:1163–1180

van Heerden SW, Wingfield MJ (2001) Genetic diversity of *Cryphonectria cubensis* isolates in South Africa. Mycol Res 105:94–99

van Heerden SW, Wingfield MJ (2002) Effect of environment on the response of *Eucalyptus* clones to inoculation with *Cryphonectria cubensis*. For Pathol 32:395–402

van Staden V, Erasmus BFN, Roux J, Wingfield MJ, van Jaarsveld AS (2004) Modelling the spatial distribution of two important South African plantation forestry pathogens. For Ecol Manage 187:61–63

van Zyl LM, Wingfield MJ (1998) Ethylene production by *Eucalyptus* clones in response to infection by hypovirulent and virulent isolates of *Cryphonectria cubensis*. S Afr J Sci 94:193–194

van Zyl LM, Wingfield MJ (1999) Wound response of *Eucalyptus* clones after inoculation with *Cryphonectria cubensis*. Eur J For Pathol 29:161–167

Vanclay JK (1995) Growth models for tropical forests: a synthesis of models and methods. For Sci 41:7–42

VanderSchaaf C, McNabb K (2004) Winter nitrogen fertilization of loblolly pine seedlings. Plant Soil 265:295–299

VanderSchaaf CL, South DB (2003) Effect of planting depth on growth of open-rooted *Pinus elliottii* and *Pinus taeda* seedlings in the United States. South Afri For J 198:63–73

Varelides C, Kritikos T (1995) Effects of site preparation intensity and fertilisation on *Pinus pinaster* survival and height growth on three sites in northern Greece. For Ecol Manage 73:111–115

Varelides C, Varelides Y, Kritikos T (2005) Effects of mechanical site preparation and fertilisation on early growth and survival of a black pine plantation in northern Greece. New For 30:21–32

Varmola M, Salminen H (2004) Timing and intensity of precommercial thinning in *Pinus sylvestris* stands. Scand J For Res 19:142–151

Velaquez-Martinez A, Perry DA, Bell TE (1992) response of aboveground biomass increment, growth efficiency, and foliar nutrients to thinning, fertilization, and pruning in young Douglas-fir plantations in the central Oregon Cascades. Can J For Res 22:1278–1289

Verbyla DL, Fisher RF (1989) An alternative approach to conventional soil-site regression modeling. Can J For Res 19:179–184

Vergara R, White TL, Huber DA, Shiver BD, Rockwood DL (2004) Estimated realized gains for first-generation slash pine (*Pinus elliottii* var. *elliottii*) tree improvement in the southeastern United States. Can J For Res 34:2587–2600

Villar-Salvador P, Ocaña L, Peñuelas J, Carrasco I (1999) Effect of water stress conditioning on the water relations, root growth capacity, and the nitrogen and non-structural carbohydrate concentrations of *Pinus halepensis* Mill. (aleppo pine) sedlings. Ann For Sci 56:459–465

Villar-Salvador P, Planelles R, Enríquez E, Rubira JP (2004a) Nursery cultivation regimes, plant functional attributes, and field performance relationships in the Mediterranean oak *Quercus ilex* L. For Ecol Manage 196:257–266

Villar-Salvador P, Planelles R, Oliet J, Peñuelas-Rubira JL, Jacobs DF, González M (2004b) Drought tolerance and transplanting performance of holm oak (*Quercus ilex*) seedlings after drought hardening in the nursery. Tree Physiol 24:1147–1155

Vogt K, Asbjornsen H, Ercelawn A, Montagnini F, Valdés M (1997) Roots and mycorrhizas in plantation ecosystems. In: Nambiar EKS, Brown AG (eds) Management of soil, nutrients and water in tropical plantation forests. Monograph no 43. Australian Centre for International Agricultural Research Canberra, pp 247–296

Volker PW (2002) Quantitative genetics of *Eucalyptus globulus*, *E. nitens* and their F_1 hybrid. PhD thesis, University of Tasmania

von Wuehlisch G, Muhs H-J, Geburek T (1990) Competitive behaviour of clones of *Picea abies* in monoclonal mosaics vs. intimate clonal mixtures. A pilot study. Scand J For Res 5:397–401

Wagner RG (1993) Research directions to advance forest vegetation management in North America. Can J For Res 23:2317–2327

Walter C, Carson SD, Menzies MI, Richardson T, Carson M (1998) Review: application of biotechnology to forestry–molecular biology of conifers. World J Microbiol Biotechnol 14:321–330

Walters BB, Cadelina A, Cardano A, Visitacion E (1999) Community history and rural development: why some farmers participate more readily than others. Agric Syst 59:193–214

Walters BB, Sabogal C, Snook LK, de Alemeida E (2005) Constraints and opportunities for better silvicultural practice in tropical forestry: an interdisciplinary approach. For Ecol Manage 209:3–18

Wang JR, Simard SW, Kimmins JP (1995) Physiological responses of paper birch to thinning in British Columbia. For Ecol Manage 73:177–184

Wang S-Y, Lin C-J, Chiu C-M (2003) Effects of thinning and pruning on knots and lumber recovery of Taiwania (*Taiwania cryptomerioides*) planted in the Lu-Kei area. J Wood Sci 49:444–449

Wang YP, Jarvis PG, Taylor CMA (1991) PAR absorption and its relation to above-ground dry matter production of Sitka spruce. J Appl Ecol 28:547–560

Wang Y, Raulier F, Ung C-H (2005) Evaluation of spatial predictions of site index obtained by parametric and nonparametric methods—A case study of lodgepole pine productivity. For Ecol Manage 214:201–211

Ward SC, Pickersgill GE, Michaelsen DV, Bell DT (1985) Responses to factorial combinations of nitrogen, phosphorus and potassium fertilizers by saplings of *Eucalyptus saligna* Sm., and the prediction of the response by DRIS indices. Aust For Res 15:27–32

Wardlaw TJ, Kile GA, Dianese JC (2000) Diseases of eucalypts associated with viruses, phytoplasmas, bacteria and nematodes. In: Keane PJ, Kile GA, Podger FD, Brown BN (eds) Diseases and pathogens of eucalypts. CSIRO, Melbourne, pp 339–352

Waring KM, O'Hara KL (2005) Silvicultural strategies in forest ecosystems affected by introduced pests. For Ecol Manage 209:27–41

Washusen R (2002) Tension wood occurrence in *Eucalyptus globulus* Labill. II. The spatial distribution of tension wood and its association with stem form. Aust For 65:127–134

Washusen R, Evans R (2001) Prediction of wood tangential shrinkage from cellulose crystallite width and density in one 11-year-old tree of *Eucalyptus globulus* Labill. Aust For 64:123–126

Washusen R, Clark N (2005) Integration of sawn timber and pulpwood production. In: Nambiar S, Ferguson I (eds) New forests. CSIRO, Melbourne, pp 185–207

Washusen R, Blakemore P, Northway R, Vinden P, Waugh G (2000a) Recovery of dried appearance grade timber from *Eucalyptus globulus* Labill. grown in plantations in medium rainfall areas of the southern Murray-Darling Basin. Aust For 63:277–283

Washusen R, Waugh G, Hudson I, Vinden P (2000b) Appearance product potential of plantation hardwoods from medium, rainfall areas of the southern Murray-Darling Basin. Green product recovery. Aust For 63:66–71

Washusen R, Ades P, Vinden P (2002) Tension wood occurrence in *Eucalyptus globulus* Labill. I. The spatial distribution of tension wood in one 11-year-old tree. Aust For 65:120–126

Watt MS, Whitehead D, Mason EG, Richardson B, Kimberley MO (2003) The influence of weed competition for light and water on growth and dry matter partitioning of young *Pinus radiata*, at a dryland site. For Ecol Manage 183:363–376

Watt MS, Kimberley MO, Richardson B, Whitehead D, Mason EG (2004) Testing a juvenile tree growth model sensitive to competition from weeds, using *Pinus radiata* at two contrasting sites in New Zealand. Can J For Res 34:1985–1992

Watt MS, Downes GM, Whitehead D, Mason EG, Richardson B, Grace JC, Moore JR (2005a) Wood properties of juvenile *Pinus radiata* growing in the presence and absence of competing understorey vegetation at a dryland site. Trees 19:580–586

Watt MS, Moore JR, McKinlay B (2005b) The influence of wind on branch characteristics of *Pinus radiata*. Trees 19:58–65

Waugh G, Rosza A (1991) Sawn products from regrowth *Eucalyptus regnans*. In: Kerruish CM, Rawlins WHM (eds) The young eucalypt report. CSIRO, Melbourne, pp 178–209

Webb DP, von Althen FW (1980) Storage of hardwood planting stock: effects of various storage regimes and packaging methods on root growth and physiological quality. N Z J For Sci 10:83–96

Webley OJ, Geary TF, Rockwood DL, Comer CW, Meskimen GF (1986) Seasonal coppicing variation in three eucalypts in southern Florida. Aust For Res 16:281–290

Weih M (2004) Intensive short rotation forestry in boreal climates: present and future perspectives. Can J For Res 34:1369–1378

Weih M, Nordh N-E (2005) Determinants of biomass production in hybrid willows and prediction of field performance from pot studies. Tree Physiol 25:1197–1206

Weiner J, Thomas SC (1986) Size variability and competition in plant monocultures. Oikos 47:211–222

West PW (1983) Comparison of stand density measures in even-aged regrowth eucalypt forest of southern Tasmania. Can J For Res 13:22–31

West PW (1991) Thinning response and growth modelling. In: Kerruish CM, Rawlins WHM (eds) The young eucalypt report. CSIRO, Melbourne. pp 28–49

West PW (2004) Tree and forest measurement. Springer, Berlin Heidelberg New York

West PW, Osler GHR (1995) Growth response to thinning and its relation to site resources in *Eucalyptus regnans* F. Muell. Can J For Res 25:69–80

West PW, Jackett DR, Borough CJ (1989) Competitive processes in a monoculture of *Pinus radiata* D. Don. Oecologia 81:57–61

West RJ (1989) Cone depredations by red squirrel in black spruce stands in Newfoundland: implications for commercial cone collection. Can J For Res 19:1207–1210

Westfall JA, Burkhart HE, Allen HL (2004) Young stand growth modeling for intensively-managed loblolly pine plantations in southeastern U.S. For Sci 50:823–835

Wheeler NC, Jech KS, Masters SA, O'Brien CJ, Timmons DW, Stonecypher RW, Lupkes A (1995) Genetic variation and parameter estimates in *Taxus brevifolia* (Pacific yew). Can J For Res 25:1913–1927

White DA, Kile GA (1991) Thinning damage and defect in regrowth eucalypts. In: Kerruish CM, Rawlins WHM (eds) The young eucalypt report. CSIRO, Melbourne, pp 152–177

White TL, Hodge GR (1989) Predicting breeding values with applications in forest tree improvement. Kluwer, Dordrecht

Whitehead D, Beadle CL (2004) Physiological regulation of productivity and water use in *Eucalyptus*: a review. For Ecol Manage 193:113–140

Whitehead D, Jarvis PG, Waring RH (1984) Stomatal conductance, transpiration, and resistance to water uptake in a *Pinus sylvestris* spacing experiment. Can J For Res 14:692–700

Whiteman A, Aglionby J (1997) The use of socio-economic data in conservation management planning: a case study from Danau Sentarum Wildlife Reserve, West Kalimantan, Indonesia. Commonw For Rev 76:239–245

Whittock SP, Greaves BL, Apiolaza LA (2004) A cash flow model to compare coppice and genetically improved seedling options for *Eucalyptus globulus* pulpwood plantations. For Ecol Manage 191:267–274

Whyte AGD (1998) Radiata pine silviculture in New Zealand: its evolution and future prospects. Aust For 51:185–196

Whyte AGD, Woollons RC (1990) Modelling stand growth of radiata pine thinned to varying densities. Can J For Res 20:1069–1076

Wikström P (2001) Effect of decision variable definition and data aggregation on a search process applied to a single-tree simulator. Can J For Res 31:1057–1066

Wikström P, Eriksson LO (2000) Solving the stand management problem under biodiversity-related considerations. For Ecol Manage 126:361–376

Wilkes P, Bren LJ (1986) Radiata pine pruning technology. Aust For 49:172–180

Wilkins AP (1986) Nature and origin of growth stresses in trees. Aust For 49:56–62

Wilkins AP (1990) Influence of silvicultural treatment on growth and wood density of *Eucalyptus grandis* grown on a previous pasture site. Aust For 53:168–172

Wilkins AP, Kitahara R (1991a) Relationship between growth strain and rate of growth in 22 year-old *Eucalyptus grandis*. Aust For 54:95–98

Wilkins AP, Kitahara R (1991b) Silvicultural treatments and associated growth rates, growth strains and wood properties in 12.5-year-old *Eucalyptus grandis*. Aust For 54:99–104

Wilkins RN, Marion WR, Neary DG, Tanner GW (1993) Vascular plant community dynamics following hexazinone site preparation in the lower coastal plain. Can J For Res 23:2216–2229

Wilkinson GR, Neilsen WA (1995) Implications of early browsing damage on the long term productivity of eucalypt forests. For Ecol Manage 74:117–124

Willebrand E, Ledin S, Verwijst T (1993) Willow coppice systems in short rotation forestry: effects of plant spacing, rotation length and clonal composition on biomass production. Biomass Bioenergy 4:323–331

Williams ER, Matheson AC, Harwood CE (2002) Experimental design and analysis for tree improvement. CSIRO, Melbourne

Williams K, Nettle R, Petheram RJ (2003) Public response to plantation forestry in southwestern Victoria. Aust For 66:92–99

Wilson J, Mason PA, Last FT, Ingleby K, Munro RC (1987) Ectomycorrhiza formation and growth of Sitka spruce seedlings on first-rotation forest sites in northern Britain. Can J For Res 17:957–963

Wilson JS, Oliver CD (2000) Stability and density management in Douglas-fir plantations. Can J For Res 30:910–920

Wingfield MJ (1999) Pathogens in exotic plantation forestry. Int For Rev 1:163–168

Wingfield MJ (2003) Increasing threat of diseases to exotic plantation forests in the southern hemisphere: lessons from *Cryphonectria* canker. Aust Plant Pathol 32:133–139

Witcosky JJ, Schowalter TD, Hansen EM (1986) The influence of time of precommercial thinning on the colonization of Douglas-fir by three species of root-colonizing insects. Can J For Res 16:745–749

Wollenweber GC, Wollenweber FG (1995) Forest wind damage risk assessment for environmental impact studies. In: Coutts MP, Grace J (eds) (1995) Wind and trees. Cambridge University Press, Cambridge, pp 404–423

Wonn HT, O'Hara KL (2001) Height:diameter ratios and stability relationships for four northern rocky mountain tree species. West J Appl For 16:87–94

Woodcock DW, Shier AD (2002) Wood specific gravity and its radial variations: the many ways to make a tree. Trees 16:437–443

Woollons RC (1985) Problems associated with analyses of long-term *Pinus* fertilizer × thinning experiments. Aust For Res 15:495–507

Woollons RC, Haywood WJ (1984) Growth losses in *Pinus radiata* stands unsprayed for *Dothistroma pini*. N Z J For Sci 14:14–22

Woollons RC, Haywood A, McNickle DC (2002) Modeling internode length and branch characteristics for *Pinus radiata* in New Zealand. For Ecol Manage 160:243–261

Wright JW (1976) Introduction to forest genetics. Academic, New York

Xiang B, Li B, Isik F (2003a) Time trends of genetic parameters in growth traits of *Pinus taeda* L. Silvae Genet 52:114–121

Xiang B, Li B, McKeand S (2003b) Genetic gain and selection efficiency of loblolly pine in three geographic regions. For Sci 49:196–208

Xie C-Y (2003) Genotype by environment interaction and its implications for genetic improvement of interior spruce in British Columbia. Can J For Res 33:1635–1643

Yang JL, Baillères H, Okuyama T, Muneri A, Downes G (2005) Measurement methods for longitudinal surface strain in trees: a review. Aust For 68:34–43

Yang J-L, Fife D, Waugh G, Downes G, Blackwell P (2002) The effect of growth strain and other defects on the sawn timber quality of 10-year-old *Eucalyptus globulus* Labill. Aust For 65:31–37

Yang J-L, Ilic J (2003) A new method of determining growth stress and relationships between associated wood properties of *Eucalyptus globulus* Labill. Aust For 66:153–157

Yang J-L, Waugh G (2001) Growth stress, its measurement and effects. Aust For 64:127–135

Yang RC (1998) Foliage and stand growth responses of semimature lodgepole pine to thinning and fertilization. Can J For Res 28:1794–1804

Yang YL, Waugh G (1996a) Potential of plantation-grown eucalypts for structural sawn products. I. *Eucalyptus globulus* Labill. ssp. *globulus*. Aust For 59:90–98

Yang YL, Waugh G (1996b) Potential of plantation-grown eucalypts for structural sawn products. II. *Eucalyptus nitens* (Dean & Maiden) Maiden and *E. regnans* F. Muell. Aust For 59:99–107

Yeates GW (1990) Nematodes in New Zealand forest nurseries. N Z J For Sci 20:249–256

Yoda K, Kira T, Ogawa H, Hozumi K (1963) Self-thinning in overcrowded pure stands under cultivated and natural conditions. (Intraspecific competition among higher plants XI). J Biol Osaka City Univ 14:107–129

Zaehle S (2005) Effect of height on tree hydraulic conductance incompletely compensated by xylem tapering. Funct Ecol 19:359–364

Zalesny RS, Riemenschneider DE, Hall RB (2005) Early rooting of dormant hardwood cuttings of *Populus*: analysis of quantitative genetics and genotype × environment interactions. Can J For Res 35:918–929

Zamudio F, Baettyg R, Vergara A, Guerra F, Rozenberg P (2002) Genetic trends in wood density and radial growth with cambial age in a radiata pine progeny test. Ann For Sci 59:541–549

Zeide B (2001) Thinning and growth: a full turnaround. J For 99:20–25

Zeng H, Peltola H, Talkkari A, Venäläinen A, Strandman H, Kellomäki S, Wang K (2004) Influence of clear-cutting on the risk of wind damage at forest edges. For Ecol Manage 203:77–88

Zobel BJ, Talbert J (1984) Applied forest tree improvement. Wiley, New York

Zou C, Penfold C, Sands R, Misra RK, Hudson I (2001) Effects of soil air-filled porosity, soil matric potential and soil strength on primary root growth of radiata pine seedlings. Plant Soil 236:105–115

Zwolinski J, Bayley AD (2001) Research on planting stock and forest regeneration in South Africa. New For 22:59–74

Appendix 1 Glossary

Abiotic—Something which does not involve living organisms.

Agroforestry—Combining forestry activities in close proximity to other agricultural activities on the one farm property.

Allele—A pair of **genes** with similar functions. Each member of the pair is carried on one of the pair of **chromosomes** which make up a **chromatid**.

Bacteria—A large group of primitive (in evolutionary terms), single-celled organisms which do not have a nucleus in the cell. They are round, spiral, or rod-shaped and often either aggregate in colonies or move using flagella (hairlike processes which project from a cell). They live in soil, water, organic matter or the bodies of plants and animals. They are important in breaking down **organic matter** and as disease-causing organisms (adapted from the dictionary of the Encyclopædia Britannica 2004).

Bark—A layer of mainly dead tissue which surrounds the stem, branches or woody roots of **trees**. It protects the thin layers of living tissues immediately beneath it.

Basal area—Cross-sectional area at **breast height** of a **tree** stem (see also **stand basal area**).

Basic density—The **oven-dry weight** of **wood** per unit green volume.

Biodiversity—The range and variety of living organisms in an **ecosystem**.

Bioenergy—**Biomass** used to make energy, usually by conversion to ethanol or burning to generate electricity.

Biomass—The weight of a living organism. It may include the water in the organism, when it is referred to as fresh biomass. Often, the **oven-dry biomass** is considered, where the water is removed by drying at 60–80ºC until the dry weight becomes constant.

Biotic—Something which involves living organisms.

Branchiness—The total cross-sectional area, at their bases, of the branches on a **tree** stem, expressed as a proportion of the total surface area of the length of the stem along which they occur.

Breast height—A height of 1.3 m (or 1.4 m in some countries and 4′6″ in the USA) above ground from the base of a **tree**. If the **tree** is growing on sloping ground, it is measured from the highest ground level at the base of the **tree**.

Breeding value—For an individual organism, it is a measure of the extent to which its offspring show a **gain** in a **trait** which a breeding programme is aiming to influence.

Broad sense heritability—The extent to which the **phenotype** of an individual is determined by its **genotype** (see also **narrow sense heritability**).

Cambium—A thin layer of live tissue surrounding stem, branch or root **wood**. It is positioned between the **phloem** and the **xylem** and its **cells** divide to produce new **phloem** and **xylem** tissue.

Canker—A spreading wound of dead tissue or a plant disease characterized by such wounds.

Canopy—The foliage and branches of an individual **tree** or a forest (see also **crown**).

Cell—The smallest part of a living organism capable of functioning independently; the bodies of higher plants and animal are made up of many billions of cells of many different types. Cells are microscopic in size and contain both the **DNA** essential for reproduction and protoplasm (a complex solution of chemical substances in water in which the chemical reactions of life occur). Cells of animals are enclosed within a fragile membrane, whilst those of plants are enclosed within a much stronger wall, made principally of **cellulose**.

Cellulose—A chemical substance which is a complex carbohydrate, consisting of many sugar molecules strung together in a long sequence.

Chromatid—A pair of **chromosomes** which carry **genes** with similar functions.

Chromosome—A body found in the nucleus of **cells** of living organisms, consisting principally of a long strand of **DNA** which carries **genes**.

Clone—The offspring of an individual organism which has been reproduced in such a way that it is genetically identical to, that is, has exactly the same **genotype** as, its parent.

Concentration—The proportion something makes up of the whole of which it is part. For example, the concentration of a **nutrient** element in a plant would be the weight of that element in the plant as a proportion of the oven-dry **biomass** of the plant.

Coppice—The resprouting of **trees** from their cut stump. Coppice is a form of **epicormic shooting**.

Crown—The foliage and branches of an individual **tree** (see also **canopy**).

Current annual increment—The present growth rate of a **stand** at any particular age. It is often abbreviated as CAI.

Density—Of **wood**, see **basic density**: Of **tree** stands, see **stand density**: Based on **tree** numbers, see **stocking density**.

Deoxyribonucleic acid—A chemical substance, consisting of a long string of molecules known as nucleotides, which provide the genetic code to determine how an organism functions. Its name is usually abbreviated as DNA. The structure of DNA allows it to be duplicated readily within the plant or animal body whenever its **cells** divide during growth or development.

Disease—An impairment to the normal functioning of a plant or animal, caused either by another living organism or by something in the non-living part of the **environment**.

DNA—See **deoxyribonucleic acid**.

Dominant height—Average height of a prescribed number per unit area of the tallest or largest-diameter **trees** in a **stand**.

Earlywood—The less dense, lighter-coloured portion of a **tree** ring, which develops when conditions for growth are favourable cf. **latewood**. It is sometimes referred to as springwood.

Ecology—The relationship between living organisms and their **environment**.

Economics—Study of the way people use the scarce resources of the earth which they consider essential for their well-being.

Ecosystem—An assemblage of plants and animals living together at a **site**.

Environment—The other living or inanimate things amongst which a living organism grows and reproduces.

Enzyme—A protein produced by living **cells** which facilitates a particular chemical reaction in the **cell**.

Epicormic shoots—Shoots which arise from buds in the **tree** stem or branch and develop ultimately into normal leaves, leaf-bearing branches or stems.

Exotic—A species planted in a location outside its natural range of occurrence.

Forestry—The use and **management** of forests to provide goods and services to people.

Fungi—(sing. fungus) A group of organisms which make up one of the five kingdoms of living organisms on earth. They are multicellular organisms with a branched, filamentous growth form, with a **cell** wall surrounding the filaments (which are called hyphae). They obtain their food by breaking down **organic matter**.

Gain (in **genetics**)—The difference between the average of a **trait** in the offspring, bred by mating some of the individuals in a population of parents, and its average in the parent population.

Gene—A specific section of the length of a **DNA** molecule which makes up a **chromosome**. Each gene provides a code for the manufacture of a **protein**, which determines the chemical functioning of a **cell**. Genes are the functional units of inheritance and are passed from parents to offspring during reproduction.

Genetic engineering—Artificial manipulation of the **genes** that an organism contains to modify the **traits** the organism displays.

Genetics—The study of how traits are passed from parents to offspring and why individuals of any species vary one from the other.

Genotype—The set of **genes** which an individual organism contains and which it inherits from its parents.

Hardwood—**Tree** species which are flowering plants, in which the seeds develop enclosed in an ovary (cf. **softwood**).

Height (of a **tree**)—The vertical distance from ground level to the highest green point on the **tree**.

Heritability—See **narrow sense heritability** (but note also **broad sense heritability**).

Hormone—A chemical substance produced in one part of a plant or animal and which moves to another part and produces a specific effect on **cell** activity there.

Knotty core—The central part of a **tree** stem in which knots occur because branches had not yet fallen from, or been pruned from, the stem at the time the **wood** in that central part was formed.

Latewood—The denser, darker-coloured portion of a **tree** ring, which develops when conditions for growth are less favourable, cf. **earlywood**. It is sometimes referred to also as summerwood.

Leaching—The loss from soil of **nutrients**, dissolved in water in the soil, as water moves through the soil.

Leaf area index—The area of the leaves of a **forest canopy**, expressed per unit area of the ground they cover. Leaf area is defined as the area of the shadow which the leaves would cast if laid flat and lit vertically from above.

Management (of forests)—Activities undertaken in a forest to achieve the provision of the goods and services which are desired from it.

Mean annual increment—The average rate of production (of **wood**, **biomass, basal area**, etc.) to any particular age of a **stand**. It is often abbreviated as MAI.

Metabolism—All that occurs in the **cells** of living organisms and which constitutes life itself. It involves chemical changes within **cells** which provide the energy and the materials for growth, maintenance and reproduction of living organisms.

Mixed-species plantation—A forest **plantation** consisting of two or more different **tree** species growing in mixture.

Molecular marker—A segment of a **chromosome** used to identify a **quantitative trait locus**.

Mycorrhiza—**Fungi** which form a symbiotic relationship with plant roots. They aid uptake of water and nutrients by roots and obtain food for their own metabolism from the plant.

Narrow sense heritability—Generally known simply as **heritability**, it is the proportion of the variation in a **phenotype**, amongst the individuals of some population, which can be attributed to **genes** transmitted from parents to offspring (see also **broad sense heritability**).

Native forest—Forest which has regenerated, usually from seed, following disturbance of the forest (such as fire, storm or logging by man), and has been allowed to develop more or less as would happen naturally without intervention by man.

Nutrient—Any one of 15 chemical elements which are essential for plants and which play a wide variety of roles in their **metabolism**. In land plants, they are mostly taken up by the roots from the soil.

Organic matter—Dead tissue of or from living organisms.

Oven-dry biomass—The weight of plant tissue after drying (usually at 60–80°C) until its weight becomes constant.

Parasite—A living organism which lives in, with or on another organism and derives benefit in some way from the other organism, without killing it

Parasitoid—A living organism which lives in association with another organism, derives benefit in some way from the other organism and eventually kills it.

Pathogen—A living organism which causes a **disease**.

Pest—A living organism which damages a **tree** and affects its growth or development in some way.

Phenotpye—The appearance or performance of a living individual. It is any **trait** that can be seen, detected or measured on the individual. An individual's phenotype is determined both by its genetic makeup (its **genotype**) and by the effects of the **environment** which influence its growth and development.

Phloem—A thin layer of living tissue surrounding stem, branch or root **wood** and lying immediately below the bark.

Photosynthesis—The process of chemical conversion by plants of carbon dioxide, taken into their leaves from the air, and water to sugars, which provide energy to the plant for other **metabolic** processes. Light absorbed by the leaves from the sun provides the energy required in this chemical process. Oxygen is released from the leaves as part of it.

Plantation—Forest created by man, where seeds or seedlings have been planted, usually at a regular spacing.

Predator—An animal which consumes all or part of another living organism.

Protein—A chemical substance consisting of a long string of smaller chemical substances known as amino acids. There are a large number of different types of proteins in living organisms, each differing in the number and sequence of the amino acids it contains. Each protein is folded into a particular three-dimensional shape which determines its function. Proteins can be used as part of the structure of a living organism. They act also by facilitating chemical reactions within **cells**, when they are known as **enzymes**; the proteins are not changed themselves by the reactions.

Provenance—The place where an individual organism lives in its natural circumstances.

Pruning—The removal of live or dead branches for some distance up the stem of a **tree**.

Pulplogs—Small logs, cut from **tree** stems or branches, of a size appropriate for chipping to be used for making paper.

Quadratic mean diameter—If the average of the stem cross-sectional areas at **breast height** of the **trees** in a **stand** has been determined, quadratic mean diameter is the diameter that corresponds to this average stem cross-sectional area.

Qualitative trait—A characteristic of a living organism which is determined by the interaction of several of the **alleles** from its genetic makeup.

Quantitative trait—A characteristic of a living organism which is determined by a single **allele** only from its genetic makeup.

Relative addition rate—The proportional increase, per unit time, in the amount of a **nutrient** element in a growing plant.

Relative growth rate—The proportional increase, per unit time, in the **biomass** of a growing plant.

Rotation—The period between establishment and final felling of a **plantation**.

Sawlog—A log cut from a **tree** stem and large enough to be sawn into one or more of the many types of sawn **wood** used for building and many other purposes.

Silviculture—The tending of **trees** in forests to achieve some desired objectives of management.

Site—An area of land which can be managed homogenously and will produce a more or less constant **wood yield** across it from a particular plantation species.

Site index—A measure of **site productive capacity**, defined as the **dominant height** of the **trees** in a **stand** at a particular, but arbitrarily chosen, age.

Site productive capacity—The total **stand biomass** produced, up to any particular stage of development, of a plantation growing on a particular site, when it uses fully the resources necessary for **tree** growth which are available from the site.

Stand—A more or less homogeneous group of **trees** in a forest in which an observer might stand and look about him or her.

Softwood—**Tree** species which do not have flowers and in which the seeds develop without the protection of an ovary. Often these 'naked' seeds are protected by the scales of a cone (cf. **hardwood**).

Stand basal area—Stem cross-sectional area at breast height, summed over all the **trees** in a **stand** and expressed per unit ground area.

Stand density—The degree of crowding of the **trees** in a **stand**.

Stocking density—The number of **trees** per unit area in a **stand**.

Stomata—Specialised, microscopic structures on the surfaces of the leaves of land plants. They may open or close to allow exchange of gases between open spaces within the leaf and the atmosphere, most noticeably carbon dioxide, oxygen and water vapour. They are sometimes called stomates.

Strain (engineering)—The change in dimensions of an object caused by a **stress** (engineering).

Stress (engineering)—The force applied per unit area of an object when it is stretched, compressed or twisted.

Thinning—Deliberate removal of some **trees** from a plantation, from time to time during its life.

Timber—**Wood** cut from **tree** stems into sizes appropriate for its final use.

Trait—A characteristic of a species which might be amenable to modification by breeding.

Transpiration—The process of transport of water from the roots to the leaves of plants.

Tree—A woody plant with a distinct stem or stems and with a mature height of several metres.

Understorey—A layer of vegetation growing beneath the main **canopy** of a forest.

Virus—Any member of a large group of submicroscopic, primitive (in evolutionary terms) agents that can be considered either as very simple microorganisms or as extremely complex molecules. They are capable of growth and multiplication only in living **cells**. Some cause important diseases of plants or animals (adapted from the dictionary of the Encyclopædia Britannica 2004).

Weed—A plant growing where it is not desired.

Wood—A strong material forming the greater part of the stem, branches and woody roots of **trees**. It consists mainly of dead tissue.

Wood density—See **basic wood density**.

Xylem—The tissue in plants through which water is transported from roots to shoots. In **trees**, it is the **wood**.

Yield—The amount of some characteristic of a plantation, such as plant **biomass** or stem **wood** volume, produced by the plantation. It is often expressed as an amount per unit area of the plantation.

Appendix 2 Conversion Factors

Abbreviations used commonly are shown in parentheses

Metric–imperial conversion factors

1 centimetre (cm)=2.538 inches (in.)
1 metre (m)=3.2808 feet (ft)=1.094 yards (yd)
1 hectare (ha)=2.471 acres
1 kilogram (kg)=2.205 pounds (lb)
1 tonne (t)=0.9842 tons
1 kilometre (km)=0.6214 miles
1 litre (l)=0.2120 gallons (UK)=0.2642 gallons (USA)
1 millilitre (ml)=0.0352 fluid ounces (fl oz)

Conversions within the metric system

1 cm=10 millimetres (mm)
1 m=100 cm=1,000 mm
1 km=1,000 m
1 ha=10,000 m^2
1 t=1,000 kg
1 kg=1,000 grams (g)
1 g=1,000 milligrams (mg)
1 l=1,000 cm^3=1,000 ml

Conversions within the imperial system

1 ft=12 in.
1 yd=3 ft
1 chain=100 links=22 yd
1 furlong=10 chains
1 mile=8 furlongs=1,760 yd=5,280 ft
1 acre=10 $chain^2$=4,840 yd^2
1 lb=16 ounces (oz)
1 ton=2,240 lb
1 gallon=4 quarts=8 pints
1 super foot=1/12ft^3
1 cord=128 ft^3
1 cunit=100 ft^3

Index

Printing: Krips bv, Meppel
Binding: Stürtz, Würzburg